Tomorrow's Healthcare by Nano-sized Approaches

A Bold Future for Medicine

Paula V. Messina

Department of Chemistry
Universidad Nacional del Sur, INQUISUR-CONICET
Buenos Aires, Argentina

Luciano A. Benedini

INQUISUR-CONICET
Buenos Aires, Argentina

Damián Placente

Department of Chemistry
Universidad Nacional del Sur, INQUISUR-CONICET
Buenos Aires, Argentina

CRC Press
Taylor & Francis Group
Boca Raton   London   New York

CRC Press is an imprint of the
Taylor & Francis Group, an **informa** business

A SCIENCE PUBLISHERS BOOK

Cover figure provided by the authors of the book.
Figure legend: Manipulation of the invisible. Nano robot attacking bacteria. Transmission Electronic Microphotograph (TEM) of Pseudomona aeruginosa.

CRC Press
Taylor & Francis Group
6000 Broken Sound Parkway NW, Suite 300
Boca Raton, FL 33487-2742

© 2020 by Taylor & Francis Group, LLC
CRC Press is an imprint of Taylor & Francis Group, an Informa business

No claim to original U.S. Government works

Version Date: 20200131

International Standard Book Number-13: 978-0-367-02301-0 (Hardback)

Library of Congress Cataloging-in-Publication Data

Names: Messina, Paula V., author. | Benedini, Luciano A., 1978- author. | Placente, Damián, 1990- author.
Title: Tomorrow's healthcare by nano-sized approaches : a bold future for medicine / Paula V. Messina, Luciano A. Benedini, Damián Placente.
Description: Boca Raton : CRC Press, [2020] | Includes bibliographical references and index.
Identifiers: LCCN 2020003588 | ISBN 9780367023010 (hardback)
Subjects: MESH: Nanomedicine | Nanostructures
Classification: LCC R857.N34 | NLM QT 36.5 | DDC 610.28--dc23
LC record available at https://lccn.loc.gov/2020003588

Visit the Taylor & Francis Web site at
http://www.taylorandfrancis.com

and the CRC Press Web site at
http://www.crcpress.com

Preface

Human Dreams, Science Fiction or Nanomedicine

Without knowing it, we have evolved with nanotechnology. To obtain evidence, we just have to go back to ancient India, 600 years BC, where carbon nanotubes and nanowires were already in use for the manufacture of steel. People in those times already benefited from the singular properties of manipulating matter at the nano-metric scale. Around 460-370 BC, The Greek philosopher Democritus proposed the first atomic model and, with it, he implanted the seed of understanding the fundament of matter. Obviously, associated with this idea, the embryonic desire of "manipulating matter at the atomic level" was engendered in human beings. As we learned more, fantastic portraits fused with scientific facts and prophetic ideas gave rise to a new vision that moved away from the supernatural, which was governed by rational or pseudo-rational principles. This was also the foundation for "science fiction". Nevertheless, let us move forward to Richard Feynman's lecture "There's Plenty of Room at the Bottom", where the first conception of matter reduction at the atomic level and the "mechanical surgeons" were introduced, followed by Norio Taniguchi's conference and K. Eric Drexler's book "Engines of Creation: The Coming Era of Nanotechnology", where the notion of "nanotechnology", "molecular engineering" and "nanoscale assembler" as active, nano-size concepts progressed.

Thus, the human dream of accessing the matter, to manipulate it on an atomic level and to enjoy the extraordinary properties associated is not original. The novelty, so to speak, is that we now have the talent to comprehend the performance of matter at an atomic level and the equipment to handle it with discretion, thus being able to access the promises that it entails. Now, in the year 2020 AC, dreams and science fiction have become a tangible reality. We ride at the Era of Nanotechnology, where the idea of the manipulation of matter at the nanoscale is completely installed and has changed the whole vision of making and learning science. The matrimony of nanotechnology and medicine gave birth to nanomedicine. This field is recognized as a global challenge, and countless worldwide research and business initiatives for the application of nanotechnology to medical care are being developed; the central themes of this book.

Presently, the nanotechnology facilities at the service of medicine preserve and improve human health through a better diagnosing, treating of chronic illness or traumatic injury, relieving pain, providing specific and targeted drug delivery carriers, artificial antibodies, and much more. Nevertheless, the ambition is immense and goes further. It attempts towards the construction of nanobots that will capture chemical and physical information from a disease site, bring it to an implanted chip where analysis is carried out and communicate wirelessly with doctors outside the body who monitor and guide healing. This will revolutionize the way we detect and treat diseases in the human body. Furthermore, nanomedicine drives towards the construction of nano- smart missile that selectively seeks out and annihilates tumour cells, virus and/or bacteria. In addition, it attempts towards the manipulation of DNA in order to solve degenerative or congenital diseases, towards the induction of precise cellular responses to regenerate damaged or missing tissue, and even more, towards the construction of artificial white, red and platelet blood cells.

In the eleven chapters of this book, we try to summarize the world of exciting possibilities that the application of nanotechnology to medicine offers us. How it has the potential to eliminate all common diseases of the 20th century, to revolutionize individual and population-based health in the 21st century and how it contributes to the extension of human physical and mental capacities.

This brings us to the question of the suitability or of the unscrupulous use of nanomedicine that leads to the controversy and fear of its application. As a double-edged weapon, the dangers and negative consequences of its application for destructive purposes should not be taken for granted. Nevertheless, in the not too distant future, we could visualize a world where medical nano-devices are routinely inserted or even inoculated into the bloodstream to screen health and, automatically, play a part in the repair and maintenance of tissue and organs to give us a superior quality of life. Seeing its potential, government initiatives will support this technology and its access to all people around the world. In this scenario, nanomedicine's public health applications will include rapid and portable diagnoses, vaccines that are more effective, and the eradication of poverty-related diseases (PRDs). Diseases of poverty (also known as poverty-related diseases) are diseases that are more prevalent in low-income populations. They include infectious diseases, as well as diseases related to malnutrition and poor health behaviors.

The technology has a great potential and there is nothing malicious in it, if it is properly applied. The solution lies in our hands. Like we once dreamt, understood and effectively manipulated matter at the nanoscale to apply its benefits in medical care, similarly, it is our responsibility to build a safe way to use and to control it, addressing social, moral and religious concerns.

This is a challenge for us and for the future generations who will read this book.

Paula V. Messina
Luciano A. Benedini
Damián Placente

Contents

The Merger of Nanotechnology and Medicine: A Little History

"I would like to describe a field, in which little has been done, but in which an enormous amount can be made… "
"I want to talk about the problem of manipulating and controlling things on a small scale…"
"Although it is a very wild idea, it would be interesting in surgery if you could swallow the surgeon…"

Richard P. Feynman "There's Plenty of Room at the Bottom"
Annual meeting of the American Physical Society at Caltech
Adapted from (Feynman 1992)

1.1 An Ancient Science, but a Still Maturing Technology

Even though nanotechnology has received a relatively recent expansion in the scientific area, the development of its central principles happened long ago. Since the first atomic model proposed by Democritus in around 460-370 BC, chemists explored and tried to understand the elements and their interactions. Based on their findings, engineers created and proposed new materials to improve our lives. In addition, physicists demonstrated that even atoms are divisible, and we seen the armed forces freed the power of the atomic nucleus. Nowadays, the scientific community has collected colossal information required for the understanding of the matter and exercise an increasing control over its fundamental units. In this new era, science and modern technology have advanced beyond our more ambitious dreams, and today, in the young field of nanotechnology, scientists and engineers are taking control of individual atoms and molecules, manipulating and driving them with an extraordinary degree of precision to attain specific uses. The importance that nanotechnology has for our future depends on the understanding that we make of it, and to adopt responsible and ethical approaches in its development and implementation. In this chapter, scientific basis and definitions of nanoscience, nanotechnology and nanomedicine are analysed, following with a brief summary of the temporal evolution of the nanotechnology applied to medical science. It will highlight the main findings from the creation of nanotechnology to present days and, how these contributions gave rise to the present nanomedicine.

1.2 Scientific Lexicon at the Nano-Size

Many definitions of nanoscience, nanotechnology and nanomedicine have been used in the vast number of conferences, reports, papers and presentations given each year on this subject (Grieneisen and Zhang 2011, Klaessig et al. 2011). In fact, the multidisciplinary nature of this scientific area almost invites a comparable multiplicity of definitions as each specialty (or scientific discipline) adjusts to the dynamic research efforts. The same dynamism leads to ambiguity in meanings and to the uncertainty in the overall impact of this field when the products are commercialized. Translational research, including fund allocation, patents, drug regulatory, ethical and approvals assessment processes, clinical trials and public acceptance are subject to the definition of nanomedicine. Therefore, a clear and consistent definition of nanomedicine would significantly increase the trust among diverse applicants, including the public, while minimizing the risk of miscommunication and unnecessary fear. Having an honest and transparent communication among nanoscience, nanotechnology and nanomedicine fields and through the understanding of how we must use these terms, the potential benefits and harms should have a better perception.

1.2.1 Nanoscience and Nanotechnology

We will begin the discussion with the scientific meaning of the prefix "nano-". Following the International System of Units (SI: from the French name of the system, Système International d'Unités), the morpheme "nano" is usually defined as a prefix meaning "billionth part of …" e.g. a nanometre means "the billionth part of a meter" (Boholm 2016). However, this analysis does not contain the full range of "nano" uses. The semantic contribution of prefix "nano" to each word is not always the same; as an example, nanotechnology is typically not interpreted as "billionth part of technology". Such reflections added to the widespread use of "nano" in many spheres of society, including science, politics and popular culture, show that the use and meaning of "nano" needs a more systematic description. According to the Oxford English Dictionary (OED), the linguistic form of "nano" originates from the classical Latin "*nanus*" or its ancient Greek etymon "*nanos*" (νάνος), meaning "dwarf" (Boholm 2016). In 1958, "nano", together with "giga", "tera", and "pico", were adopted in the newly formed SI (Burdun 1960). In 1974, Norio Taniguchi introduced the term "nanotechnology" at an engineering conference in Tokyo (Taniguchi 1974). The idea of nanotechnology was later popularized and disseminated, among others, by Eric Drexler (Drexler 1986); we will discuss about them later. In 2000, the National Nanotechnology Initiative (NNI) in the United States accepted a single definition that incorporates both science and technology: "*Nanotechnology is the understanding and control of matter at dimensions between approximately 1 to 100 nanometres, where unique phenomena enables novel applications"* (NNI 2014). Although at present, "nano-" terms are appropriate for materials with any dimension below a few hundred nano-meters, "nano" prefixed words are commonly applied to materials much larger than 100 nm throughout dozens of journals whose scope of coverage is limited to studies they define as "nanoscience" and "nanotechnology" (Grieneisen and Zhang 2011). Therefore, the 1–100 nm criterion is convenient, but simplistic to reveal the precise reality of size-dependent characteristics among all materials and disciplines. Accordingly to ISO and ASTM definitions (ASTM 2012, ISO 2015), the two major standards developing organizations, "nano-objects" are those that exhibit dimensions at the nanoscale, 1–100 nm, and such a definition includes their structural configuration, morphology, form or functionality. "Nanomaterial" is a material with any external dimension in the nanoscale or having an internal or surface structure in the nanoscale; this generic term is inclusive of "nano-object" and "nanostructured material". Nanoscience, likewise, is the understanding and control of matter and processes at the nanoscale, typically, but not exclusively, below 100 nanometres in one or more dimensions, and nanotechnology is the manipulation and control of nanoscale materials to create improved devices and systems that exploit properties and phenomena existing at that manometer scale. The Royal Society and the Royal Academy of Engineering contain similar requirements, defining "Nanoscience" as the study of phenomena and manipulation of materials at atomic, molecular and macromolecular scales, where properties differ significantly from those at a larger scale. Meanwhile, "nanotechnologies" are the design, characterization, production and application of structures, devices and systems by controlling shape and size at the nanometre scale (Enderby and Dowling 2004). For the purpose of food regulations, the recommendation is that the term nanoscale has an upper boundary of 1,000 nm, rather than the ISO and ASTM definitions. A similar tendency to favour a larger concept for nanoscale, up to 1,000 nm, was adopted by biological sciences. The European Union (EU), in a recent legislation regarding cosmetics labelling, maintains 100 nm as the upper boundary; nevertheless, those materials of general unspecified size that contain nanoscale components are also included (Klaessig et al. 2011). Scientific Committee on Emerging and Newly Identified Health Risks (SCENIHR), an advisory body to the European Commission, evaluates a metric replacement for identifying nanoscale materials based on a specific surface area greater than 60 $m^2 g^{-1}$ (Klaessig et al. 2011). On the other hand, those concerned primarily with materials science pursue smaller sizes, < 30 nm, to refer to the unique, novel and unexpected properties to be associated with nanoscale materials, especially those associated with quantum confinement (Yuan et al. 2018). A compilation of recommended upper limits suggested by different standards and non-standards developing organizations can be acquired in literature (Garboczi 2009, Linsinger et al. 2012, Boverhof et al. 2015). Some of them are summarized in Table 1.1.

Table 1.1. Dimensional limit for the term "nanoscale" according to different Standards Developing Organizations (SDOs) and non-SDOs. (T) Terminology; (M) Measurements; (EHSE) Environmental Health and Safety Effects; (E) Education Adapted from (Klaessig et al. 2011)

Upper limit/nm	Source	Topic	Identifier	Title
				SDOs
100	ISO	T	ISO/TS 80004-1:2010. ISO/TS 80004-2: 2015	Nanotechnologies. Vocabulary. Part 1: Core terms. (ISO 2010) Part 2: Nano-objects (ISO 2015)
		M	ISO/TR 13014:2012	Nanotechnologies. Guidance on physico-chemical characterization of engineered nanoscale materials for toxicological assessment. (ISO 2012)
		EHSE	ISO/TS 12901-1:2012	Nanotechnologies. Occupational risk management applied to engineered nanomaterials. Part 1: Principles and approaches. (ISO 2012)
100	ASTM International	T	ASTM E2909-13	Standard Guide for Investigation/Study/Assay Tab-Delimited Format for Nanotechnologies (ISA-TAB-Nano): Standard File Format for the Submission and Exchange of Data on Nanomaterials and Characterizations. (ASTM 2013)
		M	ASTM E2490-09(2015)	Standard Guide for Measurement of Particle Size Distribution of Nanomaterials in Suspension by Photon Correlation Spectroscopy (PCS). (ASTM 2015)
		EHSE	ASTM E2524-08(2013)	Standard Test Method for Analysis of Haemolytic Properties of Nanoparticles. (ASTM 2013)
		E	ASTM E2996-15	Standard Guide for Workforce Education in Nanotechnology Health and Safety. (ASTM 2015)

(Contd.)

			Non-SDOs
100	Royal Society	T; M; EHSE; E	Nanoscience and nanotechnologies: opportunities and uncertainties. (Enderby and Dowling 2004)
100	EC-SCENIHR	T; M; EHSE	Opinion on the scientific aspects of the existing and proposed definitions relating to products of nanoscience and nanotechnologies. (EC-SCENIHR 2007)
100	EC-SCCP	EHSE	Opinion on safety of nanomaterials in consumer products. (EC-SCCP 2007)
100	ETC Group	T; M; EHSE	Nano-scale technologies. (ETC-Group 2005)
100	Swiss Reinsurance Company	T; M; EHSE	Nanotechnology small matter, many unknowns. (Hett 2004)
200	Soil Association	T; M; EHSE	Nanotechnologies and food evidence. (Harnad 2013)
200	Department for Environment Food and Rural Affairs	T; M; EHSE	UK voluntary reporting scheme for engineered nanoscale materials. (DERFA 2006)
300	Chatham House	EHSE	Securing the promise of nanotechnologies. (Breggin et al. 2009)
300	Friends of Earth	T; M; EHSE	Nanotechnology in Food & Agriculture. (Miller and Senjen 2008)
500	Swiss Federal Office of Public Health	EHSE	Precautionary matrix for synthetic nanomaterials. (Höck et al. 2008)
1000	House of Lords Science Committee	T; M; EHSE	Nanotechnologies and Food. (Robson 2010)

1.2.2 Nanomedicine

Applications of nanoscale science to the field of medicine have resulted in the on-going development of the "nanomedicine" subfield. With regard to the concept of "nanomedicine", there is also a lack of agreement on its meaning. Robert A. Freitas (Freitas 1999) defines nanomedicine as *"the monitoring, repair, construction and control of human biological systems at the molecular level, using engineered nanodevices and nanostructures."* For the United States' National Institutes of Health Roadmap for Medical Research in Nanomedicine (NIH 2006), "nanomedicine" is *"an offshoot of nanotechnology, [which] refers to highly specific medical interventions at the molecular scale for curing disease or repairing damaged tissues, such as bone, muscle, or nerve"*, while the European Science Foundation (ESF) defines in November of 2004 *"the field of "nanomedicine" as the science and technology of diagnosing, treating and preventing disease and traumatic injury of relieving pain, and of preserving and improving human health, using molecular tools and molecular knowledge of the human body"* (ESF 2005). This last description is complete and has grown into the medical field; however, it is not appropriately differentiated from the molecular medicine, which is based on a conservative biochemical approach. To appreciate the fact that therapeutic properties of nanoscale materials specifically designed for medical context applications are, by definition, a function of their specific size and scale, a critical distinction between nanomedicine and other medical research fields has to be done. Therefore, the European Technology Platform on Nanomedicine, ETPN, defines "nanomedicine" as the application of nanotechnology to health (Tomellini et al. 2005). In addition, the English Oxford Dictionary designates "nanomedicine" as *"the branch of medicine with the use of nanotechnology"* (Stevenson 2010). Leaving the academic meaning and going to the practical one, "nanomedicine" is currently classified under *"advanced medical technologies"* by medical industry considering that such definition is focused on looking for novel therapies regardless of the technology involved (Boisseau and Loubaton 2011). The International Journal of Nanomedicine (IJN) emphasizes (Webster 2006) *"...nanomedicine research in which significantly changed medical events are elucidated only by concentrating on nanoscale events..."*

In 2006, during the ESF Scientific Forward Look on Nanomedicine (Tibbals 2010) the concept of "scientific nanomedicine" was distinguished from a more general definition of "medical nanotechnology". Nanomedicine was described as medicine that *"uses nano-sized tools for the diagnosis, prevention and treatment of disease and to gain increased understanding of the complex underlying pathophysiology of disease."* Nanotechnology applied to medicine is related to *"the use of analytical tools and devices to bring a better understanding of the molecular basis of disease, patient predisposition and response to therapy, and to allow imaging at the molecular, cellular and patient levels. It is also related to the design of nano-sized multifunctional therapeutics and drug delivery systems to yield more effective therapies."*

Many pharmaceutical formulations are actually referred to as "nanomedicines", denoting drug delivery systems by means of encapsulation or incorporation of active principles into nanoparticles. In addition, "nanomedicine" is also frequently used to mention the use of nanoparticles to enhance imaging, especially focused on the concept of theranostic. Nanomedicine is also associated with polymer therapeutics, regenerative medicine, active implants, and nano-drugs such as gene delivery nano-probes (Tibbals 2010). To summarize, "nanomedicine" has been defined in many ways and not all agree, but most emphasize a direct control and manipulation of cellular-level processes on the nanoscale, applied for diagnostics and healing.

1.3 Before the Discovery

Any understanding of the implications of nanotechnology and therefore of nanomedicine must begin by sorting out its history. Although the first steps towards the understanding of nanoscale events to create materials with tailored properties go back as far back as 460–370 BC when Greek philosopher Democritus proposed the first atomic model (Liddell and Scott 1897), nanotechnology's applications were in use centuries before. The earliest evidence of the use and applications of nanotechnology goes back to ancient India in around 600 BC. Already then, manufactured microstructure of "wootz" steel contained carbon nanotubes and cementite nanowires (Sanderson 2006). Later, many other applications have taken place without a full understanding of the events that give rise to them. A very good example of these is the Lycurgus Cup, nowadays in the British Museum. It is one of the outstanding achievements of the Roman glass industry; created in the 4th century AD, it is an architype of dichroic glass, displaying two different colours by undergoing a colour change in certain lighting conditions. The glass allows it to look opaque green when lit from outside but translucent red when light shines through the inside. The dichroic effect is caused by colloidal metal precipitated into the glass silicate matrix, specifically silver-gold alloy (silver/gold ratio of about 7/3, containing about 10% copper) nanoparticles of 50–100 nm diameter. In addition to these metallic particles, the glass was shown to contain numerous nanoparticles (15–100 nm diameter) of sodium chloride probably originating from the mineral salts used to supply the alkali during the glass assembly (Freestone et al. 2007). In the Mesopotamia of the 9th century AD, our modern nanoparticles were also used by the master craftsmen to create a glittering effect on the surface of pots (Sattler 2010, Khan 2011). Middle Ages' and Renaissance's ceramics retain a distinct gold or copper colour metallic glitter which is caused by a metallic film containing silver and copper nanoparticles homogeneously distributed into the glassy matrix of the ceramic glaze. These nanoparticles were created by the artisans by mixing together copper and silver salts, oxides, vinegar, ochre, and clay on the surface of previously-glazed pottery. The technique has its origins in the Muslim world; since Muslims were not allowed to use gold in artistic representations, they sought a way to create a similar effect without using real gold. They found a solution by applying lustre to the transparent surface of a glaze (Khan 2011, Rawson 2011). In European cathedrals from the 6th–15th centuries, the vibrant stained glass windows owed their rich colours to gold chloride and other metal oxides nanoparticles. As an extra benefit, gold nanoparticle acted as photocatalytic air purifiers. Gold nanoparticles as an inorganic dye to induce red colour was also found in Ming dynasty pottery (14–17th century) (Varadan et al. 2010). The famous 16–18th century "Damascus" steel blades were made of an ultrahigh-carbon steel formulation that gave them strength, resilience, the ability to hold a keen edge, and the visible moiré pattern in the steel that gives the blades their name. The art of producing such amazing pieces was lost long ago, but recently, researchers have established strong evidence supporting the theory that the distinct surface patterns on these blades result from a carbide-banding phenomenon produced by the micro-segregation of slight amounts of forming elements present in the "wootz" ingots from which the blades were forged. Further, it is probable that "wootz" Damascus blades with damascene patterns could have been produced simply from "wootz" ingots provided from those regions of India having the appropriate impurity-containing ore deposits (Reibold et al. 2006). It was not until the beginnings of the 19[th]century that the atomistic theory became linked to strong experimental evidence, leading to a deep study of the nature of matter, its fundamental units and the manipulation of it, the basis of our actual nanotechnology. Some contributors were John Dalton (1803), Michael Faraday (1832, 1857), J. Plucker (1859), James Clark Maxwell (1865), Dmitri Mendeleev (1869), G.J. Stoney (1874), Sir William Crookes' (1879), Johann Jakob Balmer (1885), E. Goldstein (1886), Wilhelm Roentgen (1895), Henri Becquerel (1896), Joseph John "J.J." Thomson (1897), Ernest Rutherford (1898, 1911), Marie Sklodowska Curie (1898), Max Planck (1900), Frederick Soddy (1900), Hantaro Nagaoka (1903), Richard Abegg (1904), Albert Einstein (1905), Hans Geiger (1906), Niels Bohr (1913), H.G.J. Moseley (1914), Francis William Aston (1919), Irving Langmuir (1919), Louis de Broglie (1924), Wolfgang Pauli (1925), Werner Heisenberg (1927) and Erwin Schrodinger (1930). We will not stop at their careers and discoveries, which are well known, but they deserve recognition because they laid the foundations for the perception of those who subsequently embodied the first ideas on nanotechnology. Information about them can be found in the following references (Pullman 2001, Niaz 2009, Peacocke 2013).

Increased erudite scientific understanding was followed by instrumentation and experimentation improvements, driving to the findings and advances that enabled our modern era of nanotechnology. In 1857 Michael Faraday demonstrated that nanostructured gold produced diverse coloured solutions under specific lighting conditions; he called it "ruby" gold (Thompson 2007). However, the concept of a "nanometer" to characterize a particle size was first proposed by Richard Adolf Zsigmondy (Austrian-

German chemist, 1865–1929). He studied colloidal systems including gold sols and was the first to measure the size of gold colloids using a microscope, winning the Nobel Prize Laureate in chemistry in 1925 (Zsigmondy 1909). In 1936, Erwin Müller, working at Siemens Research Laboratory, invented the field emission microscope, allowing to observe near-atomic-resolution images of materials. In 1951, he created the field ion microscope taking images of tungsten atoms structures that were published for the first time in the journal Zeitschrift für Physik (Müller 1951).

Irving Langmuir and Katharine Burr Blodgett collaborated at the nanometre scale; they studied films that were just one molecule thick. Their accomplishments were anything but minuscule; Langmuir received a Nobel Prize for his research in 1932. In 1939, Blodgett invented the non-reflecting glass by building a monolayer of 44 molecules barium stearate onto crystal. While never as acclaimed as Langmuir, Katharine Burr Blodgett received numerous awards throughout her career, including the Francis Garvan Medal from the American Chemical Society in 1951. The Age of Information came from the hand of John Bardeen, William Shockley, and Walter Brattain in 1947. Working at Bell Labs, they discovered the semiconductor transistor and greatly expanded scientific knowledge of semiconductor interfaces, laying the foundation of electronic devices. Victor La Mer and Robert Dinegar developed the theory and the process for practical growing monodisperse colloidal materials in 1950; the controlled ability to fabricate colloids opened the door of numerous industrial applications such as specific papers, paints, thin films and even the beginnings of dialysis treatments. The "molecular engineering" concept applied to dielectrics, ferroelectrics, and piezoelectrics was introduced by Arthur von Hippel at Massachusetts Institute of Technology in 1956. In 1958, Jack Kilby of Texas Instruments designed and constructed the first integrated circuit for which he received the Nobel Prize in 2000.

Until then, and despite of the advances made, the scientists did not consider the possibility of modifying the matter on an atomic scale. Till somebody did…

1.4 Theoretical Beginning

The earliest systematic discussion of nanotechnology is considered to be a speech given by Richard Feynman, the 1965 Nobel Prize Laureate in Physics for "fundamental work in quantum electrodynamics". On the evening of 29 December of 1959, during the American Physical Society meeting at Caltech, he presented a lecture titled, "There's Plenty of Room at the Bottom", in which he introduced the concept of matter manipulation at the atomic level (Feynman 1992). Feynman suggested several routes for exploration that would delineate our current nanotechnology (Keiper 2003), for example the reduction of computer machines to make them smaller and faster, or the possibility of making "mechanical surgeons" that could travel to trouble spots and heal the body from inside, the first conception of what we today call "Nubot" or the "nucleic acid robot" (Seeman 2005). Even though the degree to which Feynman induced the rise of nanotechnology was debated, his lecture is considered by the scientific community as a pivotal event in the history of the nano-science era. 25 years after his original talk, during the weeklong seminar held at the Esalen Institute in October of 1984, Feynman was invited to give an updated version of "There's Plenty of Room at the Bottom". The talk called "Tiny Machines" discuss the technological advances that researches in all fields have made since Feynman first outlined the vision of the nano-world (Feynman 1993), some of them detailed below.

In 1969, Gordon Moore, co-founder of Intel, described in the *Electronics* magazine one tendency actually known as "Moore's Law" (Moore 1965). He described the density of transistors on an integrated chip (IC), postulated about chip sizes, their costs, functionality and how its effect would transform peoples' lives and work. The basic trend that Moore envisioned has continued for 50 years, to a large extent due to the semiconductor industry's increasing reliance on nanotechnology because ICs and transistors have come close to atomic dimensions.

Investigations in the direction suggested by Feynman didn't begin immediately; he himself didn't use the word "nanotechnology" in his talk. Almost 15 years after Feynman's lecture, a Japanese scientist, Norio Taniguchi of the Tokyo University of Science, was the first to suggest the term "nanotechnology" to describe semiconductor processes occurring with precision at the level of about one nanometre. He promoted that nanotechnology involved the processing, separation, consolidation, and deformation of materials at the scale of one atom or one molecule (Taniguchi 1974). In 1981, Gerd Binnig and Heinrich Rohrer at IBM's Zurich laboratory created a sophisticated new tool that was powerful enough to allow them to see single atoms with unprecedented clarity: the Scanning Tunneling Microscope (Binnig et al. 1982). For their invention, they won the Nobel Prize in Physics in 1986. Later, Gerd Binnig, Calvin Quate, and Christoph Gerber (Binnig et al. 1986) invented the atomic force microscope, which has the capability to view, measure, and manipulate materials down to fractions of a nanometer in size, including measurement of various forces intrinsic to nanomaterials. The first dramatic demonstration of this power occurred in 1990 when the physicists Don Eigler and Erhard Schweizer at IBM's Almaden Research Center, employed 35 individual atoms of xenon to spell out the letters "IBM". In 1991, the same research team built an "atomic switch," an important step in the development of nanoscale informatics evolution.

Simultaneously, Russian physicist Alexei Ekimov discovered nano-crystalline, semiconducting quantum dots (QDs') embedded in a glass matrix and conducted pioneering studies of their electronic and optical properties (Ekimov and Onushchenko 1981). Late in 1982, the American chemist Louis E. Brus detected an analogous phenomenon in colloidal solutions (Rosetti and Brus 1982). By the end of the 1980s, the QDs' attractive optical properties and the development of the synthetic technology from them were greatly promoted (Li and Zhu 2013). This decade was stated as the beginning of the golden era of nanotechnology. Another scientific revolution arose from the discovery of new shapes of carbon molecules, the quintessential element of life. In 1985, researchers of the Rice University, Harold Kroto, Sean O'Brien, Robert Curl, and Richard Smalley revealed the Buckminsterfullerene (C60), usually identified as the "buckyball", which is a molecule resembling a soccer ball in shape and composed entirely of carbon, as are graphite and diamond (O'brien et al. 1986). For their roles in this discovery, the team was awarded the Nobel Prize in Chemistry in 1996. This investigation drive led to the discovery of a chemically and morphologically analogous molecule known as "carbon nanotube" (CNT) by Sumio Iijima of NEC Laboratories in 1991. Nanotubes have been demonstrated to be about 100 times stronger than steel just with a sixth of their weight. They displayed unusual heat and conductivity characteristics that guaranteed they would be important to high technology in the coming years (Iijima and Ichihashi 1993). In 1992, C.T. Kresge and his colleagues (Kresge et al. 1992) at the Mobil Oil Corporation discovered the nanostructured catalytic materials MCM-41 and MCM-48, at present widely used in the crude oil refining process as well as in polluted water treatments; other applications even propose them as drug delivery systems (Vallet-Regi et al. 2001). By the end of 20th century, Chad Mirkin at Northwestern University invented the dip-pen nanolithography® (DPN®) (Piner et al. 1999), a scanning probe lithography technique where an atomic force microscope (AFM) tip is used to create patterns directly on a range of substances with a variety of inks. This technology led to a tailored and reproducible "writing" of electronic circuits, in addition to patterning of biomaterials for cell biology research, nanoencryption, and other applications that we know today.

In the 90s, the first nanotechnology companies began to operate, for example Nanophase Technologies (1989), Helix Energy Solutions Group (1990), Zyvex (1997), Nano-Tex (1998), among others. By the early 2000s, nanotechnological consumer products appeared in the marketplace. To mention some, they developed lightweight nanotechnology-enabled automobile bumpers that resist denting and scratching, golf balls that fly straighter and tennis rackets that are stiffer (therefore, the ball rebounds faster). In addition, baseball bats with better flex and "kick", nano-silver antibacterial socks, clear sunscreens, wrinkle- and stain-resistant clothing, deep-penetrating therapeutic cosmetics, scratch-resistant glass coatings, faster-recharging batteries for cordless electric tools, and improved displays for televisions, cell phones, and digital cameras were designed and distributed in the market.

These nano-machines and nano-objects might soon find a practical implementation, especially in the field of medical research and diagnostics, giving rise to what we currently know as "nanomedicine".

1.5 Molecular Nanotechnology and the Start of Nanomedicine

As stated before, nanotechnology has its roots in the physical science; however, it has always-important correlations with biology, both at the rhetorical and practical levels. At the beginning of the 20th century, Paul Ehrlich, the founder of chemotherapy, was the first to propose the concept of the "magic bullet" for the application of chemotherapy to treat cancer patients (Ehrlich 1908). By the end of the 1960s, Peter Paul Speiser developed the first nanoparticles which can be used for targeted drug therapy, and in the 1970s Georges Jean Franz Köhler and César Milstein succeeded in producing monoclonal antibodies (Milstein 1996). Since then there has been intensive research into the possible syntheses and uses of various nano-carrier systems and the physicochemical functionalization of their surface structure (Qiao et al. 2018). Cancer treatment based on targeted transport of active substances can moreover take advantage of the EPR (enhanced permeability and retention) effect described in 1986 by Yasuhiro Matsumura and Hiroshi Maeda. They defined a pathway to target nanoparticles and to deposit them into the tumours to a greater degree than in healthy tissue (Matsumura and Maeda 1986, Duncan and Seymour 2007).

After Feynman ventured into a new field of research and stimulated the interest of many scientists, two directions of thought emerged to describe the numerous possibilities to produce nanostructures: on the one hand, the top-down approach agreeing to Feynman's interpretations of reducing in size the already existing machines and instruments, and on the other hand the bottoms-up approach that turns around to the construction of nanostructures atom by atom using physical and chemical methods that control and tailor the self-organizing forces of atoms and molecules, as happens in nature. This last "molecular engineering" approach became popular in 1986 after the publication of the book titled, *"Engines of Creation: The Coming Era of Nanotechnology"* by K. Eric Drexler of Massachusetts Institute of Technology (MIT). In his first and controversially discussed book, E. Drexler fused the concepts from Feynman and Taniguchi, and proposed the idea of a nanoscale "assembler" which would be able to build a copy of itself and of other items of arbitrary complexity. The possible uses of such "nanobots" or "assemblers" in medicine are described by K. Eric Drexler, Chris Peterson and Gayle Pergamit in their book *"Unbounding the Future, The Nanotechnology Revolution"* (Drexler et al. 1991) in which the term "nanomedicine" was apparently used for the first time. The term became finally established and has been used in technical literature after the book "Nanomedicine, Volume I: Basic Capabilities" by R. A Freitas was published in 1999 (Freitas 1999).

At the beginning of the 21st century, nanotechnology was already installed in our world and several government programmes were launched in support of those who advocated for their application in new discoveries.

1.6 Nanoscience of the 21st Century

1.6.1 The Global Perspective of Nanotechnology

Almost all industrialized nations created nanotechnology initiatives, leading to a worldwide proliferation of nanotechnology activities. In 2000, the National Nanotechnology Initiative (NNI) (NNI 2014) to coordinate Federal R&D efforts and promote U.S. competitiveness in nanotechnology was launched by President Clinton. The U.S. Congress funded the NNI for the first time in Fiscal Year 2001 (FY2001) and Nanoscale Science, Engineering, and Technology (NSET) Subcommittee of the National Science and Technology Council (NSTC) was selected to coordinate the planning, budgeting, programme implementation, and examination of the NNI. The Department of Defence has identified three nanoscience challenges for focused science and technology (S&T) investments: nanoelectronics, magnetics and photonics nanomaterials and nanobiodevices (Murday et al. 2005). In 2003, the U.S. Congress sanctioned the 21st Century Nanotechnology Research and Development Act (P.L. 108–153). The act provided a statutory foundation for the NNI, established programmes, assigned agency responsibilities, authorized funding levels, and promoted research to address key issues. In Asia, there are growing programmes in Japan, as well as in China, South Korea, Taiwan and Singapore (Roco 2001). Nanotechnology Research Center, RIKEN; the Institute of Nanomaterials, Tohoku University, Sendai; the Nanomaterials Laboratory, National Institute of Materials Science, Tsukuba and the Silicon Nanotechnology Center, Tsukuba established in 2001 are examples of centres supported primarily by government sources in Japan (Roco 2001). China's National High Technology Plan has included a series of projects for nanomaterial applications. From the period between 1990 and 2002, nearly 1000 of such projects were implemented. In addition, during this period, the National Natural Science Foundation of China (NSFC) approved nearly 1000 grants for small-scale projects in related areas. In 2000, the National Steering Committee for Nanoscience and Nanotechnology was created to oversee national policy and planning in these fields. The committee was set up, among other organizations, by Ministry of Science and Technology (MOST), the State Development and Planning Commission, the Ministry of Education, Chinese Academy of Sciences (CAS), the Chinese Academy of Engineering, and the NSFC. A number of interdisciplinary research centres have been established to promote and facilitate collaborations between various institutions in a particular region by sharing of resources. The National Center for Nanoscience and Technology (NCNST) in Beijing, initiated 2003 (Bai 2005) and the National Center for Nanoengineering in Shanghai, construction started in mid-2005 (Ties et al. 2005), are important additions to the list.

Later, in 2004, The European Commission proposed the institutionalization of European nanoscience and nanotechnology R&D efforts within an integrated and responsible strategy, "Towards a European

Strategy for Nanotechnology," COM (2004). This spurred European action plans and ongoing funding for nanotechnology R&D. In the same year, the Britain's Royal Society and the Royal Academy of Engineering published "Nanoscience and Nanotechnologies: Opportunities and Uncertainties" (Society 2004) encouraging the requirement to adopt novel strategies for health, environmental, social, ethical, and regulatory issues linked with nanotechnology. The first official NNI Strategy for Nanotechnology-Related Environmental, Health, and Safety (EHS) Research was published in 2008 and updated in 2011. In 2012, the NNI launched two more Nanotechnology Signature Initiatives (NSIs): Nanosensors and the Nanotechnology Knowledge Infrastructure (NKI), bringing the total to five NSIs. The next year, 2013, the NNI started the next round of Strategic Planning, starting with the Stakeholder Workshop. It was updated in 2014 with the Progress Review on the Coordinated Implementation of the NNI 2011 Environmental, Health, and Safety Research Strategy. On 10 December 2013, the European Commission adopted the "Horizon 2020, H2020, Work Programme". Its slogan was *"Leadership in enabling and industrial technologies. Nanotechnologies, advanced materials, biotechnology and advanced manufacturing and processing."* Horizon 2020 is the biggest EU Research and Innovation programme ever with nearly €80 billion of funding available over 7 years (2014 to 2020). It promises more breakthroughs, discoveries and world-firsts by taking great ideas from the lab to the market (EC 2013).

Several Latin American countries have set the development of nanotechnology policies to increase their global competitiveness. In Brazil, it effectively started in 2001 with an important effort of the Conselho Nacional de Desenvolvimento Científico e Tecnológico (CNPq) to develop nanotechnology in terms of human resources and funding (Kay and Shapira 2009). Towards 2004, there were eleven nanotechnology research groups at three universities and two research institutes, working primarily in new materials development in Mexico. Argentina, the third-ranked nanotechnology producer in Latin America, has also implemented nanotechnology policy measures including the creation of Fundación Argentina de Nanotecnología (FAN) in 2005 (D380/2005 2005). Several networks for nanoscience research have been established and are sponsored by the Agencia Nacional de Promoción Científica y Tecnológica (ANPCyT) and Consejo Nacional de Ciencia y Tecnología (CONICET). Between the years 1999 and 2006, broader programmes to improve the Chilean science and technology (S&T) systems were funded to support selected nanotechnology initiatives at different universities in the areas of physics, biology, and materials science. In 2002, the Cuban Academy of Sciences and the Ministry of Science, Technology, and Environment (CITMA) pointed out the need for creating national capabilities in nanotechnology and establishing goals to work in related areas like mathematics, physics, chemistry, IT, and new materials (OCCyT 2002). In 2004, funded by the Costa Rica-United States of America Foundation for Cooperation, the Costa Rican Ministry of Science and Technology, and the Pro-National Center for High Technology Pro-CeNAT

Foundation, the National Laboratory for Nanotechnology, Microsensors and Advanced Materials was opened, which is the first centre of this type in Central America (Kay and Shapira 2009). Although national agencies across Latin America indicate increased interest in promoting nanotechnology, public investment in nanotechnology R&D remains relatively modest in comparison with U.S and Europe. It is comparable to the estimated levels for China and greater than India. Research and development at the nanoscale require a large degree of integration, from convergence of research disciplines in new fields of enquiry to new linkages between start-ups, regional actors and research facilities; nowadays there are many nanotechnology networks in the world to promote collaboration (Balanta 2016).

1.6.2 The Global Perspective of Nanomedicine

In the early 2000s, the enthusiasm of scientists and the challenges that could be addressed though nanotechnology application have merged and promoted the governmental science and funding organisations to undertake a strategic analysis of the status of nanomedicine. Their main purposes were to assess the potential opportunities for a better healthcare, to analyse the risk-benefit of these new technologies, and to determine priorities for future funding. Thereby, the European Science Foundation launched its *"Forward look"* on nanomedicine in 2003, the first foresight study focused on medical applications of nanoscience and nanotechnology. The *"Forward Look"* had a participation of over 100 leading European experts. It allowed to determine the status of nanomedicine and to raise debates on strategic policy issues. In that meeting, the increasing excitement in the application of nanotechnology to medicine considering the significant advances in the diagnosis and treatment of diseases that it would provide was demonstrated. Recommendations of the *"Forward Look"* were published on 23rd February 2005 (ESF 2005). In 2003, the UK Government authorized the Royal Society, the UK National Academy of Science, the Royal Academy of Engineering, and the UK National Academy of Engineering, to carry out an independent study of expected advances of nanotechnology and of the necessity of advanced ethical, health, safety or social issues not actually enclosed in the current regulation. The final report was published in July 2004 with 21 suggestions to apply for a sure, safe and responsible development of nanotechnology (Society 2004). In the same year, the Commission of the European Communities presented the *"European Strategy for Nanotechnology"* (EC 2004) and the High Level Group European Technology Platform Nanomedicine was launched under their initiative. In September 2005, the Vision Paper and Basis for a Strategic Research Agenda for Nanomedicine was announced as a first step towards setting up a European Technology Platform on Nanomedicine, aiming to consolidate the Europe position and to expand the quality of life and health care of European citizens.

Beyond the Atlantic Ocean, The National Cancer Institute (NCI), as part of USA National Institutes of Health (USA-NIH), launched the Cancer

Nanotechnology Plan, a strategic initiative to transform clinical oncology and basic research through the directed application of nanotechnology. The NCI Alliance for Nanotechnology in Cancer is engaged in the efforts to harness the power of nanotechnology that can radically change the way to diagnose, treat and prevent cancer. This alliance is a comprehensive, systematized initiative encompassing the public and private sectors, designed to accelerate the application of nanotechnology to cancer (Ferrari et al. 2005). In the following years, 2005 and 2006, USA-NIH established a national network of Height Nanomedicine Development Centres, which served as the intellectual and technological attraction of the NIH Nanomedicine Roadmap Initiative. The objective of the Common Fund's Nanomedicine programme, as part of the National Health Institutes' Nanomedicine Roadmap is to determine how cellular machines operate at the nanoscale level and then use these design principles to develop and engineer new technologies and devices for repairing tissue or preventing and curing disease (NIH 2006).

Since 2000, in Japan, besides the advanced manufacturing technologies for Japanese companies, the investment in nanomedicine (Rediguieri 2009), such as drugs delivery systems (DDS), became recognized. In 2001, a great number of corporations in Japan created a nanotechnology fund dedicated to business and R&D in this area. Mitsubishi Corporation and Mitsui & Co, were the two largest ones, who launched a fund of 150 million dollars for the development of business in nanomedicine for five years. Nanomedicine has transformed itself into a priority within the Japanese policy of Science and Technology. The government started a coordinated effort in 2003, following a study conducted by the Ministry of Economy, Trade and Industry (METI) about nanobiotechnology trends and their application in industry. The METI immediately launched different nanobiotechnological projects (2003–2006) conducted by Japanese companies such as Olympus, Toray, NEC, Shimazu and others under the *"Focused 21 Program"*.

In 2005, The Nanoscience African Network (NANOAFNET) was established. With about 27 participating countries within the continent, NANOAFNET was created to enhance the global visibility and contribution of Africa in nanoscience while developing cost-effective nanotechnologies to address urgent continental societal needs in the water, energy and health sectors (Chang et al. 2015). In 2007, the European Foundation for Clinical Nanomedicine was established in Basel (Switzerland). This foundation is a non-profit institution aimed at advancing medicine for the benefit of individuals and society through the application of nanoscience (Boisseau and Loubaton 2011). The central focus of the Foundation was the recognition of the large impact of nanoscience on medical future and the accurate advance of its applications. The Foundation completed its objectives by providing support for clinical research and encouraging the exchange of a continuous flow of information, among doctors, researchers, the general public and other interested parties, guiding the prevention, diagnosis and therapy in nanomedicine, as well as exploration of its implications. In 2013, World Health Organisation (WHO) released a second report on the Priority

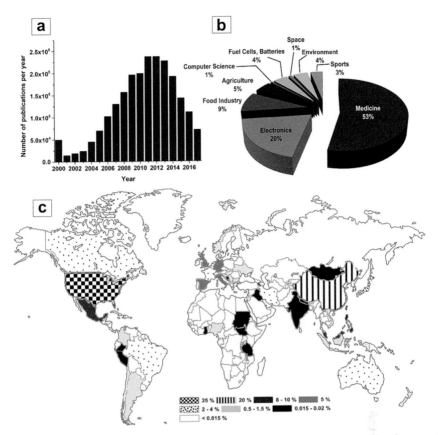

Figure 1.1. (a) Temporal evolution of the scientific interest in nanotechnology and their applications through the years 2000–2018, including all documents type and patents. (b) Different fields of nanotechnology application. (c) Main geographic areas of publications in nanomedicine. Database: Scopus; Google Scholar. Search description: nanotechnology; nanomaterials; nanoparticles; nanomedicine.

Medicines for Europe and the World with the aim to bridge the gap between public health needs and the research and development priorities. As part of the 7th framework programme, which extended over the years 2007–2013, the European Commission has financed a wide range of projects to impulse research activities generation knowledge and prototypes. Their objectives are to: (i) make active principles by challenging physical-chemical properties available to the patient (e.g. low solubility); (ii) optimise drug delivery by active targeting and controlled release into the diseased tissue; (iii) use the physical properties of the nanomaterial for the development of therapeutic and diagnostic tools; (iv) engineer tissues and (v) combine diagnostic and therapeutic tools allowing a direct monitoring of the therapeutic progress (Bremer-Hoffmann et al. 2015). The complete translational process from product design to the market is long and requires a number of approval steps with increasing data requirements. Translation of innovation and time-to-market reduction are central challenges for the currently ongoing European framework programme: *"Horizon 2020"*, H2020, 2014–2020. It places a strong emphasis on the translational aspect covering further developments in a preclinical setting before demonstrating the relevance of the product in the clinical environment (Horizon 2014).

1.6.3 A Multidisciplinary Research

Now, nanotechnology has emerged as an interdisciplinary field that combines biology, chemistry and engineering. The focus of investigation across the world continues to be nanoscale properties and the synthesis of novel materials, followed by their application to create useful devices and processes having social and economic benefits.

Figure 1a shows the temporal evolution of the scientific interest in nanotechnology and their applications through the years 2000–2018; nanomedicine corresponds to the 53.4% of the researchers' reports (Figure 1b). According to the Scopus database, United States of America is at the vanguard in nanomedicine; out of a total of 146,507 documents published and 3,478 patents during the years 2010–2018 in nanomedicine, this country produced 25% of the first and 86% of the second (Figure 1c). Throughout the book, we will provide specific detail of the areas where the multidisciplinary contributions of nanotechnology research were applied to medicine. Here are few examples that, in our opinion, stand out. In 2003, Naomi Halas, Jennifer West, Rebekah Drezek, and Renata Pasqualini at Rice University developed gold nanoshells with tuneable size to modulate the absorption of near infrared light. Their discoveries pushed diagnosis and treatment of breast cancer, leaving aside invasive biopsies, surgery, or systemically destructive radiation or chemotherapy (Loo et al. 2004). In 2005, Erik Winfree, Nick Papadakis and Paul Rothemund from the California Institute of Technology advanced in the DNA-based computation theories and proposed their "algorithmic self-assembly", embedding computational sciences with the process of DNA nanocrystal growth (Rothemund et al. 2004). During the years 2009 and 2010, Nadrian Seeman and colleagues at New York University created several nanomaterials based on DNA assembly (Seeman 2010). In 2006, James Tour and colleagues at Rice University built a nanoscale car made of oligo(phenylene ethynylene) with alkynyl axles and four spherical C60 fullerene (buckyball) wheels that was moving on a gold surface as a result of the buckyball wheels turning, as in a conventional car, in response to increased temperature (Shirai et al. 2006). Additionally, in 2007, Angela Belcher and colleagues at MIT built a lithium-ion battery using a non-harmful virus through a low-cost and environmentally benign process. The batteries were demonstrated to have the same energy capacity and power performance as modern rechargeable batteries used to impulse plug-in hybrid cars (Nam et al. 2006). In 2010, IBM presented a powerful patterning methodology for generating nanoscale patterns and structures as small as 15 nanometres using a silicon tip of few nanometres at its apex. This methodology greatly reduced cost and complexity to create nanoscale 3D relief maps, opening up new sceneries for electronics', optoelectronics', and medicine' fields. Writing in Nature Nanotechnology, Robert Langer and his colleagues examined different nanodevices applied to activate particular processes for tissue growth, and offered their view on the principal challenges and future prospects (Dvir et al. 2010). In 2014, Stanford researchers developed the first carbon nanotube computer and in 2015, Gang Zheng and colleagues at the University of Toronto published in Nature Nanotechnology a different approach to the classical drug delivery method by using ultrasound to implode microbubbles into nanoparticles (Huynh et al. 2015). Emiliano Cortés, Prineha Narang, Sebastian Schlücker and co-workers presented, in 2017, an experimental and theoretical investigation related to hot-electron-driven reactivity mapping with nanometre resolution. Results would improve efficiency in hot-carrier extraction science and nanoscale regio-selective surface chemistry (Cortés et al. 2017). Numerous inorganic nanoparticles have been explored for drug delivery systems or cancer therapy. However, the clinical application of inorganic formulations has often been hindered by their toxicity and failure to biodegrade. Huo et al. (2017) fabricated a sequential nanocatalyst based on natural glucose oxidase (GOD, enzyme catalyst) and ultra-small Fe_3O_4 nanoparticles integrated into large pore-sized and biodegradable dendritic silica nanoparticles. GOD could effectively deplete glucose into the tumour cells, and produce, in response to mild acidic tumour microenvironment, a considerable amount of highly toxic hydroxyl radicals through sequential catalytic reactions causing apoptosis and death of tumour cells. The next year, Markus B. Raschke et al. provided a facile approach to harness exciton properties in low-dimensional semiconductors offering new strategies for quantum optoelectronics (Park et al. 2018).

1.6.4 Nanotechnology Education

Nanotechnology is the promising answer to many types of problems in all fields of work. Finally, yet importantly, in the 21st century, there is a growing recognition of the education of future scientists and engineers into this emerging field, as well as to address safety and health aspects of nanomaterials (Ernst 2009, Cingolani 2013). The first programme involving nanotechnology was offered by the University of Toronto's Engineering Science programme, where nanotechnology could be taken as an option (Wikipedia 2010). The National Research Council of Canada, the University of Alberta, and the Government of Alberta established the National Institute of Nanotechnology (NINT) in 2001, located at the University of Alberta main campus, in Edmonton, Alberta, Canada. In June 2006, the institute was moved to its present 200,000 square metre facility. It was designed to be one of the world's largest buildings for nanotechnological research. Currently, there are a maximum of two or three other institutes worldwide matching the new building in scale and capacity. The National Science Foundation, in 2002, established the Network for Computational Nanotechnology (NCN), an alliance of universities' partners and the National Nanotechnology Initiative. Likewise, NCN created an exclusive cyberinfrastructure to support its website, nanoHUB.org. Through this platform, researchers, educators, and professionals collaborate, share resources, and solve real nanotechnology problems. The objective was to bring computational tools online, making the tools easy to use, and supporting the tools with educational materials. In 2007, nanoHUB.org served more than 56,000 users from 172 countries (Klimeck et al. 2008).

In 2004, the first college-level education programme in nanotechnology was launched in the United States in the College of Nanoscale Science and Engineering in SUNY Albany. In August 2006, the bachelor's programme in Nanotechnology and Molecular Engineering at the Americas Puebla University, Mexico was started. This undergraduate programme was the first one in Mexico and Latin America (Wikipedia 2010). In 2014, the programme, Remotely Accessible Instruments in Nanotechnology (RAIN), was inserted. RAIN was a network of nineteen universities and community colleges of U.S. seeking to bring free access to advanced technologies in educational settings ranging from K-12, undergraduate science courses and technical degree programmes (Ashcroft et al. 2017).

Actually, nanoscience and nanotechnology programmes are being increasingly offered by more and more institutions all around the world, developing educational programmes across the full spectrum of educational levels from K-12 to postgraduate studies. These cross-disciplinary programmes provide students with a robust scientific understanding in the field of nanotechnology, combined with a "hands-on" practical and translational focus (Jackman et al. 2016). Recognizing education as an important factor for growing the fields of nanoscience and nanotechnology, such programmes solidify and expand the role of nanotechnology and nanomedicine in the global economy.

1.7 The Actual Verisimilitude of Nano-sized Technologies

The genesis of nanotechnology can be traced from Richard P. Feynman, who showed us a promise of revolutionary advances beyond communications, technology and medicine, to this day, where scientists and engineers are effectively taking control of individual atoms and molecules to add innovative functionalities to any product or device. Numerous nanoscale technological and scientific developments were the result of biological experiments, and then nanotechnology has been rapidly applied to the investigation of biological phenomena and the demands of the medical area. World investigation on nanotechnology generated new tools to be applied in medical practice that gave rise to a new discipline from which the understanding and instruction of health and disease can be enhanced: the "nanomedicine". Later, "nanomedicine" became a serious mission, sponsored by major initiatives from governments, industries, and academies.

As the translation of Feynman's and Drexler's visions of nanoscale robots, which would travel through the body, search and destroy a disease focus, and subsequently repair organs and cells, were still in a distant future, nanomedicine concentrates on the possibilities of controlling and manipulating cell processes. In essence, through rearranging the atoms, patients can be treated with individually and personalized tailor-made medicines. Organs are likely to be regenerated or re-grown, and by using sophisticated diagnostic tools, cancer and other diseases can be prevented and treated early.

Nanomedicine has induced a health care revolution that might dwarf all other developments in the history of medical technology. However, whereas great ideas advance constantly, there are concerns around safety, environment and religion. For example, nanotechnology holds the possibility of completely changing a person's physical appearance or enhancing their physical and mental abilities. Scientists involved must engage the public to demonstrate, in enough detail, how their precise endeavours will deliver concrete economic and societal benefits to allay apprehensions and build co-operation. Successful translation requires revising the meaning of nanotechnology and nanomedicine, understanding their limitations, identifying the micro-conceptions and facing inconvenient truths. The analysis of such topics will be considered in the following chapters of this book.

References

Ashcroft, J. M., A. O., Cakmak, J. Blatti, E. Bautista and V. Wolf. 2017. It's RAINing: Remotely accessible instruments in nanotechnology to promote student success. Current Issues in Emerging eLearning 4(1): 4.

ASTM. 2012. ASTM E2456-06(2012). Standard Terminology Relating to Nanotechnology.

ASTM. 2013. ASTM E2524-08(2013). Standard Test Method for Analysis of Hemolytic Properties of Nanoparticles.

ASTM. 2013. ASTM E2909-13. Standard Guide for Investigation/Study/Assay Tab-Delimited Format for Nanotechnologies (ISA-TAB-Nano): Standard File Format for the Submission and Exchange of Data on Nanomaterials and Characterizations. Journal of Microelectromechanical Systems.

ASTM. 2015. ASTM E2490-09(2015). Standard Guide for Measurement of Particle Size Distribution of Nanomaterials in Suspension by Photon Correlation Spectroscopy (PCS).

ASTM. 2015. ASTM E2996-15. Standard Guide for Workforce Education in Nanotechnology Health and Safety.

Bai, C. 2005. Ascent of nanoscience in China. Science 309(5731): 61-63.

Balanta, D. 2016. Sitio web Nanoandes. http://www.nanoandes.org/redes.html.

Binnig, G., H. Rohrer, Ch. Gerber and E. Weibel. 1982. Surface studies by scanning tunneling microscopy. Phys. Rev. Lett. 49(1): 57.

Binnig, G., C. F. Quate and C. Gerber. 1986. Atomic force microscope. Phys. Rev. Lett. 56(9): 930-933.

Boholm, M. 2016. The use and meaning of nano in American English: Towards a systematic description. Ampersand 3: 163-173.

Boisseau, P. and B. Loubaton. 2011. Nanomedicine, nanotechnology in medicine. C. R. Phys. 12(7): 620-636.

Boverhof, D. R., Bramante, C. M., Butala, J. H., Clancy, S. F., Lafranconi, M., West, J., et al. 2015. Comparative assessment of nanomaterial definitions and safety evaluation considerations. Regul. Toxicol. Pharmacol. 73(1): 137-150.

Breggin, L., Falkner, R., Jaspers, N., Pendergrass, J. and R. Porter. 2009. Securing the promise of nanotechnologies: Towards transatlantic regulatory cooperation. Report. CHATAM HOUSE.

Bremer-Hoffmann, S., V. Amenta and F. Rossi. 2015. Nanomedicines in the European translational process. Eur. J. Nanomed. 7(3): 191-202.

Burdun, G. 1960. International system of units. Meas. Tech. 3(11): 913-919.

Cingolani, R. 2013. The road ahead. Nat. Nanotechnol. 8: 792.

Cortés, E., W. Xie, J. Cambiasso, A. S. Jermyn, R. Sundararaman, P. Narang, et al. 2017. Plasmonic hot electron transport drives nano-localized chemistry. Nat. Commun. 8: 14880.

Chang, E. H., J. B. Harford, M. A. W. Eaton, P. M. Boisseau, A. Dube, R. Hayashi, et al. 2015. Nanomedicine: Past, present and future – a global perspective. Biochem. Biophys. Res. Commun. 468(3): 511-517.

D380/2005. 2005. Aplicación y desarrollo de micro y nanotecnologías. http://servicios.infoleg.gob.ar/infolegInternet/anexos/105000-109999/105874/norma.htm.

DERFA, U. 2006. UK voluntary reporting scheme for engineered nanoscale materials. Chemicals and Nanotechnologies Division, Defra Google Scholar.

Drexler, K. E. 1986. Engines of creation, Anchor.

Drexler, E., C. Peterson, G. Pergamit and S. Brand. 1991. Unbounding the Future. William Morrow, New York.

Duncan, R. and L. Seymour. 2007. Hiroshi Maeda—defining the pathway to targeted cancer therapy. J. Drug Targeting 15(7-8): 456-456.

Dvir, T., B. P. Timko, D. S. Kohane and R. Langer. 2011. Nanotechnological strategies for engineering complex tissues. Nat. Nanotechnol. 6: 13-22.

EC-SCCP. 2007. European Commision. Scientific Committe on Consumer Products to the European Commission. Opinion on safety of nanomaterials in consumer products.

EC-SCENIHR. 2007. European Commission. Scientific Committee on Emerging Newly Identified Health Risk. Opinion on the scientific aspects of the existing and proposed definitions relating to products of nanoscience and nanotechnologies, European Commission Health Consumer Protection Directorate-General Brussels.

EC. 2004. Communication from the Commission – Towards a European strategy for nanotechnology. Brussels, Commission of the European Communities.

EC. 2013. Entrepreneurship 2020 Action Plan: Reigniting the entrepreneurial spirit in Europe, European Commission Brussels.

Ehrlich, P. 1908. Experimental researches on specific therapy: Chemotherapeutic studies on trypanosomes, Third Harben Lecture.

Ekimov, A. I. and A. A. Onushchenko. 1981. Quantum size effect in three-dimensional microscopic semiconductor crystals. Jetp. Lett. 34(6): 345-349.

Enderby, J. and A. Dowling. 2004. Nanoscience and nanotechnologies: Opportunities and Uncertainties. The Royal Society & The Royal Academy of Engineering Report, London.

Ernst, J. V. 2009. Nanotechnology education: Contemporary content and approaches. JOTS 35(1): 3-8.

ESF. 2005. European Science Foundation. Nanomedicine: An ESF–European Medical Research Councils (EMRC) Forward Look Report. Strasbourg cedex, France.

ETC-Group. 2005. A tiny primer on nano-scale technologies and 'the little bang theory'. Winnipeg: ETC Group.

Ferrari, M., A. D. Barker and G. J. Downing. 2005. A cancer nanotechnology strategy. NanoBiotechnology 1(2): 129-131.

Feynman, R. P. 1992. There's plenty of room at the bottom [data storage]. J. Microelectromech. Syst. 1(1): 60-66.

Feynman, R. P. 1993. Infinitesimal machinery. J. Microelectromech. Syst. 2(1): 4-14.

Freestone, I., N. Meeks, M. Sax and C. Higgitt. 2007. The Lycurgus cup—A roman nanotechnology. Gold Bull. 40(4): 270-277.

Freitas, R. A. 1999. Nanomedicine, volume I: Basic Capabilities, Landes Bioscience Georgetown, TX.

Garboczi, E. 2009. Concrete nanoscience and nanotechnology: Definitions and applications. Nanotechnology in Construction 3, 81-88. Springer.

Grieneisen, M. L. and M. Zhang. 2011. Nanoscience and nanotechnology: Evolving definitions and growing footprint on the scientific landscape. Small 7(20): 2836-2839.

Harnad, S. 2013. Harnad Evidence to House of Lords Science and Technology Select Committee on Open Access. House of Lords Science and Technology Committee on Open Access, pp. 119-123.

Hett, A. 2004. Nanotechnology: Small Matter, Many Unknowns. Swiss Reinsurance Co., Zurich, Switzerland.

Höck, J., T. Epprecht, H. Hofmann, K. Höhner, H. Krug, C. Lorenz, et al. 2008. Guidelines on the precautionary matrix for synthetic nanomaterials. Federal Office for Public Health and Federal Office for the Environment, Berne.

Horizon. 2014. The EU Framework Programme for Research and Innovation.

Huo, M., L. Wang, Y. Chen and J. Shi. 2017. Tumor-selective catalytic nanomedicine by nanocatalyst delivery. Nat. Commun. 8(1): 357.

Huynh, E., B. Y. Leung, B. L. Helfield, M. Shakiba, J. A. Gandier, C. S. Jin, et al. 2015. In situ conversion of porphyrin microbubbles to nanoparticles for multimodality imaging. Nat. Nanotechnol. 10(4): 325.

Iijima, S. and T. Ichihashi. 1993. Single-shell carbon nanotubes of 1-nm diameter. Nature 363(6430): 603.

ISO. 2010. TS 80004–1: 2010. Nanotechnologies—Vocabulary—Part 1: Core terms.

ISO. 2012. ISO/TR 13014: 2012. Nanotechnologies: Guidance on Physico-chemical Characterization of Engineered Nanoscale Materials for Toxicologic Assessment, ISO.

ISO. 2012. ISO/TS 12901-1: 2012. Nanotechnologies-occupational Risk Management Applied to Engineered Nanomaterials. Part 1: Principles and Approaches ISO Geneva, Switzerland.

ISO. 2015. ISO/TS 80004-2: 2015. Nanotechnologies—Vocabulary—Part 2: Nano-objects.

Jackman, J. A., D. J. Cho, J. Lee, J. M. Chen, F. Besenbacher, D. A. Bonnell, et al. 2016. Nanotechnology education for the global world: Training the leaders of tomorrow. ACS Nano 10(6): 5595-5599.

Kay, L. and P. Shapira. 2009. Developing nanotechnology in Latin America. J. Nanopart. Res. 11(2): 259-278.

Keiper, A. 2003. The nanotechnology revolution. The New Atlantis 2: 17-34.

Khan, F. A. 2011. Biotechnology Fundamentals. CRC Press, Boca Raton, FL 33487-2742, USA.

Klaessig, F., M. Marrapese and S. Abe. 2011. Current perspectives in nanotechnology terminology and nomenclature. Nanotechnology Standards, 21-52. Springer.

Klimeck, G., M. McLennan, S. P. Brophy, G. B. Adams III and M. S. Lundstrom. 2008. nanohub.org: Advancing education and research in nanotechnology. Comput. Sci. Eng. 10(5): 17-23.

Kresge, C., M. E. Leonowicz, W. J. Roth, J. C. Vartuli and J. S. Beck. 1992. Ordered mesoporous molecular sieves synthesized by a liquid-crystal template mechanism. Nature 359(6397): 710-712.

Li, J. and J.-J. Zhu. 2013. Quantum dots for fluorescent biosensing and bio-imaging applications. Analyst 138(9): 2506-2515.

Liddell, H. G. and R. Scott. 1897. A Greek-English Lexicon. New York: American Book Company.

Linsinger, T., G. Roebben, D. Gilliland, L. Calzolai, F. Rossi, N. Gibson, et al. 2012. Requirements on measurements the European Commission definition of the term "nanomaterial".

Loo, C., A. Lin, L. Hirsch, M. H. Lee, J. Barton, N. Halas, et al. 2004. Nanoshell-enabled photonics-based imaging and therapy of cancer. Technol. Cancer Res. Treat. 3(1): 33-40.

Matsumura, Y. and H. Maeda. 1986. A new concept for macromolecular therapeutics in cancer chemotherapy: Mechanism of tumoritropic accumulation of proteins and the antitumor agent smancs. Cancer Res. 46(12 Part 1): 6387-6392.

Milstein, C. 1996. Georges Jean Franz Kohler: A personal tribute. Immunol. Today 3(17): 103.

Miller, G. and R. Senjen. 2008. Out of the laboratory and onto our plates: Nanotechnology in food and agriculture, Friends of the Earth (FoE).

Moore, G. E. 1965. Cramming more components onto integrated circuits. Electronics 38: 8.

Müller, E. W. 1951. Das feldionenmikroskop. Zeitschrift für Physik 131(1): 136-142.

Murday, J. S., B. D. Guenther, C. G. Lau, C. R. K. Marrian, J. C. Pazik and G. S. Pomrenke. 2005. Overview of the Nanoscale Science and Technology Program in the Department of Defense. Defense Applications of Nanomaterials, American Chemical Society. 891: 2-14.

Nam, K. T., D. W. Kim, P. J. Yoo, C. Y. Chiang, N. Meethong, P. T. Hammond, et al. 2006. Virus-enabled synthesis and assembly of nanowires for lithium ion battery electrodes. Science 312(5775): 885-888.

Niaz, M. 2009. Foundations of Modern Atomic Theory: Thomson, Rutherford, and Bohr. Critical Appraisal of Physical Science as a Human Enterprise: Dynamics of Scientific Progress, 75-95.

NIH. 2006. National Institutes of Health. National Institute of Health roadmap for medical research: Nanomedicine. Bethesda, Maryland: HIH.

NNI. 2014. National Nanotechnology Initiative. Official website of the United States National Nanotechnology Initiative.

O'brien, S. C., J. R. Heath, H. W. Kroto, R. F. Curl and R.E. Smalley. 1986. A reply to "magic numbers in Cn+ and Cn– abundance distribution" based on experimental observations. Chem. Phys. Lett. 132(1): 99-102.

OCCyT. 2002. Observatorio Cubano de Ciencia y Tecnologia. Nanotecnología, Proyecto. Elementos Iniciales para el Análisis sobre la Nanotecnología en Cuba, OCCyT.

Park, K.-D., T. Jiang, G. Clark, X. Xu and M.B. Raschke. 2018. Radiative control of dark excitons at room temperature by nano-optical antenna-tip Purcell effect. Nat. Nanotechnol. 13(1): 59-64.

Peacocke, T. A. H. 2013. Atomic Theory and Structure of the Atom: Atomic and Nuclear Chemistry. Elsevier. Pergamon Press Ltd., Headigton Hill Hall, Oxford, U.K.

Piner, R. D., J. Zhu, F. Xu, S. Hong and C.A. Mirkin. 1999. "Dip-pen" nanolithography. Science 283(5402): 661-663.

Pullman, B. 2001. The Atom in the History of Human Thought. Oxford University Press, USA.

Qiao, Y., J. Wan, L. Zhou, W. Ma, Y. Yang, W. Luo, et al. 2018. Stimuli-responsive nanotherapeutics for precision drug delivery and cancer therapy. Wiley Interdiscip. Rev.: Nanomed. Nanobiotechnol. 11(1): e1527.

Rawson, P. 2011. Ceramics. University of Pennsylvania Press, 3905 Spruce St, Philadelphia, PA 19104, USA.

Rediguieri, C. F. 2009. Study on the development of nanotechnology in advanced countries and in Brazil. Braz. J. Pharm. Sci. 45(2): 189-200.

Reibold, M., P. Paufler, A. A. Levin, W. Kochmann, N. Pätzke and D. C. Meyer. 2006. Materials: Carbon nanotubes in an ancient Damascus sabre. Nature 444(7117): 286.

Robson, A. 2010. Nanotechnologies and Food: 1st Report of Session 2009-10, Vol. 2: Evidence. House of Lords – Science and Technology Committee.

Roco, M. C. 2001. International strategy for nanotechnology research. J. Nanopart. Res. 3(5-6): 353-360.

Rosetti, R. and L. Brus. 1982. Electron-hole recombination emission as a probe of surface chemistry in aqueous CdS colloids. J. Phys. Chem. 86: 4470-4472.

Rothemund, P. W., N. Papadakis and E. Winfree. 2004. Algorithmic self-assembly of DNA Sierpinski triangles. PLoS Biology 2(12): e424.

Sanderson, K. 2006. Sharpest cut from nanotube sword. Nature News 15.

Sattler, K. D. 2010. Handbook of Nanophysics: Nanotubes and Nanowires. CRC Press, Boca Raton, FL 33487-2742, USA.

Seeman, N. C. 2005. From genes to machines: DNA nanomechanical devices. Trends Biochem. Sci. 30(3): 119-125.

Seeman, N. C. 2010. Nanomaterials based on DNA. Annu. Rev. Biochem. 79: 65-87.

Shirai, Y., A. J. Osgood, Y. Zhao, Y. Yao, L. Saudan, H. Yang, et al. 2006. Surface-rolling molecules. J. Am. Chem. Soc. 128(14): 4854-4864.

Royal Society. 2004. Nanoscience and Nanotechnologies: Opportunities and Uncertainties: Summary and recommendations. Royal Society.

Stevenson, A. 2010. Oxford Dictionary of English. Oxford University Press, USA.

Taniguchi, N. 1974. On the basic concept of nano-technology. Proc. Intl. Conf. Prod., British Society of Precision Engineering.

Thompson, D. 2007. Michael Faraday's recognition of ruby gold: The birth of modern nanotechnology. Gold Bull. 40(4): 267-269.

Tibbals, H. F. 2010. Medical Nanotechnology and Nanomedicine. CRC Press, Boca Raton, FL 33487-2742, USA.

Ties, B. D., J. Zemin and G. B. E. Don. 2005. A Chinese nano-society? Nature Materials 4(5): 355.

Tomellini, R., U. Faure and O. Panzer. 2005. Nanotechnology for health. Vision paper and basis for a strategic research agenda for NanoMedicine. European Technology Platform Nanomedicine, 1-35.

Vallet-Regi, M., A. Ramila, R. P. Del Real and J. Pérez-Pariente. 2001. A new property of MCM-41: Drug delivery system. Chem. Mater. 13(2): 308-311.

Varadan, V. K., A. S. Pillai, D. Mukherji, M. Dwivedi and L. Chen. 2010. Nanoscience and Nanotechnology in Engineering, World Scientific Publishing Co., Inc., NJ 07601, USA.

Webster, T. J. 2006. Nanomedicine: What's in a Definition? Dove Press, Macclesfield, U.K.

Wikipedia. 2010. Under entry "Nanotechnology Education".

Yuan, F., T. Yuan, L. Sui, Z. Wang, Z. Xi, Y. Li, et al. 2018. Engineering triangular carbon quantum dots with unprecedented narrow bandwidth emission for multicolored LEDs. Nat. Commun. 9(1): 2249.

Zsigmondy, R. 1909. Colloids and the Ultramicroscope. John Wiley & Sons, Inc., New York.

The Emergence of Nanomedicine: A Vast Industry

"If we can reduce the cost and improve the quality of medical technology through advances in nanotechnology, we can more widely address the medical conditions that are prevalent and reduce the level of human suffering."

Ralph C. Merkle – Researcher and Speaker
on Molecular Nanotechnology and Cryonics
http://www.merkle.com/

2.1 Can Nano-Size Technology Boost your Health and Life?

Through the convergence of nanotechnology and medicine, nanomedicine is set to offer solutions to many of modern medicine's unsolved dilemmas (Morigi et al. 2012, Etheridge et al. 2013). The scale-length reduction that has been attained through molecular nano-synthesis (bottom-up technology) and nano-mechanical processing (top-down technology) has the potential to interact with the biological world, in a way we have never imagined before (Juanola-Feliu et al. 2012). The nanotechnologies, that operate at the interface between organized nanostructures and biomolecules, are key control routes for achieving new breakthroughs in pharmaceuticals (Kumar et al. 2013, Placente et al. 2018), medical imaging, diagnosis (Lv et al. 2016), cancer treatment (Duncan 2006, Shi et al. 2017), implantable materials (Kendall and Lynch 2016, Sartuqui et al. 2018), tissue regeneration (Zhang and Khademhosseini 2015, Santos et al. 2016) and even offer multifunctional platforms combining several of these approaches into packages of a fraction the size of a cell (Cai et al. 2015, Zhang et al. 2017).

The field is recognized as a global challenge, and countless worldwide research and business initiatives are in place to obtain a significant market position. Even though business experts find it difficult to estimate the volume and growth rates of the nanomedicine market, unquestionably, it is a billion dollar market, expected to grow faster (Bosetti and Vereeck 2011, Venugopal and Ramakrishna 2016). Nevertheless, nanomedicine has its place among those emerging areas for which business development methods have not been established yet and its economic, social and health impacts still remain unknown (Morigi et al. 2012).

While a first generation of products has been successfully commercialized and has significantly contributed to enhance patient's life, recent advances in material design and the emergence of new therapeutics methods are contributing to the development of more sophisticated systems (Ragelle et al. 2017). As the field matures, it is important to understand the challenges related to the commercial development of nanomedicine products for a more efficient and predictable translation of their use in clinic.

In this chapter, we will analyse the actual situation of nanomedicine industry and describe the financial and strategic decisions that lead to nanomedicine start-ups to reach a successful place, to obtain a satisfactory market share, and to build and maintain a competitive defendable position. Dispensing nanomedicine products from the hands of the inventor to those of the doctor is not a simple task; we will explore the technical transfer process, which connects laboratories or research institutions to the marketplace. We will provide a comprehensive and contemporary list of nanomedicine products currently marketed and in clinical trials. Finally, we will discuss the principal challenges for their commercial development, both from a manufacturing and a regulatory perspective, to gain knowledge of the translational pathways.

The application of nanotechnology to medicine leads to progress in the treatments of a number of diseases, and its offshoot products have already established a substantial presence in today's markets. The present analysis merely highlights the infancy of the field; a large portion of the identified nanomedicine applications are still in the R&D stage and should lead to a truly revolutionary advance foreseen in medicine. Now is the time to put in place an effective cooperation between members of the investment community and those involved in the scientific research in order to lead nanomedicine-based inventions to a successful business position that will improve our lives.

2.2   The Need for a New Technology

2.2.1 The General Perspective

All living organisms are nanoscale engineered materials par excellence. Cells control billions of molecules everywhere, at every second, in order to communicate with each other, spread, differentiate, grow, attack intruders or heal after injury. Even though this chaotic activity is succeeded at a macro-scale level directed by key-role organs, it always relies on the rules of physics and chemistry, which are coded within biomolecules that pertain to the nanoscale. Accordingly, the manipulation of the matter on nanoscale applied to human health should allow for novel medical treatments resulting, if they are conscientiously applied, in an enhancement of therapies, a decrease of patient's side effects and in an innovative public health breakthrough. Some relevant examples are discussed below (Zuo et al. 2007, Pautler and Brenner 2010).

2.2.1.1 Advanced Image Processing Techniques and Applications for Biological Objects

A basic need for medicine, which was remedied by the application of nanotechnology, responds to the improvement of cellular and molecular dynamics observation in living cells. High-resolution analysis of active cells is limited by the need to use non-invasive methods. Devices such as scanning electron, transmission electron, and atomic force microscopes are not useful because the required processing or the harsh contact with the system probe disturbs or destroys the cellular environment. Optical methods are non-invasive, but usually they are diffraction limited, so their resolution is limited to approximately one µm. To overcome these restrictions, fluorescent semiconductor nanocrystals (also known as quantum dots) have been tested in most biotechnological applications, including DNA array technology (Dubertret 2005) and immunofluorescence assays for cellular or animal tissues analysis (Dubertret et al. 2002, Medintz et al. 2005). In 2003, the scanning near-field optical microscope (SNOM) was developed and utilized to study the membrane activity in a live cell sample (Sasaki 2003). Near-field optical microscope can detect tiny vertical movements on the cell membrane in the range of one nm or less, a resolution about three orders of magnitude superior than conventional optical microscopes. The extreme sensitivity of SNOM added to the expansion of specific quantum dots allowed measurements that are impossible with any other method on living biomaterial, paving the way for a broad range of novel studies and applications. This methodology will open a new approach to investigate live samples providing, for example, an accurate, real-time and high-throughput screening of tumour cells. These rapid nano-optical techniques may play an important role in advancing early detection, diagnosis, and therapy (Gao et al. 2004, Choi et al. 2007, Yao et al. 2018). Advanced imaging methods and new, targeted nanoparticle contrast agents for early characterization of an exact pathology at the cellular and molecular levels might represent the next frontier to reach; combining imaging and rational drug delivery, a personalized medicine will be possible. This, for example, should be the solution for a central and unresolved problem in cardiovascular diseases (CVDs): the *in vivo* detection of atherosclerotic syndrome and the evaluation of atherosclerotic disease activity, which is the first cause of death globally according to World Health Organization (WHO) (WHO 2017).

Quantum dots, gold and magnetic nanoparticles, used in molecular imaging, also improved the practice of dermatology. High-resolution dermoscopy, microscopy, and spectrometry now offer dermatologists the opportunity to advance into a non-invasive, immediate and *in vivo* examination of the skin, improving diagnostic and therapeutic modalities.

The development of high-resolution imaging of optical systems, which conducted a precise and repeatable targeting of single cell units in living samples added to the incomparable precision of laser ablation, improved assisted surgery. While the construction of nano-surgeons that could be introduced into the body through the vascular system remains into the empire of science fiction, laser nano-surgery is a current widespread approach in several laboratories. This technology can be applied to the study of axonal regeneration, to understand the contribution of single neuron within a neural circuit with a reductionist approach, or to develop scaled-down models of brain injury (Soloperto et al. 2016).

2.2.1.2 Targeted Delivery Methods

Efficient and selective delivery of genes and drugs to the vasculature of tumours is another desirable therapeutic goal. Specific nanoparticle designs have the potential to fulfil that function; from them, vehicles can be created to protect DNA from degradation, exhibit long circulation times, efficiently bind target cells and deliver DNA to the nucleus (Ferrari et al. 2005, Blanco et al. 2015, Wilhelm et al. 2016). Induction of undesirable gene expression inhibition or the expression of therapeutic proteins by delivery of genetic materials (DNA, RNA, and oligonucleotides) is another area of considerable current attention. As a result, the possibility of gene-nanoparticle complexes that might have unique physical and biological properties to exhibit minimal side effects but high potency and efficiency has gained interest in recent years (Lee et al. 2012, Yin et al. 2014). The application of nanotechnology in the design and delivery of active principles is also a sprawling area. By the use of nanotechnology, the drug systemic distribution and delivery can be facilitated. A carrier nanoparticle can be designed by itself from the therapeutic molecule and many drugs discovered in the past, discarded because there was a lack of a suitable method of delivery, can now be applied thanks to the use of nanotechnology. It also helps to identify and characterize molecular targets of nutrient activity and biomarker effects (Hosnedlova et al. 2018), so as to generate a personalized nutrition.

2.2.1.3 Improved Tissue Regeneration

Current medical practice suffers a growing crisis, as the demand for organ transplants continues to far exceed the available supply (Perán et al. 2013). Tissue engineering and regenerative medicine seem to propose different promising options to replace organ transplantation. Regenerative medicine aims to reduce demand for organs by stimulating the body's own repair mechanisms, while the closely related field of tissue engineering promises to deliver "of-the-self" organs grown from patients' own stem cells. To fulfil the promises of regenerative medicine and tissue engineering, a reliable means of generating complex tissues is necessary. Thus, greater attention must be paid to nanoscale cues present in a cell's microenvironment, the extracellular matrix (ECM). Currently, different nanotechnology approaches are being incorporated into the design and manufacture of scaffolding to mimic the tissue's native ECM. It is expected that these can replace the natural ECM until the host cells can repopulate and rebuild the original biological matrix (Laurencin and Nair 2014). Recent advances in nanotechnology created nanostructured microspheres, including nanocomposite and nanofibrous microspheres, as injectable cell carriers scaffold to mimicking natural ECM (Zhang et al. 2016). This technology exceeds the scaffolding concept and should be applied to repair irregularly shaped tissue defects with a minimal invasive procedure leading to a reduction of postoperative times and a better patient recovery.

Nanomedicine application to functional tissue restoration also reaches neurodegenerative disorders. Neuropathology processes that contribute to central nervous system ischemia, trauma, and degenerative conditions can be limited and even reversed by promoting neural regeneration and achieving neuroprotection. Nanotechnology approaches applied to neurodegenerative disorders should facilitate: (i) a rapid detection using nanotechnology tools, in particular, those that focus on chemically functionalized semiconductor quantum dots; (ii) neuroprotection by the delivery of drugs and small molecules across the blood–brain barrier for directly interacting, recording, and/or stimulating neurons at a molecular level, or limiting the damaging effects of free radicals generated after injury (iii) the functional regeneration of the nervous system by the development of nano-engineered scaffolds that support and promote neurite and axonal growth.

Although the *in vivo* application of nanotechnology is momentarily associated to particular and still unsolved safety and legislative concerns, which makes some medical nanotechnology concepts to be only theoretical, its future prospects for the development of a personalized medicine and a top-quality health care seem to be excellent.

2.2.2 The Scope of Poverty Related Diseases

Beyond a general perspective of nanomedicine applied to enhance all areas of health, there are specific needs for the request of a new technology; an example that we would like to highlight is its impact on the poverty related diseases (PRD).

At the global level, the three primary PRDs which account for nearly 18 percent of the disease burden in the poorest countries are tuberculosis (TB), AIDS/HIV and malaria (Stevens 2017). The need is significant – according to the World Health Organization (WHO) 2014 Global TB report, an estimated 9 million people developed tuberculosis in 2013 and 1.5 million people died (up from 1.3 million estimated in 2012) (Zumla et al. 2015). The 2014 WHO also reported an estimated of 480,000 new cases of multidrug-resistant tuberculosis (MDR-TB) in 2013. Furthermore, 9% of people with MDR-TB are estimated to have extensive drug resistant tuberculosis (XDR-TB) and nearly 50,000 people worldwide have a form of the disease that, at present, cannot be treated. A similar picture can be obtained for malaria: despite the progress made in malaria control during the past decade, there remain an estimated 3.2 billion people in 97 countries and territories at risk of malaria infection. About 214 million malaria cases were estimated to occur in 2015, leading to 438,000 deaths, and key challenges exist to sustain and improve on recent gains (Hemingway et al. 2016). The plasticity of the mosquito and the *Plasmodium* parasite has led to increasing resistance to medicines and insecticides. For example, resistance to artemisinin-based combination therapies (ACTs) has been detected in five countries in Southeast Asia and in Africa, resistance has been detected against two or more insecticides in two-thirds of countries where malaria is endemic. Up to 80% of infections are asymptomatic, and *Plasmodium* vivax parasites remain dormant for months or even years after initial infection. The spread of these strains to Africa or the Indian subcontinent could be catastrophic. Current field tests are not sensitive enough to pick up the low density of parasites in low-transmission areas, thus continued R &D investment and a rapid deployment of new tools are needed.

Moreover, ignored tropical *leishmaniasis* affects some of the poorest people on earth. It is caused by the protozoan *Leishmania* parasites, which are transmitted by the bite of infected female phlebotomine sandflies, and there are three main forms: (i) visceral (LV, also known as *kala-azar*, it is the most serious form of the disease), (ii) cutaneous (LC, the most common) and, (iii) mucocutaneous (LMC). This pathology is associated with malnutrition, population displacement, poor housing, a weak immune system and lack of financial resources; an estimated of 700,000 to 1 million new cases and 20,000 to 30,000 deaths occur annually. (WHO 2018b) Effective vaccines against this disease are still under development, the available drugs can be quite toxic and costly, and there is a reported increase in *Leishmania sp.* resistance to conventional drugs.

In terms of HIV, Sub-Saharan Africa still bears the largest global share of the burden, with the highest number of people living with HIV as well as the highest number of new HIV infections, AIDS-related deaths and the highest adult HIV prevalence, which will be 70% of the global number by 2017 according to the WHO (WHO 2018a). HIV infection weakens the immune system and additionally exposes the patient to other infectious diseases such as TB, malaria and leishmaniasis leading to a closed circle of PRD prevalence.

Nano-based-platforms (nanomedicine) can be a promising line of attack to overcome the problem associated with conventional TB, malaria and HIV therapy. Emerging nanomedicine approaches enhance drug stability and bioavailability, modulate its pharmacokinetics by prolonged blood circulation time and, allow a localized delivery of high amounts of medicine, simultaneously being able to decrease its toxicity and side effects (Giardiello et al. 2016, Garg et al. 2017, Kumar et al. 2017, Gao et al. 2018, Smith et al. 2018). Despite the great need of novel diagnosis tools, almost no examples of concepts based on nanomedicine have been reported related to *leishmaniasis* or other PRDs (Gutiérrez et al. 2016).

Owing to its perceived advantages, worldwide attention impulses the use of nano-platforms in the war against poverty related diseases. As we have mentioned in chapter 1, in 2005 "The Nanoscience African Network" (NANOAFNET) was established in Africa, which focused primarily on the improvement of therapies for infectious diseases related to poverty, including

TB, malaria and HIV, by the application of nanomedicine concepts. In 2011, the CSIR (Council for Scientific and Industrial Research) Nanomedicine Platform in South Africa hosted the first "International Workshop on Nanomedicine for Infectious Diseases of Poverty". Importantly, the CSIR Nanomedicine Platform in South Africa collaborated actively with the "Pan-African Centre of Excellence in Nanomedicine" to discuss the application of nanomedicine research and training, in partnership with industry and academia, to significantly enhance the development of therapeutic compounds for infectious diseases of poverty (Saravanan et al. 2018). Nanomedicine research on PRDs also get funding from public private partnership (PPP) like the Bill and Melinda Gates Foundation (BMGF). The BMGF provided about 50% of formal PPP funding in 2007. In addition to the BMGF, PPPs received funding from large donors like the U.S. NIH, the Rockefeller Foundation, from the U.S. Agency for International Development (USAID), and from western European countries like U.K., Germany, and Denmark. Particular PPPs receive its major support from big pharmaceutical companies, like Merck or Johnson & Johnson (Woodson 2016).

Despite the great path that remains to be followed, the application of nanomedicine seems to be a promising approach to respond to WHO's requirements against poverty-related diseases.

2.3 Nanomedicine Market

2.3.1 Market Snapshot

The need for reduced side effects and for therapies with superior cost/effective ratios, in particular for cancer (WHO 2018c), motivates an expansion of novel nanotechnology-based drugs. Nano-cocktail formulations displaying a tailor-made pharmacokinetics triggered release, localized control of a tumour in combination with radiotherapy, and *in vivo* targeted activation of stem cell production, are anticipated to drive R&D revenue generation in the coming years. The application of similar technologies to improve contrast agents used to diagnose and to monitor the side effects of novel drugs will contribute to drive the growth of the nanomedicine market request, in an unprecedented short timescale. The success and growth of the project that influence the industry of nanomaterials, is based on a greater demand for biodegradable grafts with longer lives that allow tissue restoration, the use of DNA origami and nano-robotic engineering, among others. Furthermore, an increase in out-licensing of nano-drugs and the growth of healthcare facilities in emerging economies provide numerous opportunities for the nanomedicine market expansion. Biopharmaceutical and medical devices companies are enthusiastically involved in the expansion of new products as is revealed by the progressively emergent partnerships between leading enterprises and nanomedicine start-ups. Some of the key players operating in the global nanomedicines market are CombiMatrix Corporation, Abbott Laboratories, GE Healthcare, Johnson & Johnson, Sigma-Tau Pharmaceuticals,

Inc., Mallinckrodt plc, Nanosphere, Inc., Merck & Company, Inc., Celgene Corporation, Pfizer, Inc., UCB (Union chimique belge) S.A, and Teva Pharmaceutical Industries Ltd.

Those facts support that the global nanomedicine market accounted for US$ 111.9 billion in 2016, and that is anticipated to reach US$ 261.1 billion by 2023 and US$ 350.8 billion by 2025, registering a Compound Annual Growth Rate (CAGR) of 12.6% from 2017 to 2025 according to a new report disclosed by Grand View Research, Inc. (Grand View Research 2017, Insights 2018). Simultaneously, the global market of Nanobots is expected to reach US$ 100 billion in 2023 from US$ 74 billion in 2016 with a CAGR of approximately 21% during the forecast period 2017-2023 (Future 2018). Restrictions in the nanomedicine market include the cost of training health care staff to use this new technology and for patient care. Additionally, statistics show that nanomedicine market in Latin America in 2017 had a value amounting to 15.04 billion US$ and it is expected to rise to about 29.1 billion US$ by 2022 (Portal 2018).

In the light of the presented data, we can conclude that as familiarization of nanotechnology amongst the surgeons, nurses and other healthcare staff increases, it will favour the growth of nano-market.

2.3.2 Mapping Nanotechnology in the Current Healthcare Market

The global nanomedicine market is driven by a drug delivery emerging technology, an increase of nanomedicine adoption across several applications, a rise in government support & funding, and growth of therapies exhibiting fewer side effects and superior cost-effective ratios. Many nanomedicine products are already in use in humans, Table 2.1 (Etheridge et al. 2013, Kendall and Lynch 2016, Chan 2017, Ragelle et al. 2017, Dilnawaz et al. 2018). Here we present several interesting trends forecasting the future of nanotechnology tools currently applied in medical research. They should be interpreted as examples and are only a subset of the vast scope of nanomedicine research currently underway (Gordon and Sagman 2003).

- *Therapeutics:* This category is generally defined to include synthetic and natural active principles and vaccines that are proposed to directly alleviate a medical condition. According to their approved or intended uses, therapeutic products were grouped into a sub-classification: cancer treatment, hepatitis, other infectious diseases, anaesthetics, cardiac/vascular disorders, inflammatory/immune disorders, endocrine/exocrine disorders, degenerative disorders, and others. A similar proportion of approved products across all sub-categories was appreciated; however, about two-thirds of the R&D identified are concentrated on cancer treatment.
- *Administration and targeting:* It takes account of novel drugs based on nano-particles (drug encapsulation) and on the nano-systems (liposomes, polymer-drug conjugates, hydrogel nanoparticles and dendrimers) that

can be utilized to deliver active principles in an effective and precise manner (functional drug carriers).

- *Regenerative medicine:* This field comprises biocompatible materials that can be permanently or temporally implanted in a living organism. These materials can be used for substituting or repairing tissues (tissue repair and replacement, implant coatings, and tissue regeneration scaffolds) and as structural implantable materials (bone repair, bioresorbable materials and smart materials' replacement).
- *Diagnosis devices: in vivo approaches:* It refers to those technologies that involve nano-devices that can be implanted in live organisms. This category comprises those devices that can process local extracted medical information for diagnosing and treating purposes (implantable sensors) and those that can enhance sensory skills restoring lost hearing and sight functions (retina and cochlear implants).
- *Diagnostic tools: in vitro methods:* In this category are grouped those nano-systems that can help to identify the occurrence of a disease as soon as possible. Screenable markers to work directly on genes and genetic samples. In nuclear biology and molecular biology, a marker gene is a gene used to determine if a nucleic acid sequence has been successfully inserted into an organism's DNA. They are used in ultra-sensitive labelling and, detection technologies; high throughput arrays and multiple analyses. Contrast agents like nanoparticle labels and imaging devices that make possible graphical representations of the patient's condition by imaging observation.
- *Precise surgical supports:* The potential for the use of nanotechnology in surgery is huge. This area includes nano-tools (smart instruments and surgical robots) that can perform common surgical tasks in a very precise way, or by monitoring patient condition with a higher accuracy.

2.4 Transferring Nanomedicine from Bench to Bedside

Translational nanomedicine, often known as "bench to bedside", represents a new paradigm in the biomedical research enterprise catalysing a more patient-centric approach to medicine through the successful integration and the iterative modification of three traditional branches of medicine: research, drug development, and clinical medicine (Wei et al. 2007, Havel et al. 2016). Nanomedicine ushers in a different kind of remapping or reconfiguration of the time, place, and space of illness through precision materials and systems. As a rapidly developing field increasingly delivering on many of its promises, nanomedicine stands poised to have significant impacts not only on the practice of medicine, but also on the context in which it is administered and, in fact, cannot occur without societal complicity, either explicitly or implicitly. Indeed, societal considerations, patient and consumer views and preferences, institutions and infrastructures, come together to form a particularly important component of successful translation of biomedical innovation. In short, innovation in medicine cannot become a clinical

Table 2.1. Current market status of nanomedicines. (NP) nanoparticles; (IV) Intravenous; (SPION) Superparamagnetic iron oxide nanoparticles

Taxonomy/ Medical Area	Active Molecule	Description Nano-component	Product Name	Market Distribution Company	Status/Year	Application/ Administration
Administration and targeting	Paclitaxel	Albumin-bound paclitaxel NP	Abraxane	Celgene Co.	FDA approved/2005	Metastatic breast cancer/IV
	Daunorubicin citrate	Liposomes	DaunoXome	Galen US Inc.	FDA approved/1996	AIDS-related Kaposi's sarcoma/IV
	Doxorubicin (generic)	PEGylated liposomes	Doxil/Caelyx/ LipoDox	Sun pharmaceutical ind. ltd.	FDA approved/1995	Ovarian cancer, AIDS-related Kaposi's sarcoma, and multiple myeloma/IV
	Denileukin Diftitox	Recombinant DNA-derived protein NP	Ontak	Seragen Inc.	FDA approved/1999	T-Cell Lymphoma/IV
	Oxaliplatin	Liposomal	MBP-426	Mebiopharm Co., Ltd	Phase I/II	Solid Tumours/IV
	Doxorubicin	PEGylated liposomes	MCC-465	Mitsubishi Tanabe Pharma Corp	Phase I	Stomach Cancer/IV
	Doxorubicin	HPMA copolymer-doxorubicin-galactose NP	PK2	Pharmacia & Upjohn Inc.	Phase I	Liver Cancer
	Paclitaxel	Polymer micelle	Genexol[a]-PM	Samyang	FDA approved/2007	Metastatic breast cancer/IV

Category	Compound	Type	Product name	Company	Status	Indication/Route
Therapeutics	Verteporfin	Liposomes	Visudyne	Novartis	FDA approved/2002	Photodynamic therapy used in eye neovascularization/IV
	Amphotericin B	Lipid NP	Abelcet	TEVA PHARMA, B.V.	FDA approved/1995	Fungal infections/IV
	Amphotericin B	Lipid NP	Amphotec	Alza Corp.	FDA approved/1997	Fungal infections, cryptococcal meningitis, and visceral leishmaniasis/IV
	Propofol	Nanoemulsion	Diprivan	ASTRAZENECA	FDA approved/1989	General anesthesia/IV
	Difluprednate	Nanoemulsion	Durezol	Novartis	FDA approved/2008	Eye inflammation and uveitis/ocular
	Cyclosporine A	Nanoemulsion	Restasis	Allergan, Inc.	FDA approved/2003	Dry eye syndrome/ocular
	Ferumoxytol	Carbohydrate coated SPION	Feraheme	AMAG Pharmaceuticals	FDA approved/2009	Iron deficiency anemia associated with chronic kidney diseases/IV
Bone Substitute	Calcium-phosphate	Nanocrystals	Vitoss™ BA Bimodal	Orthovita	FDA approved/2011	Non-structural bone void filler for use in the extremities, pelvis and posterolateral spine
	Hydroxyapatite	Nanocrystals	Ostim	Osartis GmbH	FDA approved/2004	Bone cement/paste
	Hydroxyapatite (80%), tricalcium phosphate (20%)	Nanocrystals	OsSatura	Isotis Orthobiologics US	FDA approved/2003	Synthetic bone graft material

(Contd.)

Table 2.1. (*Contd.*)

Taxonomy/ Medical Area	Active Molecule	Description Nano-component	Product Name	Market Distribution Company	Market Status/Year	Application/ Administration
	Hydroxyapatite granules and an open structured engineered collagen carrier	Nanocrystals	NanOss	Angstrom Medica, Inc.	FDA approved/2005	Lumbar Spinal Fusion/ Synthetic bone graft material/Scaffold
	Hydroxyapatite	Nanocrystals	EquivaBone[a] BGS	ETEX Corporation	FDA approved/2009	Filling bone voids or defect of the skeletal system/self-setting paste
Dental Composite	Polysiloxane backbone	Organically modified ceramic NP combined with conventional glass fillers of ~1 μm	Ceram X Duo	Dentsply	FDA approved/2005	Restorations of primary and permanent teeth/ paste
	Silica, Zirconium	NP	Filtek™	3M Company	FDA approved/2008	Restorations of primary and permanent teeth/ paste
	Light cured resin based nano-composite	NP	Premise	Sybron Dental Specialties, Inc.	FDA approved/2009	Dental Composite Restorative Material
	Nanoparticulate reinforced adhesive	NP	Nano-Bond	Pentron® Clinical Technologies, LLC	FDA approved/2012	Composite, bonding, cementing and luting adhesives for dental use

Category	Material	Description	Product	Company	Status	Application
Devices coatings	Ag	NP	ON-Q SilverSoaker/SilvaGard™	I-Flow Corporation/AcryMed, Inc.	FDA approved/2005	Antimicrobial catheter
	Conductive material	NP embedded in a temperature-sensitive material	EnSeal Laparoscopic Vessel Fusion	Ethicon Endo-Surgery, Inc.	FDA approved/2005	Sealing arteries, veins, and transecting fatty tissue, small ligaments and connective tissue
	Calcium Phosphate	NP	NanoTite[a] Implant	Zimmer Biomet	FDA approved/2008	Dental implants
	Hydroxyapatite	nanothin-microporous hydroxyapatite surface	VESTAsync™	MIV Therapeutics	FDA approved/2003	Cardiac implants
In Vitro Assay	Ferro-fluid reagent Iron	Magnetic NP conjugated to antibodies	CellTracks[a]	Immunicon Co.	FDA approved/2003	Diagnostic Test
	Au	NP colloidal suspension	NicAlert	Nymox	FDA approved/2002	
	Alkaline phosphatase conjugated anti-CRP mouse monoclonal antibody	Dendrimer	Stratus[a] CS Acute Care™ CardioPhase[a] hsCRP TestPak (CCRP TestPak)	Dade Behring Inc.	FDA approved/2003	Reactive protein immunological test system
	Ferro-fluid reagent. Iron Oxide	NP magnetic core surrounded by a polymeric layer coated with antibodies targeting the Epithelial Cell Adhesion Molecule (EpCAM) antigen for capturing CTCs.	CellSearch[a] Epithelial Cell Kit	Veridex, LLC (Johnson & Johnson)	FDA approved/2004	Enumeration of circulating tumour cells (CTC)

(Contd.)

Table 2.1. (*Contd.*)

Taxonomy/ Medical Area	Active Molecule	Product		Market		Application/ Administration
		Description	Name	Distribution Company	Status/Year	
		Nano-component				
	Au	NP colloidal suspension	Verigene[a]	Nanosphere, Inc.	FDA approved/2007	Molecular diagnostics
Medical Dressing	Ag	15 nm Au NP, 70 to 100 ppm colloidal suspension	Acticoat[a]	Smith & Nephew, Inc.	FDA approved/2005	Antimicrobial barrier for the treatment of chronic or acute wounds
Dialysis Filter	Polysulfone	Nanoporous Membrane, pore size of 1.8 nm	Fresenius Polysulfone[a] Helixone[a]	NephroCare	FDA approved/1998	High-Flux Dialysers and Haemodiafilters
Tissue Scaffold	Titanized polypropylene	30 nm Ti NP coating	TiMESH	GfE Medizintechnik GmbH	FDA approved/2004	Titanium mesh implants use in reconstructive neurosurgical procedures, such as cranial flap fixation.
Hyperthermia	Iron Oxide	15 nm NP	NanoTherm[a]/ NanoPlan[a]/ Nano Activator[a]	MagForce	FDA approved/2018	Local treatment of solid tumours/Magnetic heating activation
	Gold	Core-shell NP	AuroShell™	Nanospectra Biosciences, Inc.	Phase I	Solid tumours/IV/IR Laser Heating
	Crystallized hafnium oxide (HfO_2)	50 nm NP	NanoXray	Nanobiotix	Phase I/II trials in liver cancers	Local treatment of solid tumours/X-Ray-induced electron emission

	Material	Type	Product	Company	Status	Application
In Vivo Imaging	Iron Oxide	NP	Endorem[a]	Guerbet S.A.	FDA approved/1995	Enhanced MRI Contrast
	Iron Oxide	NP	Sinerem[a]	Guerbet S.A.	European Marketing Authorisation Application/2007	Enhanced MRI Contrast
	Iron Oxide	NP	Clariscan[a]	Nycomed	Phase III	Enhanced MRI Contrast
In Vitro Imaging	Quantum Dot	NP	Qdot Nanocrystals	Invitrogen Corporation	Research Use Only	Fluorescent Emission
	Quantum Dot	NP	Nanodots	Nanoco Group PLC	Research Use Only	Fluorescent Emission
	Quantum Dot	NP	TriLite™ Nanocrystals	Crystalplex Corporation	Research Use Only	Fluorescent Emission
	Quantum Dot	NP	eFluor Nanocrystals	eBiosciences	Research Use Only	Fluorescent Emission
	Quantum Dot	NP	NanoHC	DiagNano	Research Only	Fluorescent Emission
In Vitro Cell Separation	Iron Oxide	NP	CellSearch[a] Epithelial Cell Kit	Veridex, LLC	FDA approved/2004	Magnetic Separation
	Iron Oxide	NP	NanoDX	T2 Biosystems	Research Use Only	Magnetic Separation

reality without the integration of these societal pillars (Murday et al. 2009, Bellare 2011).

2.4.1 Financing Nanomedicine

As it was presented in chapter 1, the booming nanotechnologies are being supported by massive governmental investments in many countries. Governments have historically funded R&D to attain a technology that potentially influences the national economy (Jia 2005). Universities are persistently pushing for funding to adapt basic nano-medical research into real products and eventually partner with large biotech or drug companies to make their enterprises a business of success. Nevertheless, purposes of nano-science must not only be focused on the realization of novel methods, techniques, devices and constructs at nano-size scale, but also about the practical enhancement of economic profits over conventional technology. Private industry does not see an appropriate opportunity for the nanomedicine implantation and investors are extremely cautious because positive returns occur only, if at all, over a long term. The currently limited availability of commercially viable nano-medical products added to the investors' concerns about the future restrictions of legal regulation that makes it difficult to find a partner. At this time, small and medium-sized companies and venture capitalists who invest in start-ups drive investment in nano-medical research (Hobson 2009).

On the other hand, the cost associated with the process of bringing a novel medicine, device, or diagnostic marker to the marketplace is very expensive, thus pharma and biotech companies focus on drugs that are expected to be blockbusters. However, potential blockbusters are difficult to identify because of a scarcity of available data. It has therefore been suggested that stakeholders, including research scientists, clinical investigators, health care providers, patient associations, and investors, develop a shared communication platform to facilitate communication and collaboration to bring innovative and profitable nano-medical products to market (Ventola 2012).

Despite these mentioned problems, investments in nanomedicine are expected to increase. Pharma and biotech companies are still expected to embrace nano-therapeutics and other nanomedical products, especially if they have novel properties, fulfil unmet medical needs, and offer an attractive cost–benefit ratio. Nanomedicine is also expected to create additional revenue streams for pharmaceutical companies by creating new patentable formulations of off-patent proprietary drugs, extending the life of these products. It is even expected that novel or reformulated nano-therapeutics will disrupt the generic drug market (Fornaguera and García-Celma 2017).

A single institute, corporation or even a single country cannot advance in the development of nanotechnology. A cooperative system must be developed to structure the link and foster competition between research groups of academic institutions around the world. At the same time, collaboration between government-funded researchers and private industry is critical to the nanotechnology transfer process and its commercialization. Such networking and collaborating scenario is now facilitated and encouraged by government-funded programmes in order to leverage public investment.

2.4.2 Business Strategies

2.4.2.1 Product Protection and Commercial Potential

Patent rights and trade secrets of production methods (expertise) are undoubtedly essential to protect every new technology, so it cannot be different for nanomedicine (Neuman and Chandhok 2016). As a rule, patents protect start-ups and smaller companies from intrusion of larger corporations; they also protect the clients of a patent owner because it may prevent a competitor from replicating the client's products made under license from the patentee (Bawa et al. 2005). Moreover, patents offer credibility to any inventor and safety to any public that may not fully understand the science behind the technology: backers, shareholders and venture capitalists, to mention a few. Therefore, prospective markets for nanomedicines are inextricably linked with an efficient patent system. A good number of specialists agree that a start-up should focus on obtaining a broad intellectual property portfolio that includes both patents and trade secrets that cover clusters of an emerging sector in nanomedicine. As start-ups progress and evolve, protecting trade secrets in this information age may be difficult; scarce venture capitalists are likely to support a start-up that relies on trade secrets instead of patents. For a start-up, patents are a means of validating the company's foundational technology to attract investment. Stakeholders are unlikely to invest in a start-up that has failed to construct adequate defenses around its intellectual property. In fact, patents generally precede funding from a venture capital firm because they are a guarantee for the capital invested; even after the dissolution of a poorly performing nanomedicine start-up, patents on its vital technologies can be sold to other companies, thereby providing some return for sponsors (Bowman et al. 2017). Because of the potential market value of nanomedicine-based systems, researchers, executives, and patent lawyers are all on a quest to obtain broad protection for new nanoscale materials. To be patented, an invention must meet both formal and informal criteria (Table 2.2); actually, numerous technologies and techniques pertaining to nanomedicine can be protected with a U.S. patent system (Bawa et al. 2005, Satalkar et al. 2016).

2.4.2.2 Intellectual Property and Patent Search

Searching for nanomedicine-related patents and publications is quite difficult compared to other technological areas, owing to its definition that covers a broad class of materials and systems; see chapter 1 section 1.2. Global patent classification systems are neither sufficiently defined nor descriptive enough to incorporate all properties derived from the unique characteristics

Table 2.2. Criteria for patentability according to the Title 35 of the United States Code. Full text can be found at the USPTO (USPTO 2018)

Patent Statute	Information
35 U.S.C. 101	Matter eligibility and utility requirements. The invention must concern patentable subject matter.
35 U.S.C. 102	Novelty requirement. The invention must be novel and the application for a patent on the invention must be timely.
35 U.S.C. 103	Non-obviousness requirement. The invention must be non-obvious to a person with knowledge in the field related to the invention.
35 U.S.C. 112 (a)	Finally, the invention must be sufficiently documented. Contains three separate and distinct requirements: (A) Adequate written description, (B) Enablement, and (C) Best mode.
35 U.S.C. 112 (b)	Definiteness requirement

that nanomedicine inventions exhibit (Bawa et al. 2005). The fundamental nature of nanomedicine is part of the challenge for effectively mapping the patent landscape. Many patent applications may result from a single nanomedicine invention; hence, a single patent may generate many products or markets. Published patents that are "truly" nanomedicine related may not use any specific nanomedicine related terminology and, conversely, there are business-savvy inventors and patent assignees that use keywords incorporating a "nano" prefix for the sake of marketing their inventions or concepts.

Part of the challenge in finding "truly" nanomedicine-related patents is the judicious use of key terms and class codes while searching the patent databases. This search, along with the additional filter of subject area expertise, is currently the most reliable way to find nanotechnology patents. Combing the patent landscape may be tricky, but a subject area expert can ultimately decide whether a patent pertains to nanomedicine (Bawa 2004, Bawa et al. 2005, Murthy 2012).

The Japanese Patent Office has created a classification system for nanotechnology, the Japan Platform for Patent Information, *J Plat Pat* (Japan Patent Office 2018). A similar platform can be obtained from the European Patents Office (EPO), the "*Espacenet*" (EPO 2018). From EPO, certain information referring to Asian documents can also be accessed. There is not a formal classification system equivalent to the "*J Plat Pat*" or "*ESpacenet*" for the U.S. nanomedicine patents; however, the U.S. Patent and Trademark Office (USPTO) possesses an organization scheme including patent applications filed, their pendency and technical information, and assignee legal and business material (EPO 2018, USPTO 2018). In any case, the organizations suffer from lack of effective automation tools specific to nanomedicine "prior art" searching (Bawa et al. 2005). The term "prior art" generally refers to the state of knowledge existing or publicly available before the date of an invention; it is a printed document that contains a disclosure or description that is relevant to an invention for which a patent is being sought or enforced. Typically, a "prior art" document is submitted by the inventor during prosecution of his/her patent application. Considering the U.S. nano-patents proliferation in the last years, it is clear that the USPTO views a scale down in physical dimensions patentable. In fact, current case law supports the proposition that a change in size can result in patentable subject matter because unique technical problems arise when physical dimensions are reduced (Bawa et al. 2005). The patent landscape is getting crowded and commercialization of a nanomedicine product should not be attempted without reviewing the patent literature and other relevant prior art. Various data sources and software tools can make a patent search more efficient and effective; some of them are summarized in Table 2.3 (Bawa et al. 2005).

Table 2.3. Patents and "Prior art" search databases (Modified from Bawa et al. 2005)

Database	Description
Governmental websites	US Patent and Trademark Office (www.uspto. gov), the European Patent Office (https://www. epo.org), and the Japanese Patent Office (www. jpo.go.jp)
Thomson databases	Providing various patent databases, including Derwent World Patents Index – Clarivate, Delphion Patent Collections and Searching, Thomson Innovation Patent Searching and Thomson Pharma.
IFI CLAIMS patents database	IFI CLAIMS specializes in annotating and classifying patent and scientific documentation using expert curators and intelligent technologies. Since 1950s.
STN chemical abstracts and chemical abstract service (CAS) database	The CAS databases CAplus and CAS REGISTRY offer intellectually analyzed content obtained from journal and patent literature.
IMSWorld drug patents international database	IMSworld Drug Patents International database provides access to the patent status of over 1200 molecules
EPO worldwide legal status database (INPADOC)	INPADOC contains legal status events from over 40 international patent authorities worldwide.
Japan Patent Information Organization (JAPIO)	JAPIO provides various patent information products and services including file wrapper copies, watching service and e-mail deliveries.
Engineering, technology, and scientific databases	Including INSPEC, EBSCO, EiCompendex and SCISEARCH
Markets and business databases	Including Factiva and PROMPT

2.4.3 Best Practices in Clinics

2.4.3.1 Translation Pathway, Examples and Challenges

Bringing new products to the market has always represented a great challenge, especially when it comes to highly innovative products with high risk/return ratio. Despite the numerous entry barriers of the nanomedicine market, there are some noteworthy examples of nano-based FDA-approved products that successfully reached the market, influencing medicine and anticipating a change in the healthcare ground (Table 2.1).

In light of this summary of best practices in the clinic, anti-cancer technology continues to occupy most of the nanomedicine market, in addition to publications and patents compared to the quantity of products marketed. Increasing acceptance with the public of the use of nanotechnologies in the clinic, along with popular widespread sensitivity for the aggressiveness of cancer, can be considered strong drivers for the commercial success of this group. Furthermore, the first tangible considerable returns due to commercial triumphs represent an undoubted source of attraction for investors. On their part, financiers must realize the importance of providing the substantive funds, necessary to gain the solid results and successful drugs as well as devices and therapies that the market requires (Morigi et al. 2012, Hua et al. 2018).

In order to move nano-medicine products (NMPs) from the bench to the bedside, several experimental challenges need to be addressed. NMP technology is usually far more complex in comparison to conventional formulation technology and some key issues must be taken into consideration, some of them listed in Table 2.4 (Hua et al. 2018).

2.4.3.2 Introduction to the Translation Pathway: Clinical Trials Phase I, II, III and Market Approval

Before initiating a clinical trial involving a nanomedicine product, manufacturers must present data to the proper regulatory entity for starting pre-clinical *in vitro* studies involving chemicals, human cells or tissues and animals. Among all regulation entities that give the green light with its approval to begin selling drugs, the U.S. Food and Drug Administration (FDA), the European Medicines Agency (EMEA) mainly, carry out the authorization and supervision of medicines (Science Medicines Health 2018, USDHHS 2018). It is estimated that these agencies receive about 70% of their funds from fees of pharmaceutical industry, which raises questions about their ability to act autonomously (nuevatribuna.es 2017).

When the proper regulator organism determines that nanomedicine is safe enough to introduce in humans, it gives the manufacturer permission to conduct a small study that includes about 25–100 test subjects, known as Phase I trial, to determine the maximum tolerable dose in human beings. If this stage is overcome, it is said that the product is safe enough to use in humans, and the regulator entity will allow the manufacturer to conduct larger studies of about 100–500 test subjects, which is known as the Phase

Table 2.4. Translational training of nanomedicines: key points and present-day obstacles

Key points	Current Obstacles
Nanomedicine Design	
Route of administration	Industrial manufacture
Processing complexity reductions	Large-scale production agreeing to Good
Biocompatibility and	Manufacturing Practice (GMP) standards
biodegradability	Cost and infrastructure
Physical and chemical stability	Reproducibility,
Final form to use in humans	Quality control assays
(dosage form)	Purity and stability
Preclinical Evaluation	
Validation and standardization	Inadequacy of specialized toxicology studies
Early detection of toxicity	Lack of understanding of the interaction of
Construction and analysis of	nanomedicines with tissues and cells
appropriate animal models of	Scarce structural stability of nanomedicines to
disease	succeeding *in vivo* administration
Suitable understanding of *in*	Limited degree of accumulation of
vivo behaviour: both cellular and	nanomedicines in target organs/tissues/cells
molecular interactions.	
Clinical Evaluation for Commercialization	
Optimal clinical experimental	Lack of detailed and specific regulatory
design	guidelines for nanomedicines
Minimization of time and expense	Complexity of nanomedicine patents
production pathways from	system and limited intellectual properties
invention to commercialization	information
Reduction of acute and chronic	Inadequate recognition of the biological
toxicity in humans	interaction of nanomedicine *in vivo* in patients
Assessment of therapeutic	
efficacy in patients	

II trial. During Phase II trials, the drug's efficacy and administration safety concerns are investigated. If the drug makes it past this stage, manufacturers are allowed to perform much larger studies involving about 500–3000 test subjects, and phase III trial starts to gather further data on safety and efficacy. When Phase III testing is complete, the regulator entity will examine the data to determine whether to approve the manufacturer's nanomedicine application to market it in a specific country or a region. If nanomedicine is approved, the manufacture may start selling it and may also conduct additional studies, known as Phase IV trials (or post-marketing studies), to gather additional information about safety, efficacy, dosing, side-effects, and adverse reactions.

Regulation entities, like FDA, encourage but do not require Phase IV studies. The FDA also animates doctors who prescribe the drug to report any adverse reactions or other safety concerns (Chan 2017, Hua et al. 2018).

A duly constituted institutional review board (IRB) must approve any of the above-mentioned studies. The IRB evaluated the ethical aspects of any study and the rights and welfare protection of test research subjects. In the decision of a study approval, the IRB must determine whether: (i) risks will be minimized; (ii) risks will be reasonable in relation to expected benefits to the subjects (e.g. medical therapy) or society (e.g. the knowledge gained); (iii) provisions for data and safety monitoring (if appropriate) will be adequate; (iv) informed consent will be properly sought and documented; (v) selection of subjects will be equitable; (vi) protections for vulnerable populations (if appropriate) will be adequate; and (vii) privacy and confidentiality will be protected. We think that nanomedicine does not raise any especially challenging issues for IRB criteria, but we will make some commentaries on the first four criteria in the list, which deal with the minimization, management, and communication of risks (Allon et al. 2017).

2.4.3.3 Risks and Risks Management

First, it is necessary to clarify that the understanding of the risks, for animals and humans, associated with the development of nanomedicines implies the understanding of some basic facts about nanomaterials. It is not possible to make any generalizations about the safety of nanomaterials. Each type of nanomaterial must be considered separately since: (i) there is a tremendous diversity among them; (ii) the risks may fluctuate according to the route of exposure, such as dermal, oral, respiratory, and/or intravenous; (iii) the risks of exposure to manufactured nanomaterials may be different from the risks of exposure to the naturally occurring nanoscale materials, since humans have had millions of years of evolution to adapt to natural exposures; and (iv) finally, the size, shape and physicochemical properties of nanomaterials are highly dependent on their microenvironment and may change once they enter into a particular organism.

Researchers, governmental institutions, academicians, and industries are just beginning to understand how exposure to nanomaterials activates the body's critical defense mechanisms: the inflammatory and oxidative stress responses, and innate and adaptive immunity. Because nanomaterials, by definition, have novel properties, they may affect animals and humans in unpredictable ways.

Although translation pathway will play an important role in minimizing the risks of nanomedicine, the experiments performed have significant limitations: (i) there may be differences in how animals and humans absorb, distribute, metabolize, or eliminate a substance or a material; thus the way that humans and the animal models used in preclinical testing (usually rodents) react to the same material or substance is different. (ii) Animal studies in pre-clinical research generally endure from 28–90 days and rarely investigate the long-term effects of new drugs, biologics, or medical devices; some of the harmful effects of materials may only materialize after many years of exposure. (iii) Despite the investigators gathering data on the risks of drugs, biologics, and medical devices during the four phases of clinical testing, the regulatory entity may continue to receive adverse event reports for many years after a product has been in the market. Clinical trials usually do not include enough subjects to detect rare adverse reactions.

It is necessary to follow 3,000 research subjects to have a 95% chance of detecting a rare drug reaction, but drugs often enter the market after being tested on fewer than 1,000 subjects (Resnik and Tinkle 2007). As a result, more than half of all new drugs require revisions in their safety information after they have been on the market, such as labeling changes or black box warnings.

To avoid or minimize the experiment's restrictions and to improve the understanding of the risks to human subjects exposed to a new substance or compound, investigators should conduct human cell and tissue studies to identify any potential differences between animals and humans before doing animal studies. Studies on genetic, metabolical, immunological, and physiological differences between animals and humans can help scientists to extrapolate from animal test to humans. Finally, pre-marketing clinical trials should include at least 3,000 total subjects and manufacturers should be required to conduct post-marketing studies.

Risk management in clinical trials involves the identification, assessment and balancing of risks and benefits. Research regulations require that risks to research subjects be reasonable in relation to the benefits to the subjects or to the society. All prominent international ethics guidelines, such as the Helsinki Declaration and the Council for the International Organization of Medical Science (CIOMS) Guidelines have similar requirements. To determine the reasonableness (or justification) of risks, sufficient information about both sides of the equation is required. One of the most important distinctions in the ethics and regulation of research is the distinction between research that poses no more than a minimal risk to subjects and research that poses more than a minimal risk. If the risks of a study are minimal, the benefits need only be more than minimal to justify the risks, while if the risks of a study are further than the minimal, the benefits must also be further than the minimal. Additionally, special protection for vulnerable population, such as children, foetuses, and prisoners, apply to more than minimal risk research. If a study exposes research subjects to nanomaterials, the risks will probably be more than minimal. Thus, the expected benefits to subjects or society must outweigh these risks to assure that the study meets ethical and legal requirements. Achieving a shared understanding of the application of risk management among diverse stakeholders is difficult because each stakeholder might perceive different potential harms, place a different probability on each harm occurring and attribute different severities to each harm. In relation to pharmaceuticals, a similar approach can be considered for nanomedicines. Although there are a variety of stakeholders, including patients and medical practitioners as well as government and industry, the protection of the patient by managing the risk to quality should be considered of prime importance (Food and Drug Administration 2011).

2.5 Nanomedicine under the Microscope: Industrial Revolution, Future Trends, Drawback and Controversy

Atoms and molecules compose the human body as all the matter. Therefore, the development of a novel technology that will permit the material manipulation at the same scale of atoms and molecules will dramatically increase the progress in human medical services. Rather than just an extension of "molecular medicine," nanomedicine will help to understand how the biological machinery inside living cells operates and such information can be used to address complicated medical conditions such as cancer, neurodegenerative diseases, AIDS, and aging, to mention some examples. At present, most nanomedicine approaches are centered on drug delivery to treat cancer by developing nanoscale particles to improve drug bioavailability, targeting and delivering with a cell precision. It has been also applied to medical imaging, diagnosis, creation of implantable devices and tissue regeneration. A more futuristic vision lies in the use of nanobots in medicine. By introducing these nano-sized robots into the body, damages and infections could be immediately localized and restored, holding the ability to change the traditional medical approaches.

However, nanomedicine research and the translation of nanomedicine products and technology from bench to bedside reveal a variety of issues that generate different concerns, questions, problems, requirements and interests. Those issues join different scenarios (*in vitro*, *in vivo*) and different scales (nano, micro, and macro) of action applied to different kinds of subjects (human, nonhuman) exposed.

Imposing a nanoscale gives the matter properties different from its conventional form. Every single nanoparticle has different characteristics and a combination of two or more of them may act unpredictably. Considering the fact that the naturally occurring nanoparticles may be harmful, it may be possible that the artificially made nanoparticles will be more harmful. Thus, before the full-fledged implementation shall start, an intense series of testing the features and characteristics of nanomedicine products should be done. Another reason of concern is the cost of nanotechnology applications. It is expected that the price of nanomedicine or any of its application will be quite expensive and certainly be out of reach for middle class people. New medical technologies are irrelevant for poor people if they are not accessible or affordable and, currently, most of the countries are still deprived of the basic medicinal facilities. In such a condition, the huge sponsorships and investment in nanomedicine are questioned.

Finally, nanotechnology, and particularly nanomedicine, raises many ethical and social issues that are associated with many emerging technologies, such as questions concerning risks to human beings and the environment and access to the technology, and several new questions, such as the use of nanotechnology to enhance human traits. Because the physicochemical properties of nanoscale materials have not been fully studied, clinical trials involving nanomedicine present some unique challenges related to risk minimization, management and communication involving human subjects. Although these clinical trials do not raise any truly novel ethical issues, the rapid development of nanotechnology and its potentially profound social and environmental impacts create a sense of urgency to the problems that arise. Long approval process and risks associated with nanomedicine restrain the market growth (Schillmeier 2015).

Summarizing, there is a colossal break with respect to the topic of nanomedicine applied to patient care among different fields of expertise: the viewpoint of researchers and medical experts, the business and legal professional and finally the ethical specialist. As soon as these viewpoints will be unified, the process transferring nanomedicine will be real.

References

Allon, I., A. Ben-Yehudah, R. Dekel, J. -H. Solbakk, K. -M. Weltring and G. Siegal. 2017. Ethical issues in nanomedicine: Tempest in a teapot? Med. Health Care Philos. 20: 3-11.

Bawa, R. 2004. Nanotechnology patenting in the US. Nanotech. L. & Bus. 1: 31.

Bawa, R., S. Bawa, S. B. Maebius, T. Flynn and C. Wei. 2005. Protecting new ideas and inventions in nanomedicine with patents. Nanomedicine 1(2): 150-158.

Bellare, J. R. 2011. Nanotechnology and nanomedicine for healthcare: Challenges in translating innovations from bench to bedside. J. Biomed. Nanotechnol. 7(1): 36-37.

Blanco, E., H. Shen and M. Ferrari. 2015. Principles of nanoparticle design for overcoming biological barriers to drug delivery. Nat. Biotechnol. 33(9): 941-951.

Bosetti, R. and L. Vereeck. 2011. Future of nanomedicine: Obstacles and remedies. Nanomedicine 6: 747-755.

Bowman, D., A. D. Marino and D. J. Sylvester. 2017. The patent landscape of nanomedicines. Med. Res. Arch. 5(9): 1-8.

Cai, W., C. C. Chu, G. Liu and Y. X. J. Wáng. 2015. Metal-organic framework-based nanomedicine platforms for drug delivery and molecular imaging. Small 11: 4806-4822.

Chan, W. C. 2017. Nanomedicine 2.0. Acc. Chem. Res. 50: 627-632.

Choi, H.S., W. Liu, P. Misra, E. Tanaka, J.P. Zimmer, B.I. Ipe, et al. 2007. Renal clearance of quantum dots. Nat. Biotechnol. 25: 1165-1170.

Dilnawaz, F., S. Acharya and S. K. Sahoo. 2018. Recent trends of nanomedicinal approaches in clinics. Int. J. Pharm. 538(1-2): 263-278.

Dubertret, B. 2005. Quantum dots: DNA detectives. Nat. Mater. 4(11): 797-798.

Dubertret, B., P. Skourides, D. J. Norris, V. Noireaux, A. H. Brivanlou and A. Libchaber. 2002. In vivo imaging of quantum dots encapsulated in phospholipid micelles. Science 298: 1759-1762.

Duncan, R. 2006. Polymer conjugates as anticancer nanomedicines. Nat. Rev. Cancer. 6(9): 688-701.

EPO, European Patent Oficie. 2018. Searching for Patents.

Etheridge, M. L., S. A. Campbell, A. G. Erdman, C. L. Haynes, S. M. Wolf and J. McCullough. 2013. The big picture on nanomedicine: The state of investigational and approved nanomedicine products. Nanomedicine 9(1): 1-14.

Ferrari, M., A. D. Barker and G. J. Downing. 2005. A cancer nanotechnology strategy. NanoBiotechnology 1(2): 129-131.

Food and Drug Administration. 2011. Guidance for Industry Postmarketing Studies and Clinical Trials – Implementation of Section 505 (O)(3) of the Federal Food, Drug, and Cosmetic Act. Food and Drug Administration.

Fornaguera, C. and M. J. García-Celma. 2017. Personalized nanomedicine: A revolution at the nanoscale. J. Pers. Med. 7(4): pii: E12.

Future, M. R. 2018. Nanobots Market Research Report – Global Forecast To 2023. ID: MRFR/MED/0793-HCRR.

Gao, X., Y. Cui, R. M. Levenson, L. W. Chung and S. Nie. 2004. In vivo cancer targeting and imaging with semiconductor quantum dots. Nat. Biotechnol. 22(8): 969-976.

Gao, Y., J. C. Kraft, D. Yu and R. J. Y. Ho. 2018. Recent developments of nanotherapeutics for targeted and long-acting, combination HIV chemotherapy. Eur. J. Pharm. Biopharm. S0939-6411(18)30157-7.

Garg, A., K. Bhalala, D. S. Tomar and W. Muhammad. 2017. Nanomedicine: Emerging trends in treatment of malaria. pp. 475-509. In: A. M. Grumezescu (ed.). Antimicrobial Nanoarchitectonics. Elsevier, Radarweg 29, PO Box 211, 1000 AE Amsterdam, Netherlands.

Giardiello, M., N. J. Liptrott, T. O. McDonald, D. Moss, M. Siccardi, P. Martin, et al. 2016. Accelerated oral nanomedicine discovery from miniaturized screening to clinical production exemplified by paediatric HIV nanotherapies. Nat. Commun. 7: 13184.

Gordon, N. and U. Sagman. 2003. Nanomedicine taxonomy. Canadian Institutes of Health Research (CIHR).

Grand View Research I. 2017. Nanomedicine Market Size Worth $350.8 Billion by 2025 | CAGR: 11.2%. Available at: https://www.grandviewresearch.com/press-release/global-nanomedicine-market.

Gutiérrez, V., A. B. Seabra, R. M. Reguera, J. Khandare and M. Calderón. 2016. New approaches from nanomedicine for treating leishmaniasis. Chem. Soc. Rev. 45: 152-168.

Havel, H., G. Finch, P. Strode, M. Wolfgang, S. Zale, I. Bobe, et al. 2016. Nanomedicines: From bench to bedside and beyond. The AAPS Journal 18: 1373-1378.

Hemingway, J., R. Shretta, T. N. Wells, D. Bell, A. A. Djimdé, N. Achee, et al. 2016. Tools and strategies for malaria control and elimination: What do we need to achieve a grand convergence in malaria? PLoS Biology 14: e1002380.

Hobson, D. W. 2009. Commercialization of nanotechnology. Wiley Interdiscip. Rev. Nanomed. Nanobiotechnol. 1(2): 189-202.

Hosnedlova, B., M. Kepinska, S. Skalickova, C. Fernandez, B. Ruttkay-Nedecky, Q. Peng, et al. 2018. Nano-selenium and its nanomedicine applications: A critical review. Int. J. Nanomedicine. 13: 2107-2128.

Hua, S., M. B. C. de Matos, J. M. Metselaar and G. Storm. 2018. Current trends and challenges in the clinical translation of nanoparticulate nanomedicines: Pathways for translational development and Commercialization. Front. Pharmacol. 9: 790.

Insights, C.M. 2018. Report CMI6393. Nanomedicines Market – Global Industry Insights, Trends, Outlook, and Opportunity Analysis, 2018-2026. https://www.coherentmarketinsights.com/ongoing-insight/nanomedicines-market-393.

Japan Patent Office. 2018. Japan Platform for Patent Information.

Jia, L. 2005. Global governmental investment in nanotechnologies. Curr. Nanosci. 1: 263-266.

Juanola-Feliu, E., J. Colomer-Farrarons, P. Miribel-Català, J. Samitier and J. Valls-Pasola. 2012. Market challenges facing academic research in commercializing nano-enabled implantable devices for in-vivo biomedical analysis. Technovation 32: 193-204.

Kendall, M. and I. Lynch. 2016. Long-term monitoring for nanomedicine implants and drugs. Nat. Nanotechnol. 11(3): 206-210.

Kumar, A., F. Chen, A. Mozhi, X. Zhang, Y. Zhao, X. Xue, et al. 2013. Innovative pharmaceutical development based on unique properties of nanoscale delivery formulation. Nanoscale 5: 8307-8325.

Kumar, N., B. Das and S. Patra. 2017. Drug resistance in tuberculosis: Nanomedicines at rescue. pp. 261-278. *In*: A. M. Grumezescu (ed.). Antimicrobial Nanoarchitectonics. Elsevier, Radarweg 29, PO Box 211, 1000 AE Amsterdam, Netherlands.

Laurencin, C. T. and L. S. Nair. 2014. Nanotechnology and Regenerative Engineering: The Scaffold. CRC Press, Boca Raton, FL 33487-2742, USA.

Lee, H., A. K. Lytton-Jean, Y. Chen, K. T. Love, A. I. Park, E. D. Karagiannis, et al. 2012. Molecularly self-assembled nucleic acid nanoparticles for targeted in vivo siRNA delivery. Nat. Nanotechnol. 7(6): 389-393.

Lv, G., W. Guo, W. Zhang, T. Zhang, S. Li, S. Chen, et al. 2016. Near-infrared emission CuInS/ZnS quantum dots: All-in-one theranostic nanomedicines with intrinsic fluorescence/photoacoustic imaging for tumor phototherapy. ACS Nano 10: 9637-9645.

Medintz, I. L., H. T. Uyeda, E. R. Goldman and H. Mattoussi. 2005. Quantum dot bioconjugates for imaging, labelling and sensing. Nat. Mater. 4(6): 435-446.

Morigi, V., A. Tocchio, C. Bellavite Pellegrini, J. H. Sakamoto, M. Arnone and E. Tasciotti. 2012. Nanotechnology in medicine: From inception to market domination. J Drug Deliv. 2012: 389485.

Murday, J. S., R. W. Siegel, J. Stein and J. F. Wright. 2009. Translational nanomedicine: Status assessment and opportunities. Nanomedicine 5: 251-273.

Murthy, R. S. R. 2012. Challenges and emerging issues in patenting nanomedicines. Patenting Nanomedicines. Springer, pp. 25-48.

Neuman, D. and J. N. Chandhok. 2016. Patent Watch: Nanomedicine patents highlight importance of production methods. Nat. Rev. Drug. Discov. 15(7): 448-449

nuevatribuna.es. 2017. La enfermedad, un negocio para la industria farmacéutica.

Pautler, M. and S. Brenner, S. 2010. Nanomedicine: Promises and challenges for the future of public health. Int. J. Nanomedicine 5: 803-809.

Perán, M., M. A. García, E. Lopez-Ruiz, G. Jiménez and J. A. Marchal. 2013. How can nanotechnology help to repair the body? Advances in cardiac, skin, bone, cartilage and nerve tissue regeneration. Materials 6: 1333-1359.

Placente, D., L. A. Benedini, M. Baldini, J. A. Laiuppa, G. E. Santillán and P. V. Messina. 2018. Multi-drug delivery system based on lipid membrane mimetic coated nano-hydroxyapatite formulations. Int. J. Pharm. 548: 559-570.

Portal, T.S. 2018. Nanomedicine Market Value in Latin America in 2017 and 2022.

Ragelle, H., F. Danhier, V. Préat, R. Langer and D. G. Anderson. 2017. Nanoparticle-based drug delivery systems: A commercial and regulatory outlook as the field matures. Expert Opin. Drug Delivery 14: 851-864.

Resnik, D. B. and S. S. Tinkle. 2007. Ethical issues in clinical trials involving nanomedicine. Contemp. Clin. Trials. 28(4): 433-441.

Santos, T., C. Boto, C. M. Saraiva, L. Bernardino and L. Ferreira. 2016. Nanomedicine approaches to modulate neural stem cells in brain repair. Trends Biotechnol. 34: 437-439.

Saravanan, M., B. Ramachandran, B. Hamed and M. Giardiello. 2018. Barriers for the development, translation, and implementation of nanomedicine: An African perspective. J. Interdiscip. Nanomed. 3(3): 106-110.

Sartuqui, J., C. Gardin, L. Ferroni, B. Zavan and P.V. Messina. 2018. Nanostructured hydroxyapatite networks: Synergy of physical and chemical cues to induce an osteogenic fate in an additive-free medium. Mater. Today Commun. 16: 152-163.

Sasaki, H. 2003. Near field optical microscope. Google Patents.

Satalkar, P., B. S. Elger and D. M. Shaw. 2016. Defining nano, nanotechnology and nanomedicine: Why should it matter? Sci. Eng. Ethics. 22: 1255-1276.

Science Medicines Health, S. 2018. European Medicines Agency (EMEA)

Schillmeier, M. 2015. Caring for social complexity in nanomedicine. Nanomedicine 10: 3181-3193.

Shi, J., P. W. Kantoff, R. Wooster and O. C. Farokhzad. 2017. Cancer nanomedicine: Progress, challenges and opportunities. Nat. Rev. Cancer. 17(1): 20-37.

Smith, L., D. R. Serrano, M. Mauger, F. Bolás-Fernández, M. A. Dea-Ayuela and A. Lalatsa. 2018. Orally bioavailable and effective buparvaquone lipid-based nanomedicines for visceral leishmaniasis. Mol. Pharm. 15: 2570-2583.

Soloperto, A., M. Bisio, G. Palazzolo, M. Chiappalone, P. Bonifazi and F. Difato. 2016. Modulation of neural network activity through single cell ablation: An in vitro model of minimally invasive neurosurgery. Molecules 21(8). pii: E1018.

Stevens, P. 2017. Diseases of Poverty and the 10/90 Gap. Fighting the Diseases of Poverty. Routledge. pp. 154-168.

USDHHS, U.S.D.o.H.a.H.S., 2018. U.S. Food and Drug Administration, FDA.

USPTO, U.S.P.a.T.O., 2018. Manual of Patent Examining Procedure Latest Revision January 2018 [R-08.2017].

Ventola, C. L. 2012. The nanomedicine revolution: Part 2: Current and future clinical applications. Pharmacy and Therapeutics 37(10): 582-591.

Venugopal, J. R. and S. Ramakrishna. 2016. Nanotechnology: 21st century revolution in restorative healthcare. Nanomedicine (Lond). 11(12): 1511-1513

Wei, C., N. Liu, P. Xu, M. Heller, D.A. Tomalia, D.T. Haynie, et al. 2007. From bench to bedside: Successful translational nanomedicine: Highlights of the Third Annual Meeting of the American Academy of Nanomedicine. Nanomedicine 3: 322-331.

WHO, 2017. World Health Organization. Cardio Vascular Diseases (CVDs). Available from: http://www.who.int/news-room/fact-sheets/detail/cardiovascular-diseases-(cvds).

WHO, 2018a. HIV/AIDS. Data and statistics. http://www.who.int/hiv/data/en/.

WHO, 2018b. Leishmaniasis. Key facts. http://www.who.int/en/news-room/fact-sheets/detail/leishmaniasis.

WHO, 2018c. World Health Organization. World cancer day 2018. Available at: http://www.who.int/cancer/en/.

Wilhelm, S., A.J. Tavares, Q. Dai, S. Ohta, J. Audet, H.F. Dvorak, et al. 2016. Analysis of nanoparticle delivery to tumours. Nat. Rev. Mater. 1: 16014.

Woodson, T. S. 2016. Public private partnerships and emerging technologies: A look at nanomedicine for diseases of poverty. Research Policy 45(7): 1410-1418.

Yao, J., P. Li, L. Li and M. Yang. 2018. Biochemistry and biomedicine of quantum dots: From biodetection to bioimaging, drug discovery, diagnosis, and therapy. Acta Biomater. 74: 36-55.

Yin, H., R. L. Kanasty, A. A. Eltoukhy, A. J. Vegas, J. R. Dorkin and D. G. Anderson. 2014. Non-viral vectors for gene-based therapy. Nat. Rev. Genet. 15(8): 541-545.

Zhang, S., W. Guo, J. Wei, C. Li, X. -J. Liang and M. Yin. 2017. Terrylenediimide-based intrinsic theranostic nanomedicines with high photothermal conversion efficiency for photoacoustic imaging-guided cancer therapy. ACS Nano 11: 3797-3805.

Zhang, Y. S. and A. Khademhosseini. 2015. Seeking the right context for evaluating nanomedicine: From tissue models in petri dishes to microfluidic organs-on-a-chip. Nanomedicine 10: 685-688.

Zhang, Z., T. W. Eyster and P. X. Ma. 2016. Nanostructured injectable cell microcarriers for tissue regeneration. Nanomedicine 11: 1611-1628.

Zumla, A., A. George, V. Sharma, R. H. N. Herbert, A. Oxley and M. Oliver. 2015. The WHO 2014 global tuberculosis report-further to go. The Lancet Global Health 31(1): e10-e12.

Zuo, L., W. Wei, M. Morris, J. Wei, M. Gorbounov and C. Wei. 2007. New technology and clinical applications of nanomedicine. Med. Clin. North Am. 91(5): 845-862.

CHAPTER

3

Nano-sized Organization in Nature

"A miracle constantly repeated becomes a process of nature."

Lyman Abbott (1835–1922)
American Congregationalist Theologian

"No human technology can replace nature´s technology, perfected over hundreds of millions of years to sustain life on Earth."

Marco Lambertini
General Director of World Wildlife Fund (WWF) International

3.1 The Scale at which Biological Processes Befall

How can nanotechnology and medicine converge and, from their matrimony, conduct the birth of a new discipline: the nanomedicine? The answer to this question is simple, nanotechnology works on the same scale as biology.

Human beings like to think of themselves as the great architects of innovation but, as is usually the case, Mother Earth beats us to the knockout. Nature has refined the art of biology at the nanoscale centuries ago. It has created a plethora of inorganic and organic nanomaterials in the forms of nanotubes, nanodots, nanowires, and even nanomachines, acting as the first level of organization of any biological system where its vital properties are defined in response to a specific demand of highly functional optimized adaptations (Fratzl 2007).

A good example is the cell. The surrounding's lipid bilayer is on the order of 6 nm thick and all inner cellular structural components like microtubules, microfilaments, chromatin and organelles work at the nanoscale (Ti Tien and Ottova-Leitmannova 2000, U.S.- NNI 2015). In the visual cortex of the human brain, neurotransmission depends on the movements of transmitter-laden organelles of average diameter of 39.5 ± 5.1 nm, the synaptic vesicles (Horih 1989, Qu et al. 2009). This is also the scale of large biological macromolecules such as haemoglobin, the protein that, exhibiting 5.5 nm in diameter, carries oxygen through the body, and DNA, the molecule that in its helix of about 2 nm diameter and a short structural helical pitch of 3.4–3.6 nm, encodes the genetic information of all life forms (Bueno 2011, U.S.- NNI 2015). Likewise, several types of tissue extracellular matrix (ECM) exhibit mesh-like nanoscale structures; for example, the basement membrane, a ubiquitous ECM structure separating cell layers from interstitial ECM responsible for a number of tissue functions, including polarization and compartmentalization, shows elevations of 150 –190 nm, fibre diameters of 77 nm, and pore diameters of 72 nm organized in a general repetitive pattern spaced on the order of 100 nm; ECM fibres in the dermis have diameters of approximately 60 – 120 nm; and in bone tissue, collagen fibres have diameters between 80 and 100 nm, with small 1 – 4 nm thick and 50×25 nm^2 carbonated apatite regions. Some ECM environments are rather chaotic, while others, including the myocardium and tendons, are highly organized, exhibiting fibres of approximately 100 nm diameter strongly parallel aligned with the cell layer (Young et al. 2016).

Observation of the biological world serves to understand and to take advantage of naturally occurring phenomena when the matter is organized at the nanoscale. It is a source of inspiration for scientists to delve on the conception of novel methods that can enhance their work in many fields, including medicine. It is not a simple question of worked at smaller dimensions; rather, operating at the nanoscale enables scientists to utilize the unique physical, chemical, mechanical, and optical properties of materials that naturally occur at such scale.

Drawing on the natural nanoscale of biology, many researchers are working on designing tools, treatments, and therapies that are further accurate and personalized than conventional ones and, that can be applied preventively in the course of a disease and have fewer side-effects. Research on biosystems at the nanoscale has created one of the most dynamic science and technology fields at the confluence of physical sciences, molecular engineering, biology, biotechnology and medicine; the reasons are quite diverse.

First, there is a growing need for new types of biomaterials with specific and well-defined interaction with biological host systems. One medical example is the development of surface-directed nanobiotechnology techniques for the manipulation of molecules within cells, including the use of bioselective surfaces, control of biofouling and cell culture (Roco 2003).

Secondly, recent advances in material characterization and fabrication technologies have prompted scientists to ask how one can reformulate the structure of natural materials, which developed in the course of evolution, into biomimetic designs for engineering applications. The exploration of a radically new approach for the design of biomimetic/bio-inspired materials included the development of synthetic polymers with stimuli-responsive mechanical properties that can be exploited, for example, in biomedical implants, robotic elements or orthopedic devices (Bello et al. 2013).

Finally, it is increasingly recognized that the properties of biomolecular structures, tissues and organs, derived from nanoscale organization can be critical for their biological function and that also has impact in medicine. Some scientists are looking at ways to use nanoscale biological principles of molecular self-assembly, self-organization, and quantum mechanics to repair damage and to induce self-healing (Ozin et al. 2009, U.S.- NNI 2015). There are, indeed, many opportunities to learn lessons from the nanoscale natural world: on growth, functional adaptation, hierarchical structuring, on damage repair and self-healing (Fratzl 2007).

In this chapter, we will analyse the structure-function relations in natural configurations and how this information can be applied to the construction of novel engineering bio-inspired materials.

3.2 Nanoscale Assembly in Nature

The main principles on which nanotechnology is founded, that is, the ability to reversible hierarchically assemble atoms and molecules into objects, is what nature has already been doing in living systems and in the environment through millenniums. Rearranging matter at the nanoscale using "weak non-covalent" molecular interactions, such as van der Waal forces, hydrogen bonds, electrostatic dipoles, fluidics and several surface forces, requires low-energy consumption and allows for reversible subsequent changes. Such changes of usually "soft" nanostructures in a limited temperature range are essential for the occurrence of bioprocesses and as a consequence, biosystems are governed by nanoscale controls that have been optimized over millions of years to attain specific functional adaptations (Roco 2003).

3.2.1 The Self-assembly Concept

Molecular self-assembly is defined as the spontaneous organization of molecules, according to their non-covalent type reciprocal interactions, into spatial and/or temporal ordered aggregates (Mendes et al. 2013). Supramolecular chemistry provides the basis of self-assembly where the instructions of how to assemble into larger entities are coded in the structural motifs of individual molecules. According to Whitesides (Whitesides et al. 2005), the structural units of a molecular self-assembling system involve a group of similar or different molecules, or segments of a macromolecule that interact with one another. Their interaction begins from a less arranged state (a solution, a disordered aggregate or random coil conformation) leading to a final state (a crystal or a folded macromolecule) that is well-organized.

Two main grades of self-assembly have been considered within the above definition: (i) the static and (ii) the dynamic self-assembly. In the first one, the static self-assembly, the constituents form ordered static structures that do not temporally change without energy exchange with the environment; for example, folded globular proteins. On the other side, during a dynamic self-assembly process, ordered non-equilibrium structures are formed. They are maintained far from equilibrium through an input of energy from an external source and its subsequent dissipation into the environment; for example, biological cells (Whitesides and Grzybowski 2002).

3.2.2 Biological Self-assembly

Self-assembly and replication, the paradigms of molecular and cellular biology, greatly inspired the concept of *"machines that make themselves"* from Drexler's book *Engines of Creation* (Drexler 1986) and became the standard of tissue engineering and regenerative medicine (Furth et al. 2007, Bernstein

2011). According to the principle of "programmed chemistry" promoted by Lehn (Lehn 1995), the natural molecular components come already programmed with the information needed to bring them together in the right configuration to construct complex structures. Thus, biology is replete with examples of highly functional complex nanoscale structures formed by self-assembly. One of the greatest well-designed self-assembled example from nature is the cellular membrane; at its simplest level it is a bilayer of phospholipid molecules. Lipids have all the chemical functionality they need, a hydrophilic head group and a hydrophobic tail, to ensure assembly into the oriented array in which the hydrophobic regions are shielded from water (Ball 2002). This particularly valuable arrangement permits several different nanostructured organizations: flat lamellar stacks, closed vesicles (liposomes) with shapes ranging from spherical to disc-like or branched, and topologically complex disordered and ordered networks like those seen, respectively, in the smooth endoplasmic reticulum and in "cubosomes" (Ball 2002). The local membrane environment of these nanostructures influences dynamics and membrane receptor distributions, which in turn impacts the signal transduction and, evidently, the structure functionality (Cambi and Lidke 2012).

Another particularly refined nanoscale molecular self-assembly process in nature is the folding of a protein to its native state; here the concept of "programmed molecules" operates in one of its most remarkable manifestations. The proteins contain all the information required to acquire its correct functional configuration in the form of hydrophobicity, hydrogen-bonding capacity, electrostatic charge distribution and other properties of the amino acid residues. Even when the folding process requires the support of several molecular assistants, the linear polypeptide chain is pre-programmed to adopt and maintain its compact, three-dimensional arrangement (Ball 2002).

The aforementioned process has special relevance because proteins are nature's building blocks. They are programmed to assemble and operate as templates in the crucial step to chemical replication or to form materials that confer stiffness, organization and utility to all biological networks. The idea of using non-covalent supramolecular interactions to assist covalent synthesis is inherent in the formation of proteins on messenger ribonucleic acid (mRNA), catalyzed by the ribosome, as well as the transcription of the mRNA on deoxyribonucleic acid (DNA) and the replication of DNA itself (Ball 2002). On the other hand, the processes of template and of co-operative self-assembly between macromolecules are at the core of natural material generation and result in functional constructs possessing structural hierarchy ranging from the nano- to the macroscale, for example during biomineralization. The basis of any biomineralization process comprises a delicate and precise interplay between assemblies of macromolecules and forming mineral phases in a bottom-up approach. Macromolecules guided the nucleation and growth of the inorganic phase in a molecular level starting on atomic and molecular scales, followed by the formation of nanostructured building blocks and finalizing with their organization into complex hierarchical structures. Many organisms incorporate inorganic solids into their tissues to enhance their functionality, primarily mechanical hardening or stiffening, as support of the body, protection of the vital organs and defense against predators. These mineralized tissues, also called biominerals, are unique organo-mineral nanocomposites, organized at nano to macroscale hierarchical levels (Beniash 2011). To mention some examples, we can include the formation of silicates in algae and diatoms (Sumper and Brunner 2006), carbonates in invertebrates (Tang et al. 2003), and calcium phosphates and carbonates in the hard tissues of vertebrates (Fratzl 2007, Lotsari et al. 2018).

The self-assembly course in nature involves much more than just the self-association of molecules; it must be reversible in some specific biological processes. Let us say, it is a characteristic during cell surface signaling, neurotransmission and along the mechanisms of the immune response. For example, in molecular replication templating an eventual separation of the product from the template is required, so the bonds holding the two together while assembly takes place must eventually be broken. Moreover, if molecular recognition and self-assembly is included in some dynamical process such as ion transport, e.g. when ionophores such as valinomycin selectively bind metal (K^+) ions and carry them across a cell membrane, at that point the substrate must be able to leave the receptor as well as to bind to it. In proteins such as haemoglobin, a conformational change alters the substrate binding affinity, highlighting the relationship between rigidity and selectivity on the one hand, and flexibility and reversibility on the other (Ball 2002).

The previously mentioned examples, are only few cases that clearly show how sophisticated self-assembly can be and how nature's assembly principles are used to create supramolecular materials with unique and precise purposes. Although the specific approaches may be different from the slowly evolving living systems in aqueous medium, many concepts such as self-assembly, templating of atomic and molecular structures on other nanostructures, interaction on surfaces of various shapes, self-repair, and integration on multiple length scales can be used as sources of inspiration.

3.3 Why Copy Nature?

Even if nature has not necessarily found all the best ideas already, we have to recognize that it does have some awfully good ones. Within this context, the science of biomimicry, defined as the copying, adaptation or derivation of biology, has relatively recently gained popularity. However, the idea has been in existence for thousands of years. Since the Chinese attempted to make artificial silk around the year 200 BC, there have been many examples of humans learning from nature to design new materials and devices (Gleason 2006). Leonardo da Vinci, for example, designed ships and planes by looking at fish and birds, respectively (Kemp 2007). The Wright brothers created the first successful airplane model only after comprehending that birds do not fold their wings constantly, rather they slide along on air currents (Jakab 2014). In 1866, the Engineer Karl Culmann, during a visit of the dissecting room of anatomist Hermann von Meyer, discovered a conspicuous connection between the patterns of stress trajectories (tension and compression lines) in a loaded crane-head and the anatomical trabecular architecture of the proximal human femur (Mow and Huiskes 2005). In other words, nature has strengthened the bone accordingly to its function, precisely in a manner dictated by modern engineering (Byrom 2014).

But, how can we learn from nature? There are some remarkable differences between the strategies commonly used in engineering and those used by nature. Primarily, life uses mild conditions, enabling sustainable solutions; it does not operate at very high temperatures, in high-vacuum conditions or in organic solvents. Another difference is the selection of elements, which is far away from that of the engineer. Elements such as iron, chromium and nickel are very rare in biological tissues and certainly they are not used in the metallic form, as would be the case for steel (Fratzl 2007). Most of the structural building blocks used by nature are polymers or nanocomposites. Such supplies would usually not to be the first choice of an engineer to build strong and long-bearing loads mechanical structures. Nevertheless, nature uses them to build trees and skeletons. The last difference is the approach in which materials are made. While an engineer chooses the proper material to fabricate a piece according to an exact design and function, nature goes the opposite way and grows both the piece and the whole organism (a plant or an animal) using the principles of self-assembly providing control over the structure of the material at all levels of hierarchy. That is definitely the key point in the successful use of polymers and composites as structural building blocks.

Summarizing, nature shows that if we paid a little attention both to the biological structure and to the set of problems that such structure is designed to solve, we can take living examples as inspiration for the construction of novel and efficient materials applied to multiple requirements. These are especially useful in new nanomedicine approaches.

3.4 The Dynamic Strategy

As we have mentioned previously, in the case of an engineered machine, each of its parts are designed. Then, the raw material for their construction are selected according to knowledge and experience regarding the functional requirements, taking into account possible changes in those requirements during service (e.g. typical or maximum loads) and fatigue (and other lifetime issues) of the material. This strategy is static, since the design is established at the beginning and is maintained during the useful life of the machine, hoping that it meets all the requirements for what it was created. Life, on the other hand, involves a dynamic strategy: *growing rather than being fabricated*. It is not the exact design of the organ that is coded in the genes, but rather a recipe to build it (Fratzl 2007, Fratzl and Weinkamer 2007).

This means that the final result is rather obtained by an algorithm than by the replication of a design. This approach presents three main advantages. First, it allows flexibility at all levels, enabling a functional adaptation through the growth. For example, the bone in a healthy person or an animal will adapt to the loads under which it is placed; that is the statement of "Wolff's law", developed by the German anatomist and surgeon Julius Wolff (1836–1902) in the 19th century. If the load on a particular bone increases, the bone will remodel itself over time to become stronger to resist that sort of loading. The internal architecture of the trabeculae undergoes adaptive changes, followed by secondary changes to the external cortical portion of the bone, perhaps becoming thicker as a result. The inverse is true as well: if the loading on a bone decreases, the bone will become less dense and weaker due to the lack of the stimulus required for continued remodeling (Frost 1990).

Second, it allows the growth of hierarchical materials, where the microstructure at each position of the part is adapted to the local needs. Following the bone example, seven levels of hierarchy can be determined in the structure of bone tissue. Mineralized collagen fibrils are the basic building block of bone tissue at the first hierarchical level; they are a composite of collagen and mineral nanoparticles that closely resemble the synthetic hydroxyapatite ($Ca_{10}(PO_4)_6(OH)_2$, HA) crystals (Messina et al. 2017). Collagen proteins ~200 nm length and 2 - 3 nm in diameter self-assemble into fibrillary clusters, which is the second level of hierarchy. Then they further aggregate into superior size fibrils arrays of about 500 nm in diameter, the third level of hierarchy. TEM and neutron scattering observations revealed that the regular stacking pattern of collagen molecules in a fibre leaves small gap zones with 35 nm length and overlap zones with 32 nm length. This striped structure is occupied by plate-shaped HA nanocrystals of 10 - 20 nm in length and 2-3 nm wide, that are mostly arranged parallel to each other and to the long axis of collagen fibrils. Crystals arise at regular intervals along the fibrils, with an estimated repetitive distance of 67 nm. In addition to crystals embedded in fibrils, there is also an extra-fibrillar mineral coating the whole structure. The fourth level of hierarchy is the pattern of arrays that they are formed. These include the organization in lamellar structures that resembles a rotated plywood assembly, where the fibres are parallel within a thin sub-layer and where the fibre direction rotates around an axis perpendicular to the layers. The origin of the rotated plywood assembly could be a twisted-nematic (or cholesteric) liquid crystalline array of collagen fibres (Ruso et al. 2015). At a macroscopic level and determined by the type of bone and its associated function in the body, these mineralized fibres are arranged into larger level structures. Cylindrical structures called "osteons" are the fifth level of hierarchy, while the sixth level of bone organization is the classification of osseous tissue as either spongy (trabecular or cancellous bone) or compact (cortical bone). The final level is simply the whole structure on the macroscopic scale, incorporating all of the latter mentioned levels of hierarchy (Drouet et al. 2012). In addition to the bone tissue, other examples of structural hierarchical biological materials are trees, seashells, spider silk (Eisoldt et al. 2011), the attachment systems of geckos (Greiner et al. 2009), super-hydrophobic surfaces (Guo et al. 2011), optical microstructures (Vukusic and Sambles 2003), the exoskeleton of arthropods (Romano et al. 2007) and the skeleton of glass sponges (Tang et al. 2003).

The dual need for optimization of the part's form and the material's microstructure is well known for any engineering problem. However, in natural materials, shape and microstructure become intimately related due to their common origin, which is the growth of the organ. Hence, at every size level, in the example that we saw, bones modify the form and material simultaneously. This leads us to the third advantage of the flexibility design that nature imparts to the materials at the time of its creation. The processes of growth and "remodeling" (this is a combination of growth and removal of old material) allow a constant renewal of the material, thus reducing problems of material fatigue, enabling healing and self-repair.

At higher levels, many organisms have the capability to remodel the material. For our case in point, the bone, specialized cells (osteoclasts) are permanently removing material, while other cells (osteoblasts) are depositing new tissue. In technical terms, this would mean that a sensor/actuator system is put in place to replace damaged material wherever needed. To close, nature can heal a critically damaged tissue. In most cases, wound healing is not a one-to-one replacement of a given tissue, except for the bone tissue, which is able to regenerate completely and where the intermediate tissue (the callus) is eventually replaced by a material of the original type (Fratzl 2007).

With such examples from nature as a guide, dynamic design and the hierarchical, self-assembly construction are becoming more and more a laboratory tool for the building of mesoscopic structures with increasing levels of complexity. Such hierarchically formed architectures may exhibit unique properties and functions that are not displayed by their individual components and represent a major opportunity for biomimetic materials research.

3.5 Lessons from Nature: Nanoscale Biomimetic and Bioinspired Applications in Nanomedicine

3.5.1 Hierarchical Frameworks Applied to Tissue Regeneration

Hierarchical arranging is one of the omnipresent natural forms in biological systems and is a result of the evolution process in the conception of tissues. In this way, it is not strange that tissue engineering, generally conceived as a tool to re-establish lost functionality and morphology to a previously damaged tissue, emphasizes in detail the role of the specific organization, i.e. the hierarchy, in every part of a biomaterial synthetic design. Hierarchical structuring allows the construction of large and complex biological tissues based on much smaller, often very similar, building blocks. It also allows the adaptation and optimization of the matter at each level of hierarchy to yield outstanding performance, often relating to a specific mechanical function besides fulfilling other functional requirements. Therefore, the significance of hierarchical structuring on mechanical properties has been exploited through many currently available tissue regeneration methodologies.

As it was mentioned previously, examples of such building blocks are collagen fibrils in the bone tissue, which have units with a few hundred nanometers thickness and can be assembled to a variety of bones with very different functions (Fratzl 2007). The combined action of such structural elements at the nano- and micro-meter levels provides to the bone a superior stiffness than classic tissues, while at the same time offers a remarkably high fracture toughness and resistance to damage (Messina et al. 2017). Having a wealth of potential clinical applications from the treatment of nonunion fractures to spinal fusion and long bones damaged areas substitution, the enhancement of bone regeneration is an exciting approach to directly repair bone defects or engineer bone tissue for transplantation. The construction of hierarchical porous scaffolds from the synergic entanglement of biogenic ceramics and biopolymer components plays a pivotal role in providing both template and an extracellular environment to support regenerative cells and to induce tissue regeneration. Current challenges include the engineering of materials that can match simultaneously with the mechanical and biological context of real bone tissue matrix and support the vascularization of large tissue constructs. Scaffolds with new levels of biofunctionality that attempt to recreate the fibrillary hierarchical organization, the nanoscale topography and the biofactor cues from the extracellular environment are emerging as interesting candidates for biomimetic materials (Stevens 2008, Liu et al. 2016, Dang et al. 2018).

The inspiration from natural marine skeletons also has been applied to the use and development of functional musculoskeletal tissue-engineered constructs. Marine skeletons, such as seashells exhibiting dense lamellar structures or the sea urchin, cuttlebone and coral displaying interconnected porous structures, have tailored architectures, which give them structural support and are enriched with bioactive elements that potentially could be either effective to improve existing bone repair materials or to develop novel bone tissue materials (Espinosa et al. 2011, Nandi et al. 2015, Green et al. 2017, Neto and Ferreira 2018).

On the other hand, the well-known mechanical characteristics (Du et al. 2011) such as high elasticity, extreme dilatability and tensile strength of silk, added to its degradability in mild conditions (ambient temperatures, low pressures, and water as solvent), and high cytocompatibility and low immunogenicity *in vivo* served as inspiration for the construction of a variety of biomedical textiles and fibre-based implants. Those are routinely in clinical use for nearly five decades facilitating healing; some examples include wound dressings, vascular grafts, heart valves, sutures to wearable medical implants and polymer sensors (Li et al. 2015).

Currently, spider silk characteristics were exploited in the construction of a nerve scaffolding that assists the endogenous repair processes of extensive nerve injuries of more than five centimeters in length (Radtke et al. 2011).

This really constitutes a breakthrough because, although surgical reposition of peripheral nerve results in some axonal regeneration and functional recovery, the clinical outcome in long distance nerve defects is disappointing and research continues to utilize further interventional approaches to optimize functional recovery (Radtke et al. 2011).

3.5.2 Biomimetic Camouflage to Drug Delivery

Increasing attention is being devoted to the biomembranes' configurations, their specific assembly and their associated function, see section 3.1 (Ball 2002). Thus, it is no wonder that the amphiphilic sheet, bilayer and vesicle have become familiar motifs in biomimetic materials and structures. Biomembranes biomimicry has been exploited to the point of exhaustion during the production of drug-delivery vehicles. Given the complexity of mass transport into the vascular environment, "biomimetic camouflage" strategies gained popularity and have been proposed as potential strategies for overcoming vascular barriers to drug delivery (Blanco et al. 2015).

The therapeutic efficacy of systemic drug-delivery vehicles depends on their ability to evade the immune system, to cross the biological barriers of the body and to localize at target tissues. All these abilities were imparted, for example, to nanoporous silicon particles after their coating with cellular membranes purified from leukocytes (Parodi et al. 2012). Macrophage cell membrane (MPCM) camouflaged mesoporous silica nanocapsules (MSNCs) loaded with hydrochloride doxorubicin (DOX) was also prepared using a top-down assembly by Xuan and coworkers (Xuan et al. 2015). The MPCM has two functions, offering camouflage and providing active targeting ability with the guidance of surface proteins. These developed biomimetic drug delivery platform effectively demonstrated a high loading capacity, an immunological adjuvant, and a tumour-assisted targeting action through the amalgamation of synthetic nanoparticles and natural membrane edges.

For decades, poly(ethylene glycol) (PEG) has been widely incorporated into nanoparticles for evading immune clearance and improving the systematic circulation time. However, recent studies have reported a phenomenon known as "accelerated blood clearance (ABC)" where a second dose of PEGylated nanomaterials is rapidly dispersed when given several days after the first dose. After biomimetic coating of Fe_3O_4 nanoparticles with red blood cell (RBC) membrane, Rao and coworkers (Rao et al. 2015a) demonstrated that natural RBC membrane is a superior alternative to PEG to evade the immune clearance response through interactions with the signal regulatory protein-alpha (SIRP-α) receptor.

Interestingly, research groups have attempted to use the ubiquitous protein corona to their advantage to target specific cells and/or disease sites. Kreuter et al (Kreuter et al. 2002), using poly(butyl cyanoacrylate) nanoparticles coated with polysorbate 80, demonstrated enhanced drug delivery beyond the blood-brain barrier. Through adsorption of apolipoproteins to the polysorbate 80 surface, nanoparticles effectively act as

"Trojan horses" that mimic lipoprotein particles, traversing the blood-brain barrier through low-density lipoprotein–mediated endocytosis. Another example is the one attained by Cao et al (Cao et al. 2017); a nano-assembly (DRI-S) was created containing human serum albumin (HSA) which acted as a biomimetic nano-corona onto an IR-780 iodide (IR-780) and small interfering RNA-loaded cell-penetrating peptide. Camouflage (DRI-S@HSA) nanoparticles exhibited high accumulation and deep penetration within tumour tissues, thereby holding a great promise for synergistic therapy.

Inspired by the remarkable performance of the "cell membrane-camouflage" nanoparticles, and taking into consideration the large number of molecular recognition moieties of stem cell membranes that can be used for tumour-targeting materials, Gao et al (Gao et al. 2016a) developed a bone marrow derived mesenchymal stem cell membrane-coated gelatin nanogels (SCMGs) as highly efficient tumour-targeted drug delivery platforms. The same research group reported a biomimetic delivery of microsized capsule-cushioned leukocyte membrane vesicles (CLMVs). This system was created through the conversion of freshly reassembled leukocyte membrane vesicles (LMVs), including membrane lipids and membrane-bound proteins, onto the surface of layer-by-layer assembled polymeric multilayer microcapsules (Gao et al. 2016b).

3.5.3 Insect-inspired Optoelectronic Structures at the Service of Medical Imaging

By observing optical science and technology, it can be found that most of the basic physical ideas were already evolved from nature over millions of years ago. Accordingly, advances in micro- and nanofabrication technologies enable insect-inspired optoelectronic structures leading to new optical working principles and new optical components (Fink 2018). Insects bring extraordinary tutorials through the structures and the functions of their nanoarchitectural organizations, like versatile photonic properties, for example, wide field-of-view (FOV), negligible distortion, low aberration, and deep depth-of-field as well as highly efficient light focusing and guiding. A wide range of insect configuration from a plain topology to delicate and complicated schemes provides a comprehensive archive of natural templates and optical technologies to hire, duplicate, and adopt (Chung et al. 2018). Multilevel and hierarchical assemblies incorporating insect micro- and nanostructures designed to reproduce observed properties in nature were largely exploited for optical colour sensing, virtual reality, 3D imaging, light field cameras, robotic systems, security including surveillance and collision-free navigation of terrestrial and aerospace vehicles (Chung et al. 2018). Nanomedicine has also benefited from multifaceted eye structures in insects. The study of ultra-thin imaging systems in nature resulted in the construction of advanced camera lenses. Later they were incorporated into miniature imaging systems, improving optical performance such as FOV, resolution and aberration. Recently, such miniaturized imaging lens systems have been actively exploited for biomedical applications (Cogal and Leblebici 2017).

Inspiration from the multifunctional biophotonic nanostructures found on the transparent wings of the longtail glasswing butterfly (*Chorinea faunus*) was used to advance into the versatility of micro-optical implants whose practical use is often limited by the angle dependency of sensing and the readout processes (Narasimhan et al. 2018). A combination of bottom-up nanofabrication with top-down microfabrication processes was used to yield a nanostructured micro-optical implant that senses intraocular pressure (IOP) for diagnosis and management of glaucoma, a leading cause of irreversible blindness (Fink 2018).

3.5.4 Improving Nanomedicine via Nanoscale Biological Interfaces

Nanotechnology holds great potential for transforming the medicine concepts. One of the most important of these conceptions is the biological interaction of nanomaterials with interfacial biomolecules and surroundings (Park and Hamad-Schifferli 2010). Physical communications of soft matter under biologically relevant conditions are governed by complex interplays of generic and specific interfacial interactions, which are clearly different from those acting at the interface between hard matters (Tanaka 2013). Taking into account such interactions in nature, very significant advances can be made in the development of a new device or medical treatment using nanotechnology schemes.

A representative example is the persistent clinical need to prevent blood clotting while minimizing administration of anticoagulant drugs. This fact has led to the search for biomaterial surface coatings that can directly suppress blood clot formation. To date, the most effective approach has been the chemical immobilization of heparin on blood-contacting surfaces to reduce thrombosis and lower anticoagulant administration. Although this tactic has been widely adopted, there is a limitation due to the surface-bound heparin leaches, which results in the progressive loss of anticoagulation activity. Thus, the use of heparin-coated materials has not led to a drastic reduction in the clinical use of soluble heparin. In the search for new contact surfaces for medical devices, the researchers focused on natural designs such as those presented by gecko skin. The multifunctional features of the gecko skin provide a potential natural template for synthetic applications where specific interfacial control of biological interactions are required. The skin of the box-patterned gecko (*Lucasium* sp.) consists of dome shaped scales arranged in a hexagonal patterning. It is comprised of spinules (hairs), from several hundred nanometers to several microns in length, with a sub-micron spacing and a small radius of curvature typically from 10 to 20 nm. This micro- and nanostructure of the skin exhibited ultralow adhesion with contaminating particles. The topography also provides a super hydrophobic, anti-wetting barrier, which can self-clean by the action of low velocity rolling or impacting droplets of various size ranges from microns to several millimeters (Watson et al. 2015). It was demonstrated that gecko skin bioinspired surface coating on medical devices completely repels blood-preventing thrombosis and can be applied to the creation of innovative implantable devices (Leslie et al. 2014).

One engineering challenge solved in nature, but plagued of medical conditions, is biological fouling, commonly referred to as biofouling (Bixler and Bhushan 2012). For humans, medical biofouling, highly associated with prosthetic and dental implants, biosensors, catheters and medical equipment, is the population of surfaces from pathogenic bacteria and the formation of immovable biofilms. They cause 80% of infections in medical care facilities, by releasing toxins or interfering with the functions of a range of medical devices *in situ* and *ex situ* (Li et al. 2016). Biofouling drive to significant health risks and financial losses.

Biomimetic approaches addresses the importance of surface textures and chemistries that influence the antifouling properties. Furthermore, and according to the concept of "the race for the surface", improving biocompatibility will reduce medical biofouling as fouler will not target the implanted device (Subbiahdoss et al. 2009). Here, the inspection of assembly designs found in natural organisms provides also a powerful way of discovering new strategies and innovations to solve the problem. Naturally occurring nano-protrusion patterns have been shown to influence the contact of bacteria on surfaces. Insect wing membranes structures (Kelleher et al. 2015) and the gecko skin (Leslie et al. 2014) resist microbial attack while these types of surfaces with micro and nanostructures have been shown to be biocompatible with eukaryotic cells. These antibacterial surfaces, which are capable of exhibiting differential responses to bacterial and eukaryotic cells, represent surfaces that have excellent prospects for biomedical applications (Bhadra et al. 2015, Andrés et al. 2018, Rigo et al. 2018), including to fight bacteria on contact lenses and artificial cornea (Han et al. 2018).

The adequate superficial nanophotography can also be used to induce tissue regeneration because stem cells respond to nanoscale surface of ECM features. The interaction of nanotopographical features with integrin receptors in the cells' focal adhesions alters how they adhere to material's surfaces and defines their fate through both biochemical and morphological changes. Inspired by the hierarchically organized protein fibres in extracellular matrix (ECM) and assessing the exquisite control that they exert on stem cells response, the material scientists are beginning to understand how the behaviour of stem cells in their niche can be controlled (Dalby et al. 2014). Such information is being currently used for the development of the next generation of culture materials, of new implant interfaces, and to fuel the discovery of stem cell therapeutics to support regenerative therapies. Thus, novel biomaterial matrices are being developed to mimic the nanotopography characteristics of the ECM, including the presenting adhesion sites (Bae et al. 2015, Chen et al. 2018, Sartuqui et al. 2018).

3.5.5 Stimuli Responsive Polymers

Response to stimulus is a basic process of living systems, for example, kinetic and thermodynamic control of the coil-to-globule transition in proteins, the basic structural units of most self-assembled structures in nature, that are achieved through changes in polymer composition and topology (Jeong and Gutowska 2002). Based on the lessons from nature, scientists have been designing useful materials that respond to physical cues (such as temperature, light, voltage or magnetic field), chemical cues (such as pH, ionic strength, redox agent or competitive host–guest interactions) and biological cues (from enzymes or proteins, for example). These responses are manifested as dramatic changes in shape, surface characteristics, solubility, and the formation of an intricate molecular self-assembly or a sol-to-gel transition. Then, such responsiveness can be leveraged to engineer supramolecular biomaterials that allow for the modulation of properties in real time in response to externally applied stimuli, or that are capable of autonomously sensing and responding to environmental stimuli, thus ultimately yielding smarter therapeutics. In addition, the reversible nature of supramolecular interactions gives rise to biomaterials that can sense and respond to physiological cues, or that mimic the structural and functional aspects of biological signalling (Webber et al. 2016).

Applications of stimuli responsive, or 'smart', polymers in delivery of therapeutics, tissue engineering, bio-separations, sensors or actuators have been studied extensively. Recent reviews summarizing research progress can be found in Lu et al (2017); some examples are discussed below.

Owing to the varied roles that enzymes have in different biological processes, disease-associated enzyme deregulations have recently become an emerging target for medications. For example, ester bonds are often incorporated for targeting phosphatases, intracellular acid hydrolases and several other esterases. Amides, although relatively stable to chemical attack in physiological environments, are vulnerable to enzymatic digestion and have been used for constructing materials sensitive to hydrolytic proteases, such as prostate-specific antigens, and materials containing cleavable azo linkers can target bacterial enzymes in the colon for site-specific drug release.

At present, blood sugar monitoring and insulin injection ('open-loop' treatment) is still the primary management of type 1 and advanced type 2 diabetes. The first glucose-responsive insulin delivery system (GRIDS) was developed in 1979, and used concanavalin A (ConA), a member of the saccharide-binding lectin family. Free glucose can dock within the specific binding sites of the ConA–polymer complex, causing the dissociation of the complex and subsequent insulin release. Currently, application of glucose-responsive materials is not limited to GRIDS. For example, in a recent study, hydrogel formulations composed of poly butyl acrylate (PBA) derivatives showed potential as both the delivery carrier for protein therapeutics and substrates for 3D cell culture. The hydrogel exhibited self-healing (rapid structural recovery) based on the dynamic interaction between PBA and diols. A nanostructured surface constructed by grafting a PBA-containing brush from an aligned silicon nanowire array captured and released cells in response to changes in pH values and glucose concentration in an AND logic manner. In addition, in a recently reported microneedle-based cancer immunotherapeutic delivery system targeting melanoma, gaseous oxygen (GOx) converted blood glucose to localized acidification for triggering the sustained release of anti-PD1 antibody with enhanced retention in the tumour site.

In addition to drug-delivery systems, glucose-responsive materials are also used for long-term glucose monitoring. For example, single-walled carbon nanotubes (SWNTs) functionalized with a glucose analogue were developed for glucose sensing. The SWNTs form aggregations in the presence of saccharide-binding ConA or PBA, quenching the fluorescence signal. By contrast, the dissociation of such aggregates owing to the competitive binding of glucose leads to the recovery of the fluorescence. Fluorescent polyacrylamide hydrogel beads fabricated from monomers containing glucose-recognition sites and fluorogenic sites have shown potential for continuous blood glucose levels monitoring.

Biological systems are exceptionally complex, and there remains a great need to develop synthetic materials capable of recapitulating their structural and functional complexity in order to create truly mimetic systems for enhanced therapeutic function. Supramolecular biomaterials can replicate aspects of structural and/or functional features of biological signal transduction. This is especially true for materials prepared from peptide or protein building blocks, whether assembled through molecular stacking or from polymeric precursors, as these systems are already prepared from the same amino acid coding elements used in native biological signaling. Biomaterials that can replace or recapitulate deficient native materials or signaling pathways could be especially useful for applications in regenerative medicine or tissue engineering. As synthetic scaffolds, supramolecular biomaterials can act as structural mimics of fibrous matrix components. For example, materials that can replicate the structure or function of native collagen (the most abundant component of native extracellular matrix) could be useful in the preparation of replacement cell scaffolds or in replicating the role of collagen in the nucleation of aligned biomineralization. Functional biomimicry of protein signaling can be achieved through controlled display of bioactive epitopes—often protein-derived signaling sequences—on supramolecular biomaterials. This strategy promotes the selective differentiation of progenitor cells and can be used to present specific cell-adhesion cues at defined densities, or to mimic potent mitogenic proteins. Some materials can simultaneously provide both structural and functional biomimicry, such as recombinant protein materials that combine elastin-like properties fused with functional cell-adhesive cues (Webber et al. 2016).

Inspired by nastic plant motions, where a variety of organs such as tendrils, bracts, leaves and flowers respond to environmental stimuli (such as humidity, light or touch) by changing their internal turgor and leading to a variety of dynamic conformations governed by the tissue composition and microstructural anisotropy of cell walls, emerging pathways incorporate materials that can respond to external stimuli, such as shape memory alloys and swellable hydrogel composites, and assemble them by methods such as four dimensional (4D) printing and self-folding origami, with the aim to mimic such dynamic architectures. Gladman et al. (Gladman et al. 2016) developed a biomimetic hydrogel composite based on cellulose fibrils alignment that can be 4D printed into programmable bilayer architectures patterned in space and time. Printed composite hydrogel architectures that are encoded with localized, anisotropic swelling behaviour controlled by the alignment of cellulose fibrils along prescribed four-dimensional printing pathways were combined with a minimal theoretical framework to attain complex and programmable plant-inspired architectures that change shape on immersion in water. According to the authors, these biocompatible and flexible ink designs open new avenues for creating designer shapes-shifting architectures for tissue engineering, biomedical devices, soft robotics and beyond (Gladman et al. 2016).

3.5.6 Lessons from Natural Molecules: Preservation of Mammalian Cells

Simply freezing or drying mammalian cells and tissues usually results in dead and obviously nonfunctional materials. However, nature employs a variety of compounds and strategies to enhance the survival of ectothermic animals during extreme environmental conditions. The study of such mechanisms by which animals tolerate environmental extremes may provide strategies for preservation of living mammalian materials. Even though, actually, how most animals survive freezing conditions is not known, adaptations may include production of high concentrations of polyols (glycerol, glucose, sorbitol, ribitol, erythritol, threitol, and/or ethylene glycol), disaccharides (particularly trehalose) and a variety of thermal hysteresis producing antifreeze compounds (Brockbank et al. 2011).

The basic research on anhydrobiotic organisms, the tardigrades, which initially did not appear to have any practical applications in biotechnology, inspired the investigation on mammalian cells preservation. As soon as the water dries up, the tardigrades dry up too, and they can persist in this dry state for decades, during which time they are remarkably resistant to environmental extremes; after rehydration, they were successfully revived. The analysis of their capacity to survive dry conditions showed that they must be dried slowly to provide time for the synthesis of the disaccharide trehalose. In the 1980s, we learned that trehalose preserves membranes and proteins in a physical state resembling fully hydrated conditions, suggesting that the hydroxyl groups of the sugar might physically interact with polar residues in these cellular components. Much evidence has accumulated in favour of this idea, known as the "water replacement hypothesis" (Clegg et al. 1982). In the dry state, trehalose and many other sugars form a highly viscous liquid, identified as a glassy state, which has been shown to be necessary for stabilization. Unlike other sugars, however, trehalose remains in this glassy state at very high temperatures and at extreme levels of desiccation, properties that may explain its particular effectiveness. Its extreme tolerance of less than ideal drying and storage conditions may explain its status as nature's "sugar of choice" for preservation (Crowe et al. 1998).

The major challenge, however, is to load cells with membrane-impermeable trehalose for intracellular protection. Beattie et al. (Beattie et al. 1998) were first to demonstrate the ability of trehalose to preserve mammalian cells during freezing. They showed that the insulin-producing cells from mammalian pancreas have a membrane lipid phase transition from the fluid to solid phases well above the freezing point. These workers exploited the inherent leakiness of membranes during the phase transition to introduce trehalose into cells, which were, subsequently, successfully frozen for extended periods. These findings are being used to develop a commercial product that may provide a stable, transplantable device for treatment of diabetes in humans. Following, research by Eroglu et al. (Eroglu et al. 2000) incorporate a clever new twist: instead of using a lipid phase transition to introduce the sugar, they use a recombinant pore-forming hemolytic protein, *a*-hemolysin. This protein had previously been engineered to enable the pore to respond to the presence of zinc. In its presence, the pore remained closed, removing the zinc, however, allowed it to open. As the pore protein spontaneously inserts into membranes, Eroglu et al. (Eroglu et al. 2000) found that they could simply incubate the cells with the recombinant protein and then deliver trehalose to the cells in the absence of zinc. Using this procedure, they were able to obtain very high rates of survival of two lines of mammalian cells in the frozen state. In a different approach, Guo et al. (Guo et al. 2000) transferred the genes for trehalose synthesis into the target cells. This is an old idea that seemed attractive from the start because trehalose synthesis requires only two enzymes, encoded by only two genes. An adenoviral vector delivered the genes and the infection of the cells with increasing quantities of adenovirus correlated with increased biosynthesis of trehalose (Crowe and Crowe 2000).

After the above mentioned pioneered approaches where transgenic cell lines have been generated expressing genes for trehalose synthesis (Guo et al. 2000), genetically engineered pore-forming proteins (Eroglu et al. 2000), thermal/osmotic shock (Puhlev et al. 2001), thermal cycling while passing membrane phase transitions (Beattie et al. 1998), fluid phase endocytosis (Wolkers et al. 2001), and electro-permeabilization (Shirakashi et al. 2002) were used as trehalose loading methods rendering cells in part desiccation tolerant. Actually, novel methods that include synthetic biopolymers which interact with membranes (Lynch et al. 2010), cell penetrating peptides (Wei et al. 2014), engineered lipophilic membrane-permeable derivatives of trehalose (Abazari et al. 2015), nanoparticle-mediated intracellular delivery of encapsulated trehalose (Rao et al. 2015b), and intracellular delivery using thermally responsive nanocapsules (Zhang et al. 2009, Stewart and He 2018) have been used to facilitate trehalose uptake in order to enhance freezing and/or drying survival. Current studies with red blood cells and platelets of Zhang and coworkers (Zhang et al. 2016) have shown that freezing these cells in the presence of high extracellular trehalose concentrations resulted in surprisingly high post-thaw survival rates. Cryosurvival was found to coincide with high post-thaw intracellular trehalose concentrations, which was assumed to be taken up by the cells during freezing and/or thawing due to a combination of freezing-induced osmotic forces and fluid-to-gel membrane phase transitions.

3.6 Using Nature to Inspire Human Innovation

Through evolution, nature has "tested" various solutions to its challenges and has improved the successful ones in a process that involves nano- to macro-scale hierarchical approaches that are transmitted from one generation to another encoded into their genes. Some of these solutions may have inspired humans to achieve outstanding outcomes. The subject was pioneered by Leonardo da Vinci (1452-1519), who was fascinated by the phenomenon of flight and produced many studies, including the *"Codex on the Flight of Birds"*, published in 1505. Then, Galileo Galilei (1564-1642) with his book *"Discourses and Mathematical Demonstrations Relating to Two New Sciences"* published in 1638, the forbearers of modern material engineering and kinematics, and D'Arcy Wentworth Thompson (1860-1948) with his book *"On Growth and Form"* published in 1917, among others, followed this mission.

Nowadays, biomimetic materials research (sometimes also signaled as material bionics or bio-inspired materials exploration) has begun to progress enthusiastically. One of the reasons is the conception of nanomedicine, the convergence between modern nanotechnology and medicine. Such field, which applies the nanoscale principles and techniques to understand and transform biosystems (living or non-living) and which uses biological principles and materials to create new medical devices and systems integrated from the nanoscale, has expanded and is expected to have a revolutionary impact on healthcare. Biomimetic materials research generates numerous opportunities for devising new strategies to build multifunctional materials for the clever use of interfaces and the development of active or self-healing materials. Interdisciplinary teams developed a portfolio of bio-inspired processes for obtaining new function by structuring and assembling of known elements.

But what else can we learn? It is evident that not all the lessons imparted by nature will be directly applicable to the design of new engineering materials. This will require cooperation between biologists and technologists/engineers as well as the establishment of such an educational path in academic institutes that, I hope, will also lead to new disciplines of biomimetic science and engineering.

The inspiration from nature is expected to continue leading to technology improvements and the impact is expected to be felt in every aspect of our lives. Some of the solutions may be considered science fiction according to our current capabilities, but as we improve our understanding of nature and with the development of better expertise our dreams may become a reality; I particularly believe that we are closer than we think.

We are the future and it is in our hands so that we can expand and transform it. Nature will most likely provide us many more useful lessons as we continue to extend our knowledge and acquire the abilities to understand them. In this way, we will apply it to the construction of functional nanomedicine approaches and devices to improve our world through medical care.

References

Abazari, A., L.G. Meimetis, G. Budin, S. S. Bale, R. Weissleder and M. Toner. 2015. Engineered trehalose permeable to mammalian cells. PLoS One 10: e0130323.

Andrés, N.C., J. M. Sieben, M. Baldini, C. H. Rodriguez, Á. Famiglietti and P. V. Messina. 2018. Electroactive Mg^{2+}-Hydroxyapatite Nanostructured Networks against Drug-Resistance Bone Infection Strains. ACS Appl. Mater. Interfaces. 10(23): 19534-19544.

Bae, W. -G., J. Kim, Y. -H. Choung, Y. Chung, K. Y. Suh, C. Pang, et al. 2015. Bio-inspired configurable multiscale extracellular matrix-like structures for functional alignment and guided orientation of cells. Biomaterials 69: 158-164.

Ball, P., 2002. Natural strategies for the molecular engineer. Nanotechnology 13: R15.

Beattie, G. M., J. H. Crowe, F. Tablin and A. Hayek. 1998. Cryopreservation of human adult and fetal pancreatic cells and human platelets. United States Patent US5827741.

Bello, O., K. Adegoke and R. Oyewole. 2013. Biomimetic materials in our world: A review. J. Appl. Chem. 5: 22-35.

Beniash, E. 2011. Biominerals-hierarchical nanocomposites: The example of bone. Wiley Interdiscip. Rev. Nanomed. Nanobiotechnol. 3: 47-69.

Bernstein, H. S. 2011. Tissue Engineering in Regenerative Medicine. Humana Press. Totowa, NJ. USA.

Bhadra, C.M., V. Khanh Truong, V. T. H. Pham, M. Al Kobaisi, G. Seniutinas, J.Y. Wang, et al. 2015. Antibacterial titanium nano-patterned arrays inspired by dragonfly wings. Sci. Rep. 5: 16817.

Bixler, G.D. and B. Bhushan. 2012. Biofouling: Lessons from nature. Phil. Trans. R. Soc. A 370: 2381-2417.

Blanco, E., H. Shen and M. Ferrari. 2015. Principles of nanoparticle design for overcoming biological barriers to drug delivery. Nat. Biotechnol. 33(9): 941-951.

Brockbank, K. G., L. H. Campbell, E. D. Greene, M. C. Brockbank and J. G. Duman. 2011. Lessons from nature for preservation of mammalian cells, tissues, and organs. In Vitro Cell Dev. Biol. Anim. 47(3): 210-217.

Bueno, O. 2011. When physics and biology meet: The nanoscale case. Stud. Hist. Philos. Biol. Biomed. Sci. 42(2): 180-189.

Byrom, R. 2014. William Fairbairn, Karl Culmann and the Origin of Wolff's Law. IJET. 84: 52-58.

Cambi, A. and D. S. Lidke. 2012. Nanoscale membrane organization: Where biochemistry meets advanced microscopy. ACS Chem. Biol. 7: 139-149.

Cao, H., L. Zou, B. He, L. Zeng, Y. Huang, H. Yu, et al. 2017. Albumin biomimetic nanocorona improves tumor targeting and penetration for synergistic therapy of metastatic breast cancer. Adv. Funct. Mater. 27: 1605679.

Clegg, J. S., P. Seitz, W. Seitz and C. F. Hazlewood. 1982. Cellular responses to extreme water loss: The water-replacement hypothesis. Cryobiology. 19: 306-316.

Cogal, O. and Y. Leblebici. 2017. An insect eye inspired miniaturized multi-camera system for endoscopic imaging. IEEE Trans. Biomed. Circuits. Syst. 11(1): 212-224.

Crowe, J. H., J. S. Clegg and L. M. Crowe. 1998. Anhydrobiosis: The water replacement hypothesis. The properties of water in foods. ISOPOW 6. Springer, pp. 440-455.

Crowe, J. H. and L. M. Crowe. 2000. Preservation of mammalian cells—learning nature's tricks. Nat. Biotechnol. 18(2): 145-146.

Chen, W., S. Han, W. Qian, S. Weng, H. Yang, Y. Sun, et al. 2018. Nanotopography regulates motor neuron differentiation of human pluripotent stem cells. Nanoscale 10: 3556-3565.

Chung, T., Y. Lee, S. P. Yang, K. Kim, B. H. Kang and K. H. Jeong. 2018. Mining the Smartness of Insect Ultrastructures for Advanced Imaging and Illumination. Adv. Funct. Mater. 1705912.

Dalby, M. J., N. Gadegaard and R. O. C. Oreffo. 2014. Harnessing nanotopography and integrin–matrix interactions to influence stem cell fate. Nat. Mater. 13(6): 558-569.

Dang, M., L. Saunders, X. Niu, Y. Fan and P. X. Ma. 2018. Biomimetic delivery of signals for bone tissue engineering. Bone Res. 6: 25.

Drexler, K. E. 1986. Engines of creation. Anchor.

Drouet, C., J. Gomez-Morales, M. Iafisco, S. Sarda, L. Rimondini, C. Bianchi, et al. 2012. Calcium phosphate surface tailoring technologies for drug delivering and tissue engineering. Chap 2, pp. 43-111. *In*: Rimondini L., C. L. Bianchi and E. Vernè (Eds.). Surface Tailoring of Inorganic Materials for Biomedical Applications. Bentham Science Publishers, Sharjah, United Arab Emirates.

Du, N., Z. Yang, X. Y. Liu, Y. Li and H. Y. Xu. 2011. Structural origin of the strain-hardening of spider silk. Adv. Funct. Mater. 21: 772-778.

Eisoldt, L., A. Smith and T. Scheibel. 2011. Decoding the secrets of spider silk. Mater. Today 14: 80-86.

Eroglu, A., M. J. Russo, R. Bieganski, A. Fowler, S. Cheley, H. Bayley, et al. 2000. Intracellular trehalose improves the survival of cryopreserved mammalian cells. Nat. Biotechnol. 18(2): 163-167.

Espinosa, H. D., A. L. Juster, F. J. Latourte, O. Y. Loh, D. Gregoire and P. D. Zavattieri. 2011. Tablet-level origin of toughening in abalone shells and translation to synthetic composite materials. Nat. Commun. 2: 173.

Fink, W. 2018. Nature-inspired sensors. Nat. Nanotechnol. 13: 437-438.

Fratzl, P. 2007. Biomimetic materials research: What can we really learn from nature's structural materials? J. R. Soc. Interface. 4(15): 637-642.

Fratzl, P. and R. Weinkamer. 2007. Nature's hierarchical materials. Prog. Mater. Sci. 52(8): 1263-1334.

Frost, H. M. 1990. Skeletal structural adaptations to mechanical usage (SATMU): 1. Redefining Wolff's law: The bone modeling problem. Anat. Rec. 226: 403-413.

Furth, M. E., A. Atala and M. E. Van Dyke. 2007. Smart biomaterials design for tissue engineering and regenerative medicine. Biomaterials 28: 5068-5073.

Gao, C., Z. Lin, B. Jurado-Sánchez, X. Lin, Z. Wu, Q. He, et al. 2016a. Stem cell membrane-coated nanogels for highly efficient in vivo tumor targeted drug delivery. Small 12: 4056-4062.

Gao, C., Z. Wu, Z. Lin, X. Lin and Q. He. 2016b. Polymeric capsule-cushioned leukocyte cell membrane vesicles as a biomimetic delivery platform. Nanoscale 8: 3548-3554.

Gladman, A. S., E. A. Matsumoto, R. G. Nuzzo, L. Mahadevan and L. A. Lewis. 2016. Biomimetic 4D printing. Nat. Mater. 15(4): 413-418.

Gleason, C. 2006. The Biography of Silk. Crabtree Publishing. St. Catharines, ON L2M 5V6, Canada.

Green, D. W., B. Ben-Nissan, K. S. Yoon, B. Milthorpe and H. -S. Jung. 2017. Natural and synthetic coral biomineralization for human bone revitalization. Trends Biotechnol. 35(1): 43-54.

Greiner, C., E. Arzt and A. Del Campo. 2009. Hierarchical gecko-like adhesives. Adv. Mater. 21: 479-482.

Guo, N., I. Puhlev, D. R. Brown, J. Mansbridge and F. Levine. 2000. Trehalose expression confers desiccation tolerance on human cells. Nat. Biotechnol. 18(2): 168-171.

Guo, Z., W. Liu and B. -L. Su. 2011. Superhydrophobic surfaces: From natural to biomimetic to functional. J. Colloid Interface Sci. 353(2): 335-355.

Han, S., S. Ji, A. Abdullah, D. Kim, H. Lim and D. Lee. 2018. Superhydrophilic nanopillar-structured quartz surfaces for the prevention of biofilm formation in optical devices. Appl. Surf. Sci. 429: 244-252.

Horih, S.I. 1989. Basic Neurochemistry Molecular, Cellular, and Medical Aspects. Neurology 39: 460-460b.

Jakab, P.L. 2014. Visions of a flying machine: The Wright brothers and the process of invention. Smithsonian Institution.

Jeong, B. and A. Gutowska. 2002. Lessons from nature: Stimuli-responsive polymers and their biomedical applications. Trends Biotechnol. 20: 305-311.

Kelleher, S. M., O. Habimana, J. Lawler, B. O'reilly, S. Daniels, E. Casey, et al. 2015. Cicada wing surface topography: An investigation into the bactericidal properties of nanostructural features. ACS Appl. Mater. Interfaces. 8(24): 14966-14974.

Kemp, M. 2007. Leonardo da Vinci: The Marvellous Works of Nature and Man. Oxford University Press. Oxford OX1 2JD, UK.

Kreuter, J., D. Shamenkov, V. Petrov, P. Ramge, K. Cychutek, C. Koch-Brandt, et al. 2002. Apolipoprotein-mediated transport of nanoparticle-bound drugs across the blood-brain barrier. J. Drug Target. 10: 317-325.

Lehn, J.-M. 1995. Supramolecular Chemistry. Vch, Weinheim Germany.

Leslie, D. C., A. Waterhouse, J. B. Berthet, T. M. Valentin, A. L. Watters, A. Jain et al. 2014. A bioinspired omniphobic surface coating on medical devices prevents thrombosis and biofouling. Nat. Biotechnol. 32(11): 1134-1140.

Li, G., Y. Li, G. Chen, J. He, Y. Han, X. Wang, et al. 2015. Silk-based biomaterials in biomedical textiles and fiber-based implants. Adv. Healthc. Mater. 4(8): 1134-1135.

Li, X., G. Cheung, G. S. Watson, J. A. Watson, S. Lin, L. Schwarzkopf, et al. 2016. The nanotipped hairs of gecko skin and biotemplated replicas impair and/or kill pathogenic bacteria with high efficiency. Nanoscale 8: 18860-18869.

Liu, Y., D. Luo and T. Wang. 2016. Hierarchical structures of bone and bioinspired bone tissue engineering. Small 12: 4611-4632.

Lotsari, A., A. K. Rajasekharan, M. Halvarsson and M. Andersson. 2018. Transformation of amorphous calcium phosphate to bone-like apatite. Nat. Commun. 9(1): 4170.

Lu, Y., A. A. Aimetti, R. Langer and Z. Gu. 2017. Bioresponsive materials. Nat. Rev. Mater. 2: 16075.

Lynch, A. L., R. Chen, P. J. Dominowski, E. Y. Shalaev, R. J. Yancey Jr. and N. K. Slater. 2010. Biopolymer mediated trehalose uptake for enhanced erythrocyte cryosurvival. Biomaterials 31: 6096-6103.

M Ruso, J., J. Sartuqui and P. V. Messina. 2015. Multiscale inorganic hierarchically materials: Towards an improved orthopaedic regenerative medicine. Curr. Top. Med. Chem. 15(22): 2290-2305.

Mendes, A. C., E. T. Baran, R. L. Reis and H. S. Azevedo. 2013. Self-assembly in nature: Using the principles of nature to create complex nanobiomaterials. Wiley Interdiscip. Rev. Nanomed. Nanobiotechnol. 5(6): 582-612.

Messina, P. V., N. L. D'Elía and L. A. Benedini. 2017. Bone tissue regenerative medicine via bioactive nanomaterials. Nanostructures for Novel Therapy. Elsevier, pp. 769-792.

Mow, V. C. and R. Huiskes. 2005. Basic Orthopaedic Biomechanics & Mechano-Biology. Lippincott Williams & Wilkins. 2001 Market Street Philadelphia, PA 19103, USA.

Nandi, S. K., B. Kundu, A. Mahato, N. L. Thakur, S. N. Joardar and B. B. Mandal. 2015. In vitro and in vivo evaluation of the marine sponge skeleton as a bone mimicking biomaterial. Integr. Biol. (Camb). 7(2): 250-262.

Narasimhan, V., R. H. Siddique, J. O. Lee, S. Kumar, B. Ndjamen, J. Du, et al. 2018. Multifunctional biophotonic nanostructures inspired by the longtail glasswing butterfly for medical devices. Nat. Nanotechnol. 13: 512-519.

Neto, A. and J. Ferreira. 2018. Synthetic and Marine-Derived Porous Scaffolds for Bone Tissue Engineering. Materials 11(9): pii: E1702.

Ozin, G. A., K. Hou, B. V. Lotsch, L. Cademartiri, D. P. Puzzo, F. Scotognella, et al. 2009. Nanofabrication by self-assembly. Mater. Today 12: 12-23.

Park, S. and K. Hamad-Schifferli. 2010. Nanoscale Interfaces to Biology. Curr. Op. Chem. Biol. 14: 616-622.

Parodi, A., N. Quattrocchi, A. L. van de Ven, C. Chiappini, M. Evangelopoulos, J. O. Martinez, et al. 2012. Synthetic nanoparticles functionalized with biomimetic leukocyte membranes possess cell-like functions. Nat. Nanotechnol. 8(1): 61-68.

Puhlev, I., N. Guo, D. R. Brown and F. Levine, 2001. Desiccation tolerance in human cells. Cryobiology 42: 207-217.

Qu, L., Y. Akbergenova, Y. Hu and T. Schikorski. 2009. Synapse-to-synapse variation in mean synaptic vesicle size and its relationship with synaptic morphology and function. J. Comp. Neurol. 514(4): 343-352.

Radtke, C., C. Allmeling, K. -H. Waldmann, K. Reimers, K. Thies, H.C. Schenk, et al. 2011. Spider silk constructs enhance axonal regeneration and remyelination in long nerve defects in sheep. PLoS ONE 6: e16990.

Rao, L., L.L. Bu, J.H. Xu, B. Cai, G.T. Yu, X. Yu, et al. 2015a. Red blood cell membrane as a biomimetic nanocoating for prolonged circulation time and reduced accelerated blood clearance. Small 11: 6225-6236.

Rao, W., H. Huang, H. Wang, S. Zhao, J. Dumbleton, G. Zhao, et al. 2015b. Nanoparticle-mediated intracellular delivery enables cryopreservation of human adipose-derived stem cells using trehalose as the sole cryoprotectant. ACS Appl. Mater. Interfaces. 7(8): 5017-5028.

Rigo, S., C. Cai, G. Gunkel-Grabole, L. Maurizi, X. Zhang, J. Xu, et al. 2018. Nanoscience-based strategies to engineer antimicrobial surfaces. Adv. Sci. 5(5): 1700892.

Roco, M.C. 2003. Nanotechnology: Convergence with modern biology and medicine. Curr. Op. Biotechnol. 14: 337-346.

Romano, P., H. Fabritius and D. Raabe. 2007. The exoskeleton of the lobster Homarus americanus as an example of a smart anisotropic biological material. Acta Biomater. 3: 301-309.

Sartuqui, J., C. Gardin, L. Ferroni, B. Zavan and P. V. Messina. 2018. Nanostructured hydroxyapatite networks: Synergy of physical and chemical cues to induce an osteogenic fate in an additive-free medium. Mater. Today Commun. 16: 152-163.

Shirakashi, R., C. Köstner, K. Müller, M. Kürschner, U. Zimmermann and V. Sukhorukov. 2002. Intracellular delivery of trehalose into mammalian cells by electropermeabilization. J. Membr. Biol. 189(1): 45-54.

Stevens, M. M. 2008. Biomaterials for bone tissue engineering. Mater. Today 11: 18-25.

Stewart, S. and X. He. 2018. Intracellular delivery of trehalose for cell banking. Langmuir. doi: 10.1021/acs.langmuir.8b02015

Subbiahdoss, G., R. Kuijer, D. W. Grijpma, H. C. van der Mei and H. J. Busscher. 2009. Microbial biofilm growth vs. tissue integration: "The race for the surface" experimentally studied. Acta Biomater. 5: 1399-1404.

Sumper, M. and E. Brunner. 2006. Learning from diatoms: Nature's tools for the production of nanostructured silica. Adv. Funct. Mater. 16: 17-26.

Tanaka, M. 2013. Physics of interactions at biological and biomaterial interfaces. Curr. Op. Colloid Interface Sci. 18(5): 432-439.

Tang, Z., N. A. Kotov, S. Magonov and B. Ozturk. 2003. Nanostructured artificial nacre. Nat. Mater. 2(6): 413-418.

Ti Tien, H. and A. Ottova-Leitmannova. 2000. Fundamental aspects of biological membranes. pp. 23-82. In: Ti Tien, H. and Ottova-Leitmannova, A. (eds.). Membrane Science and Technology. Elsevier. Radarweg 29, PO Box 211, 1000 AE Amsterdam, Netherlands.

U.S.-NNI, U.S.N.N.I., 2015. What's so special about the nanoscale?

Vukusic, P. and J. R. Sambles. 2003. Photonic structures in biology. Nature. 424: 852-855.

Watson, G. S., D. W. Green, L. Schwarzkopf, X. Li, B. W. Cribb, S. Myhra, et al. 2015. A gecko skin micro/nano structure – A low adhesion, superhydrophobic, anti-wetting, self-cleaning, biocompatible, antibacterial surface. Acta Biomater. 21: 109-122.

Webber, M. J., E. A. Appel, E. Meijer and R. Langer. 2016. Supramolecular biomaterials. Nat. Mater. 15(1): 13-26.

Wei, Y., C. Li, L. Zhang and X. Xu. 2014. Design of novel cell penetrating peptides for the delivery of trehalose into mammalian cells. Biochim. Biophys. Acta. 1838(7): 1911-1920.

Whitesides, G. M. and B. Grzybowski. 2002. Self-assembly at all scales. Science. 295: 2418-2421.

Whitesides, G. M., J. K. Kriebel and B. T. Mayers. 2005. Self-assembly and nanostructured materials. pp. 217-239. In: Wilhelm E. S. Huck (Ed.). Nanoscale Assembly. Springer, Salmon Tower Building, Midtown Manhattan, NY, USA.

Wolkers, W. F., N. J. Walker, F. Tablin and J. H. Crowe. 2001. Human platelets loaded with trehalose survive freeze-drying. Cryobiology 42: 79-87.

Xuan, M., J. Shao, L. Dai, Q. He and J. Li. 2015. Macrophage cell membrane camouflaged mesoporous silica nanocapsules for in vivo cancer therapy. Adv. Healthc. Mater. 4: 1645-1652.

Young, J. L., A. W. Holle and J. P. Spatz. 2016. Nanoscale and mechanical properties of the physiological cell – ECM microenvironment. Exp. Cell Res. 343: 3-6.

Zhang, M., H. Oldenhof, H. Sieme and W. F. Wolkers. 2016. Freezing-induced uptake of trehalose into mammalian cells facilitates cryopreservation. Biochim. Biophys. Acta. 1858: 1400-1409.

Zhang, W., J. Rong, Q. Wang and X. He. 2009. The encapsulation and intracellular delivery of trehalose using a thermally responsive nanocapsule. Nanotechnology 20(27): 275101.

Biotechnology: Tuning Nanoscale Bio-systems

"…One of the phenomena which had peculiarly attracted my attention was the structure of the human frame, and, indeed, any animal endued with life. Whence, I often asked myself, did the principle of life proceed? It was a bold question, and one, which has ever been considered as a mystery; yet with how many things are we upon the brink of becoming acquainted, if cowardice or carelessness did not restrain our inquiries…."

Extract from "Frankenstein; or, The Modern Prometheus". Mary Shelley's 'Gothic' (proto-)science fiction novel, first published in 1818 by Lackington, Hughes, Harding, Mavor & Jones, Gradifco eds. (Shelley 1818)

4.1 Tuning Nanoscale Bio-systems

In the preceding chapters, we have analysed and understood how biological and physical sciences share a common interest in nano-sized structures; how a dynamic line of work across the limits of these scientific areas was established and developed around new materials and tools obtained from the physical skills, and how new phenomena were displayed from the biological subjects. Thus, the physical sciences offer a wide range of tools for the synthesis and fabrication of novel devices designed for a better manipulation of cells and sub-cellular components features, and of materials that provide innovative tools to operate on cellular machineries and to study molecular biology rules (Whitesides 2003). On the other hand, taking into consideration that the evolution has already encountered and solved many of the challenges that nanotechnologists confront, nature gives us a peek into the most sophisticated collection of existent functional nanostructures. Even though the benefits of biomimesis in nanotechnology are widely recognized, there is no guarantee that nature's solutions can be translated to a technological setting. Nevertheless, biology does seem to be an abundant storehouse of ideas (Ball 2004). At this level, one of the boldest and most controversial plans in the fundamental biological research consists of understanding biology at the nanoscale and then turn biology into an applied, engineering science. The nanoscale structures of the cells can be mimicked and manipulated to such an extent that completely new organisms will be chemically designed to adapt to new technological objectives. Among them, complex systems-level are continuously designed to reprogram living entities' metabolic routes and that opens an avenue for the exploitation of microscopic diversity and biosynthetic potential in biofabrication of nanomedicines (Vázquez and Villaverde 2013).

In this chapter, we will examine the concept of biofabrication, i.e. the set of methodologies addressed to the production of rather complex constructs with predefined biological properties, and how bacterial and even higher organisms' processes are manipulated to obtain functional recombinant nanomedicines. The recent advances in genomics together with high throughput analysis techniques and integrative approaches will also be analysed and contrasted with the traditional chemical and physical methods. Next, we will summarize the main achievements and the current situation in the development of novel biological factories, from single cells to livestock animals, as valuable platforms for large-scale production of therapeutic proteins. Finally, the potential of these platforms for production of blood human substitutes will be discussed.

4.2 Beginning, Understanding and Running Biotechnology

Which is the point of convergence among bread, cheese, wine, stonewashed jeans, the pharmaceutical form of insulin and home pregnancy test? At a glance, it could be said that nothing, but a deeper analysis reveals that they are all wares resulting from the manipulation of biological processes, organisms or systems (Lone-Star-College 2018). Indeed, they are all biotechnology products. The term "biotechnology" was first coined in 1919 by the Hungarian engineer Karl Ereky, who referred to the production of manufactured goods from raw materials with the aid of living organisms (Fári and Kralovánszky 2006).

Today, much of modern biotechnology deals with the manipulation of DNA; however, classical biotechnology began long before we even knew about genes or chromosomes. The most primitive type of biotechnology, which stretches back over 10 000 years ago, is the cultivation of plants and the domestication of animals (Bhatia 2018). Our ancestors already knew processes for production of cheese, yogurt, bread and various alcoholic drinks such as beer and wine from microorganisms. As examples, we can cite the Egyptians that used yeasts to bake bread, the Chinese whose developed fermentation techniques for brewing and cheese making, and the Aztecs that used *Spirulina* algae to make cakes. Furthermore, in earlier times several natural products were used as substitutes for our modern medicines, such as honey, which was used to treat several respiratory ailments and as an ointment for wounds healing. Likewise, in China as far back as 600 BC, soybean curds were used to treat boils and Ukrainian farmers utilized mouldy cheese to treat infected wounds. However, what began as recipes for production of food or homemade medicines now includes responsiveness to enhance everything: from agriculture to pharmaceuticals.

Around the end of the nineteenth century, what has been called "the modern biotechnology" commenced. It is associated with the introduction of scientific evidence for fermentation process presented by Louis Pasteur (Bhatia 2018). Pasteur's efforts would contribute towards several branches of science, as the germ theory of disease (Pasteur and Lister 1996), the first vaccine against rabies (Geison 2014), the discoveries of chemical crystal asymmetry and, of course, the pasteurization process (Schwartz 2001). Meanwhile, a better understanding of fermentation led to the industrial field of zymotechnology or zymurgy (Bud 2002). Particularly focused on the production of beer, zymotechnology became very popular in the mid-19th century. In the 1860s, institutes and remunerative consultancies were dedicated to the technology of brewing. The most famous was the private Carlsberg Institute, founded in 1875, which employed Emil Christian Hansen, who pioneered the pure yeast process for the reliable production of consistent beer (Travis et al. 1992). Lesser well known were private organizations as the Zymotechnic Institute, established in Chicago by the German-born chemist John Ewald Siebel, who advised the brewing industry. (Thackray 1998). In late 19th century Germany, brewing contributed as much to the gross national product as steel, and taxes on alcohol proved to be significant sources of revenue to the government (Thackray 1998). The zenith and expansion of zymotechnology came during World War I in response to industrial needs to support the war. The late nineteenth century was known to be a milestone in biology. Agriculture and automobile industries incorporated crop rotation approaches, animal-drawn technology and, the production of acetone and paint solvents by fermentation processes. The industrial potential of fermentation was outgrowing its traditional home in brewing, and "zymotechnology" soon gave way to "biotechnology". Finally, in the nineteenth century and at the beginning of the twentieth century, it was observed that antibiotics present in moulds killed bacteria and averted the spread of infection. In 1928 Alexander Fleming extracted penicillin, the first antibiotic, from mould (Fleming 1944). By then, Mendel's work on genetics (Orel 1996) was completed and institutes for investigating fermentation along with other microbial processes had been founded by Koch, Pasteur, and Lister (Bhatia 2018).

The origins of biotechnology culminated with the birth of genetic engineering. By mid-20th century, Watson and Crick described major advances in genetics, among them, the discovery of the "double helix" structure of DNA and that it carries the genetic information (Watson and Crick 1953). Rapidly, new techniques were developed to allow DNA manipulation, like the recombinant DNA technique discovered in 1973 by Cohen, Chang and Boyer (Cohen et al. 1973, Cohen and Chang 1974) by which a section of DNA was cut from the plasmid of an *Escherichia coli* bacterium and transferred into the DNA of another. After these two key events, genetically engineered plants, microbes, animals, biotech-based medicines products like insulin and penicillin and different immunoassays were made. Further research into viral biology has led to improved vectors for delivering new genetic material (Kay et al. 2001). An explosion of enzymes for clipping, editing, ligating, and copying DNA, as well as efficient techniques for the chemical synthesis and repair of DNA (Lindahl and Wood 1999, Li and Elledge 2007, Shendure and Ji 2008), has allowed the creation of complicated new genetic constructs (Muramatsu et al. 2000, Laird 2010). Currently, engineered bacteria now create large quantities of natural proteins for medicinal use, mutated proteins for research, hybrid chimeric proteins for specialized applications, and entirely new proteins, highlighting the future of biotechnology (Langer and Tirrell 2004).

The importance of the new genetics leads to *"The Human Genome Project"*, an international project instituted by the U.S. Department of Energy and the National Institutes of Health (NIH) to "map the human genome" (Olson 1993). James Watson was elected as the first director of the Office of Human Genome Research at the NIH which aimed to identify the structure of the entire human genome, including its three billion base pairs and estimated 30-40,000 genes (Lone-Star-College 2018). The hope is that this knowledge will help scientists to identify, prevent and treat many of the illnesses resulting from genetic malfunction. As genetic discoveries have progressed, the importance of regulating how the knowledge will be used has become of primary importance. Therefore, in addition to experimental research, a portion of the budget of the human genome project was set aside to study the ethical, legal and social implications of the project and of recombinant DNA experiments. Spliced DNA has been used to create a glowing bunny rabbit, to breed a goat whose milk contains spider silk and to repair genetic defects in sick people. DNA and genetic functions are very complex, so you cannot make a giraffe with elephant tusks, but concrete benefits are accruing quickly. Thus, genetic engineering stimulated hopes for both therapeutic proteins, drugs and biological organisms themselves, such as seeds, pesticides, engineered yeasts, and modified human cells for treating genetic diseases. From the perspective of its commercial promoters, scientific breakthroughs, industrial commitment, and official support were finally coming together, and biotechnology became a normal part of business. Their message had finally become accepted and incorporated into the policies of governments and industry. Nowadays, even though it has historical roots, biotechnology still plays, and likely it will for years to come, a central role in diverse industrial fields, as well as in the development and production of numerous effective medicines.

4.3 The Splice of Life, Recombinant Technology and the Nanoscale Precision Editing of DNA

At present, biotechnology concept is rapidly associated with DNA splicing. That is the process by which one organism's DNA is cut separately and another organism's DNA is slipped into the gap. The result is recombinant DNA that includes features of the host organism modified by the trait in the foreign DNA, producing organisms with new characteristics. What would you think of a plant that glows like a firefly, a cotton crop that produces its own insecticide or of a cloned transgenic cow that produces recombinant human growth hormone (hGH) in the milk? During the prehistoric era, fermentation was perhaps first explored by chance and nobody knew how it worked; some civilizations considered fermentation to be a gift from their gods (Bhatia 2018). At that time, by means of such processes it was possible to play gods, as nowadays it is, misnamed, the recombinant DNA engineering.

Manipulation of the genetic code of living organisms was always in the minds of scientists. At the beginning, it was attained using biology's own tools of mating and crossing or, more aggressively and less controlled, by random mutagenesis with chemicals or ionizing radiation (Goodsell 2004). Today, researchers modify the genetic code rationally at the nanoscale atomic level making use of recombinant DNA technology. Recombinant DNA technology, although revolutionary in its impact, applies tools and procedures that *per se* were known and were not revolutionary. For example, one molecular biology process of the most widely studied microbe, *Escherichia coli*, whose study had been intensified at that time was the bacteriophage property (commonly referred as "phage") (Berg and Mertz 2010). Such process, known as "transduction", involves the transference of genes from one strain to another; its deep examination gave rise to the discovery and characterization of restriction proteins, scientific event that earned Werner Arber, Daniel Nathans, and Hamilton O. Smith the Nobel Prize for Physiology or Medicine in 1978 (Salmond and Fineran 2015).

Restriction enzymes are one class of the broader endonuclease group of enzymes; they are made-up by bacteria to provide a defence mechanism against viral infection. Inside a prokaryote, the restriction enzymes selectively cut up foreign DNA in a process called restriction digestion; meanwhile, host DNA is protected by a modification enzyme (a methyltransferase) that adapts the prokaryotic DNA and blocks cleavage. According to their structure and whether they cut their DNA substrate at their recognition site, or if the recognition and cleavage sites are separate from one another, restriction enzymes are commonly classified into five types (Orlowski and Bujnicki 2008). Today, over 3000 restriction enzymes have been studied in detail, and more than 600 of these are commercially available (Roberts et al. 2007). Thus restriction enzymes, originally evolved merely for their destructive capacity, are now tools for atomic-precision editing of large pieces of DNA in laboratories playing a vital function in molecular cloning.

Theoretically, any DNA sequence can be inserted into a plasmid or phage vector and then propagated indefinitely. The gene of interest and the vector are biochemically modified *in vitro*, ligated together, and the resultant recombinant DNA is used to transform a bacterial host, usually *Escherichia coli*. If a gene is readily isolated in a relatively pure form by standard techniques, then it may be cloned directly into a plasmid or phage. This approach is valid for the genes coding for ribosomal RNA (rRNA) or a very abundant messenger (mRNA). The resultant recombinant molecule is referred to as a gene-specific probe, since when it is hybridised to human DNA, it will only react with the human gene sequence which it specifies. In most cases, the eukaryotic mRNA cannot be purified easily, and the preparation of gene-specific probes requires the prior assembly of a library of DNA sequences followed by the identification of the recombinant molecule of interest (Davies 1981). At present, recombinant DNA technology has flowered. Clever researchers are continually discovering new methods for harnessing the protein production machinery of the cell in several ways. Consistent methods, often in the form of commercial kits, are available for every possible process. We can find and extract specific genes from any organism, duplicate and determine the sequence of large quantities of these genes, mutate, recombine, and splice these genes or create entirely new genes nucleotide by nucleotide. Finally, we can replace these genes into cells, modifying their genetic information to obtain organisms with very new properties. Such transgenic entities can act as factories of biomedicines or to generate diagnostic tools that improve the care of patients with infectious diseases.

4.4 The Cell Reactor, a Tactical Player in the Recombinant Nanomedicine's Line of Attack

Diabetes, growth or clotting disorders are among the spectrum of human diseases related to malfunction or protein absence. Since these pathologies cannot be yet regularly treated by gene therapy, the administration of functional proteins produced *ex vivo* is required (Kinch 2015). Now is when recombinant DNA technologies come into play and rapidly become a choice for expression and purification of the desired therapeutic recombinant protein in a large production quantity. There is no doubt that the production of recombinant proteins in microbial systems has revolutionized biochemistry and transformed the tactic by which medicines are discovered and industrialized. The days where kilograms of animal and plant tissues or large volumes of biological fluids were needed for the purification of small amounts of a given protein are almost gone.

In 1976 Herbert Boyer, a key pioneer of recombinant DNA research, had founded *Genentech*, developed a recombinant form of human insulin in 1978, partnered with Eli Lilly pharmaceutical and gained FDA approval for the first recombinant DNA product in 1982: Humulin®. Since the approval of Humulin®, the FDA as therapeutics has approved many recombinant-protein-based molecular entities (rPMEs); some of them are summarized in Table 4.1. The numbers and diversity of FDA approvals of biologics NMEs accelerated from the early 1980s through the middle of the past decade. The first generation of biotechnology drugs largely consisted of recombinant growth factors and other modulators of cell surface proteins. Over time, the field has seen the emergence and growth of recombinant enzymes and monoclonal antibodies as well, Box 4.1.

Box 4.1. Biotechnology: Timeline of key events

- 500 B.C.: In China, the first antibiotic, mouldy soybean curds, is put to use to treat boils.
- A.D. 100: The first insecticide is produced in China from powdered chrysanthemums.
- 1761: English surgeon Edward Jenner pioneers vaccination, inoculating a child with a viral smallpox vaccine.
- 1870: Breeders crossbreed cotton, developing hundreds of varieties with superior qualities.
- 1870: The first experimental corn hybrid is produced in a laboratory.
- 1911: American pathologist Peyton Rous discovers the first cancer-causing virus.
- 1928: Scottish scientist Alexander Fleming discovers penicillin.
- 1933: Hybrid corn is commercialized.
- 1942: Penicillin is mass-produced in microbes for the first time.
- 1952: The phenomenon of host-controlled restriction and modification of bacterial phage or bacteriophage was first identified in the laboratories of Salvador, Lucia, Weigle and Giuseppe Bertani.
- 1953: James Watson and Francis Crick discovered the double helix structure of DNA
- 1958: Werner Arber and Matthew Stanley Meselson showed that the restriction is caused by an enzymatic cleavage of the AND, the enzyme involved was therefore termed as "restriction enzyme". Type I restriction enzyme cleaved DNA randomly from the recognition site.
- 1961: Marshall Nirenberg sequenced the bases in each codon.
- 1970: Hamilton O. Smith, Thomas Kelly and Kent Wilcox isolated and characterized the first Type II restriction enzyme, HindII, from bacterium *Haemophilus influenzae*. They cleaved DNA at the site of their recognition sequence.
- 1971: Daniel Nathans and Kathleen Danna showed that the cleavage of simian virus 40 (SV40) DNA by restriction enzyme yields specific fragments that can be separated using polyacrylamide gel electrophoresis, showing that restriction enzymes can also be used for mapping DNA.
- 1973: Paul Berg, Herbert Boyer, Annie Chang, and Stanley Cohen of Stanford University and University of California San Francisco developed recombinant DNA technology.
- 1975: During "The Asilomar Conference" regulation, safe use of rDNA technology was discussed.
- 1976: Herbert Boyer founded Genentech Inc.
- 1978: Recombinant form of insulin was developed. H. Boyer partnered with Eli Lilly.
- 1979: Genentech Inc., announced the successful bacterial production of human Growth Hormone (hGH).

- 1982: FDA approved the first recombinant DNA-derived product, Humulin®.
- 1985: FDA discontinued the use of cadaveric hGH and led to the approval of Genentech's synthetic methionyl GH for the therapy of severe childhood GH disorders in the U.S.: Protropin; first commercialized in 1987.
- 1986: The first recombinant vaccine for humans, a vaccine for hepatitis B, was approved: Recombivax HB by Merck Sharp and Dohm. Peter T. Jones, Paul H. Dear, Jefferson Foote, Michael S. Neuberger and Greg Winter created the first humanized monoclonal antibody
- 1987: FDA approved the recombinant tissue plasminogen activator (tPA), an enzyme that dissolves blood clots: Activase by Genentech Inc.
- 1988: Interferon became the first anticancer drug produced through biotechnology: Intron A, interferon alfa-2b recombinant by Schering Co.; Referon-A, interferon alfa-2a recombinant by Hoffman – La Roche.
- 1989: The second recombinant – type hepatitis B vaccine is licensed: Engerix – B by Glaxo-Smith-Kline (GSK) Biologicals S.A. Greg Winter and Dave Chiswell founded Antibody Technology (CAT).
- 1990: G. Winter created the phage display monoclonal antibodies.
- 1991: First display and selection of human antibodies phage by Carlos F. Barbas and Richard A. Lerner from the Scripps Research Institute.
- 1992: Monoclonal antibodies market crashed following FDA's call for more information for Centocor's drug, Centoxin.
- 1993: Centoxin withdrawn from European market.
- 1994: First chimeric monoclonal antibody therapeutic approved for market.
- 1995: First monoclonal antibody drug for cancer approved in Europe.
- 1998: FDA approved Trastuzumab (Herceptin) for the treatment of metastatic breast cancer. FDA and European regulatory authorities approved the first monoclonal antibody drug for an autoimmune disease: Infliximab (Remicade).
- 2002: Adalimumab (Humira®) became the first phage display-derived antibody granted a marketing approval. Approved for rheumatoid arthritis treatment and found using CAT's technology, it is the world's top selling medicine, with sales reaching almost £ 12 billons in 2014.
- 2004: FDA approved Technetium (99m Tc) fanolesomab marketed as NeutroSpec (12/2005). A new imaging agent for detecting difficult to diagnose cases of appendicitis.
- 2006: FDA approved the recombinant vaccine Gardasil®, the first vaccine developed against human papillomavirus (HPV), and the first preventive cancer vaccine.
- 2009: FDA approved the first genetically engineered animal for production of a recombinant form of human anti-thrombin.
- 2013: European Commission approved Alemtuzumab (Lemtrada) for multiple sclerosis treatment.

(Contd.)

Box 4.1. (*Contd.*)

- 2014: FDA approved nivolumab (Opdivo®), an immune checkpoint inhibitor targeting PD1, for treating melanoma.
- 2016: FDA approved atezolizumab (Tecentriq®), an immune checkpoint inhibitor targeting PD1, for treating urothelial carcinoma, the most common form of bladder cancer. Monoclonal antibody drug for Alzheimer's disease shown to be promising in phase II clinical trials.
- 2017: Monoclonal antibody shown to effectively cut cholesterol levels, thereby preventing heart attacks and strokes. FDA approved evolocumab (Amgen's Repatha®). Public health services (NHS) in the United Kingdom made available Nivolumab (Opdivo®) for patients with advanced lung cancer. First patient treated in first clinical trial using monoclonal antibody drug to treat schizophrenia.

The knowledge and technology needed for the design of rPMEs, will be defined by the choice of the host cell whose protein synthesis machinery will produce the accurate molecule according to its required therapeutic action. In this sense, the key drive of a protein production platform is to reach an effective protein folding (Gasser et al. 2008) and post-translational modifications that enhance drug functionality, while maintaining low complexity, high flexibility of cell culture and great levels of production. The spectrum of organisms exploited as recombinant cell factories has expanded from the early prevalent platform *Escherichia coli* in the 1980s, to alternative bacteria, yeasts, filamentous fungi, unicellular algae, insect cells and especially mammalian cells, which benefit from metabolic and protein processing pathways similar to those in human cells (Sanchez-Garcia et al. 2016). Currently, 85% of all biopharmaceuticals that require post translational modifications are overwhelmingly expressed in Chinese Hamster Ovary (CHO) mammalian cells systems, while those that do not are expressed in *Escherichia coli* systems. The benefits of using *Escherichia coli* as the host organism are well known. First, it has unparalleled fast growth kinetics; in the optimal environmental conditions, it takes 20 minutes to double its population. That is very important because the expression of a recombinant protein may impart a metabolic burden on the microorganism, causing a considerable decrease in the generation time. Secondly, high cell density cultures are easily achieved. The theoretical density limit of an *Escherichia coli* liquid culture is estimated to be about 200 g dry cell weight/L or approximately 1×10^{13} viable bacteria/mL. The third point is that rich complex media can be made from readily available and inexpensive components and finally that the transformation with exogenous DNA is fast and easy. Plasmid transformation of *Escherichia coli* can be performed in about 5 minutes (Rosano and Ceccarelli 2014). *Escherichia coli* use was followed by the implementation of the yeast *Saccharomyces cerevisiae*. Such system and the associated genetic methodologies exhibit the similar unusually high versatility than *Escherichia coli* making it adaptable to different production demands (Sanchez-Garcia et al. 2016). In addition, an efficient protein secretion in absence of endotoxic cell wall components was obtained after the use of Gram-positive bacteria such as *Bacillus megaterium* (Secore et al. 2017) and *Lactococcus lactis* (Almeida et al. 2016). Investigation of insect cells as sources of recombinant protein production was performed with positive results (Maiorella et al. 1988), especially for vaccine-oriented protein achievements (Gheysen et al. 1989, Wickham et al. 1992, Cox 2012). As an example, the *baculovirus*-insect cell expression system (BEVS) can be mentioned. The BEVS production platform has been extensively explored for the production of viral, parasitic antigens, and more recently, vaccines. Five approved vaccines for human or veterinary use are currently commercially available, Table 4.1, including the more recently CERVARIX– GSK's human papilloma virus vaccine (Cox 2012).

Due to their suitability to produce conveniently glycosylated proteins, mammalian cells are, nowadays, the prevailing animal-derived cell system (Sanchez-Garcia et al. 2016). Mammalian cell hosts can correctly fold, assemble, and glycosylate monoclonal antibody (MAb) polypeptides. The latter is crucial, for example, in the case of recombinant MAbs that are designed to harness biological activities such as antibody-dependent cell cytotoxicity and complement mediated lysis *in vivo*. In addition, mammalian cell-based MAb production systems are capable of generating the multi-kilogram quantities of product that are required to support the administration of relatively clinical high doses (>100 mg), thus satisfying the critical demands on the biopharmaceutical industry since blockbuster antibodies are currently produced at a multi-ton scale per year. As it was previously mentioned, large-scale processes generally employ Chinese Hamster Ovary (CHO) cells as production vehicles, although other mammalian cells types, such as murine lymphoid cells (NS0, SP2/0), are also utilized (Dinnis and James 2005). In fact, they are the gold-standard mammalian host cells for the production of therapeutic MAb and F_c-fusion proteins that have already reached the market (Beck and Reichert 2012). However, recently, glyco-engineered antibodies with humanized glycoforms in heterologous expression systems have been obtained. Filamentous fungi (such as *Trichoderma reesei* (Bischof et al. 2016)), moss (*Physcomitrella patens* (Khan et al. 2017)) and protozoa (*Leishmania tarentolae* (Rahmati et al. 2016)) promote glycosylation patterns similar to those in mammalian proteins but are still cultured through simpler methods. Also, genetically engineered tobacco (Strasser et al. 2014), moss (Decker and Reski 2007, Decker et al. 2014), and yeast (Choi et al. 2003, Gerngross 2004) has been used. Some of these systems have made the leap to commercial manufacture, although they are still very much in their infancy. High mannose-type N-glycans contain from five to nine mannose residues and are found on antibodies produced in mammalian cells, yeast, insect cells and plants, but only at a very low level in normal human antibodies (Beck and Reichert 2012). High mannose glycans on the F_c region of therapeutic immune globulin G (IgG) antibodies increase serum clearance in humans. Several other glycoforms containing fucose or xylose moieties characteristic of mice, yeast or plant-derived glycoproteins are also highly immunogenic in

Table 4.1. Examples of therapeutic recombinant proteins FDA approved until 2018 (U.S. Department of Health and Human Services 2018)

Year	Product	Company	Clinical Indication	Platform
1982	Humulin®	Eli Lily	Humulin, the first recombinant DNA drug approved by FDA, is an intermediate-acting recombinant human insulin indicated to improve glycemic control in adults and pediatric patients with diabetes mellitus.	E. coli
1983	Recombivax HB®	Merck Sharp & Dohme Co.	Recombivax HB (Hepatitis B Vaccine, Recombinant) is a vaccine indicated for prevention of infection caused by all known subtypes of hepatitis B virus.	S. cerevisiae
1985	Muromonab CD3 Orthoclone OKT3	Centocor Ortho Biotech Inc.	Recombinant monoclonal antibody to Human CD3 is an immunosuppressant drug given to reduce acute rejection in patients with organ transplants. It is a monoclonal antibody targeted at the CD3 receptor, a membrane protein on the surface of T cells. It was the first monoclonal antibody to be approved for clinical use in humans.	Murine ascites
1986	Roferon-A®	Roche	Interferon alpha-2a recombinant has been approved for use in hairy-cell leukaemia, AIDS-related Kaposi's sarcoma, follicular lymphoma, chronic myeloid leukaemia and melanoma.	E. coli
1989	Epogen®	Amgen	Epogen® (epoetin alfa) Epoetin alfa is a 165-amino acid erythropoiesis stimulating glycoprotein manufactured by recombinant DNA technology. Authorised by the European Medicines Agency on 28 August 2007, it stimulates erythropoiesis (increasing red blood cell levels) and is used to treat anemia, commonly associated with chronic renal failure and cancer chemotherapy.	CHO

1991	Neupogen®	Amgen	Neupogen® (filgrastim) is a manufactured form of granulocyte colony stimulating factor (G-CSF), which is a substance naturally produced by the body. It stimulates the growth of neutrophils, a type of white blood cell important in the body's fight against infection. It is used to treat low blood neutrophils following chemotherapy, radiation poisoning or HIV/AIDS.	*E. coli*
1991	Leukine®	Genzyme Co.	Leukine® (sargramostim) is a recombinant human granulocyte-macrophage colony-stimulating factor (rhu GM-CSF). GM-CSF is a hematopoietic growth factor, which stimulates proliferation and differentiation of hematopoietic progenitor cells. It is indicated in acute myelocytic leukaemia	*S. cerevisiae*
1993	Pulmozyme®	Genentech	Pulmozyme® (dornase alfa) is a recombinant human deoxyribo-nuclease I (rhDNase), an enzyme that selectively cleaves DNA. It is indicated for daily administration in conjunction with standard therapies for the management of cystic fibrosis (CF) patients to improve pulmonary function.	CHO
1994	Reopro® Abciximab	Centocor Ortho Biotech Inc., Eli Lilly & Co.	Abciximab is the F_{ab} fragment of the chimeric human-murine, monoclonal antibody 7E3. It binds to the glycoprotein (GP) IIb/IIIa receptor of human platelets and inhibits platelet aggregation. It has been approved as an adjunctive therapy to prevent cardiac ischemic complications (PCI), as well as in unstable angina patients not responding to conventional medical therapy when PCI is planned within 24 hours.	Sp2/0
1996	Activase®	Genentech	Activase® (alteplase) is a tissue plasminogen activator (tPA) indicated for the treatment of acute ischemic stroke (AIS), acute myocardial infarction (AMI) for the reduction of mortality and reduction of the incidence of heart failure, and for the lysis of acute massive pulmonary embolism (PE).	CHO

(Contd.)

Table 4.1. *(Contd.)*

Year	Product	Company	Clinical Indication	Platform
1996	Retavase	Boehringer-Mannheim	Reteplase is a recombinant non-glycosylated form of human tissue plasminogen activator, which has been modified to contain 357 of the 527 amino acids of the original protein. It is used in the treatment of acute myocardial infarction	E. coli
1998	Simulect®	Novartis Pharm. Co.	Simulect® (basiliximab) is a chimeric (murine/human) monoclonal antibody (IgG1k), produced by recombinant DNA technology, that functions as an immunosuppressive agent, specifically binding to and blocking the interleukin-2 receptor a-chain (IL-2Ra, also known as CD25 antigen) on the surface of activated T-lymphocytes. It is an immunosuppressant agent, used to prevent immediate transplant rejection in kidney-transplanted patients.	Sp2/0
1998	Proleukin®	Chiron	Proleukin® (aldesleukin), a human recombinant interleukin-2 product, indicated for the treatment of metastatic renal cell carcinoma, metastatic melanoma, kidney cancer, angiosarcoma.	E. coli
2001	Aranesp®	Amgen	Aranesp® (darbepoetin alfa) is an erythropoiesis stimulating protein, closely related to erythropoietin, which stimulates bone marrow to make red blood cells. It is indicated for the treatment of anemia associated with chronic renal failure, including patients on dialysis and patients not on dialysis, and on March 23, 2006 for the treatment of anemia due to the effects of concomitant myelosuppressive chemotherapy.	CHO
2007	Lantus®	Sanofi-aventis	Lantus® (insulin glargine) is a long-acting human insulin analog indicated to improve glycemic control in adults and pediatric patients with type 1 diabetes mellitus and in adults with type 2 diabetes mellitus.	E. coli

Year	Product	Company	Description	Expression system
2009	Cervarix®	Glaxo Smith Kline (GSK)	Cervarix® is a vaccine indicated for the prevention of the following diseases caused by oncogenic human papillomavirus (HPV).	Insect cells using the BEVS.
2014	Eloctate®	Biogen Idec. Inc.	Antihemophilic rFactor VIII, F_c Fusion Protein. It is indicated for the treatment and prophylaxis of bleeding and adult patients with hemophilia A.	HEK293
2014	Alprolix®	Biogen Idec. Inc.	rFactor IX fused to a human IgGI Fc domain. It is indicated for the control and prevention of bleeding episodes and peri-operative management in patients with hemophilia B.	HEK293
2014	Ruconest®	Salix Pharm. Inc.	rC1 Esterase Inhibitor. It is indicated for the treatment of acute attacks in adult and adolescent patients with hereditary angioedema (HAE).	Transgenic rabbits' milk
2014	Afrezza®	MannKind Co.	Recombinant human insulin indicated to improve glycemic control in adult patients with diabetes mellitus.	*E. coli*
2014	Cyramza®	Eli Lilly	Cyramza® (ramucirumab) is a human vascular endothelial growth factor receptor 2 antagonist indicated for the treatment of gastric cancer.	NSO
2014	Entyvio®	Takeda Pharm.	Entyvio® (vedolizumab) is an integrin receptor antagonist indicated for adults with moderate to severely active ulcerative colitis and Crohn's disease.	CHO
2014	Sylvant®	Janssen Biotech. Inc.	Sylvant® (siltuximab) is an interleukin-6 (IL-6) antagonist indicated for the treatment of patients with multicentric Castleman's disease (MCD) who are human immunodeficiency virus (HIV) negative and human herpesvirus-8 (HHV-8) negative.	CHO

(Contd.)

Table 4.1. (*Contd.*)

Year	Product	Company	Clinical Indication	Platform
2015	Strensiq®	Alexion Pharm. Inc.	Strensiq® (asfotase alfa) is a tissue nonspecific alkaline phosphatase enzyme replacement therapy indicated for the treatment of patients with perinatal, infantile and juvenile-onset hypophosphatasia (HPP).	CHO
2015	Tresiba®	Novo Nordisk	Tresiba® (insulin degludec) is a long-acting human insulin analog indicated to improve glycemic control in adults with diabetes mellitus.	*S. cerevisiae*
2015	Repatha®	Amgen	Repatha® (evolocumab) is a PCSK9 (proprotein convertase subtilisin kexin type 9) inhibitor antibody indicated to reduce the risk of myocardial infarction, stroke, and coronary revascularization in adults with established cardiovascular disease.	CHO
2015	Unituxin™	United Therapeutics Co.	Unituxin™ (dinutuximab) is a GD2-binding monoclonal antibody indicated, in combination with granulocyte-macrophage colony-stimulating factor (GM-CSF), interleukin-2 (IL-2), and 13-cis-retinoic acid (RA), for the treatment of pediatric patients with high-risk neuroblastoma who achieve at least a partial response to prior first-line multiagent, multimodal therapy.	Sp2/0
2015	Natpara®	NPS Pharm.	Natpara® is a parathyroid hormone, indicated as an adjunct to calcium and vitamin D to control hypocalcemia in patients with hypoparathyroidism.	*E. coli*
2016	Zinplava®	Merck Sharp & Dohme Co.	Zinplava® is a human monoclonal antibody that binds to Clostridium difficile toxin B, indicated to reduce recurrence of Clostridium difficile infection (CDI) in patients 18 years of age or older who are receiving antibacterial drug treatment of CDI and are at a high risk of CDI recurrence.	CHO

2016	Lartruvo™	Eli Lilly	Lartruvo™ (olaratumab) is a prescription medicine used with a type of chemotherapy called doxorubicin to treat adult patients with soft tissue sarcoma (STS) for whom doxorubicin is appropriate and who cannot be cured with radiation or surgery.	NS0
2016	Zinbryta®	Biogen	Zinbryta® (daclizumab) is an interleukin-2 receptor-blocking antibody indicated for the treatment of adult patients with relapsing forms of multiple sclerosis (MS).	NS0
2016	Anthim®	Elusys Therapeutics, Inc.	Anthim® (obiltoxaximab) is a monoclonal antibody directed against the protective antigen of *Bacillus anthracis*. It is indicated in adult and pediatric patients for treatment of inhalational anthrax due to *B. anthracis* in combination with appropriate antibacterial drugs and for prophylaxis of inhalational anthrax when alternative therapies are not available or are not appropriate.	NS0
2017	Luxturna®	Spark Therapeutics, Inc.	Indicated for the treatment of patients with confirmed biallelic RPE65 mutation-associated retinal dystrophy.	Retinal pigment epithelial (RPE) cells
2017	HeplisaV-B™	Dynavax Technologies Co.	Hepatitis B Vaccine (Recombinant), Adjuvanted is indicated for immunization against infection caused by all known subtypes of hepatitis B virus.	*Hansenula polymorpha*
2017	Shingrix®	Glaxo Smith Kline (GSK) Biologicals	Zoster Vaccine Recombinant, Adjuvanted. Prevention of herpes zoster (shingles) in adults aged 50 years and older.	CHO
2017	cobas® Zika	Roche Molecular Systems, Inc.	The cobas® Zika for use on the cobas® 6800 and cobas® 8800 systems is a qualitative *in vitro* nucleic acid screening test for the direct detection of Zika virus RNA in human plasma.	BHK-21 cells

(Contd.)

Table 4.1. (*Contd.*)

Year	Product	Company	Clinical Indication	Platform
2017	Rebinyn®	Novo Nordisk Inc.	Rebinyn is a coagulation rFactor IX, GlycoPEGylated, indicated for the use in the treatment and control of bleeding episodes and for the perioperative management of bleeding in adults and children with hemophilia B.	CHO
2018	Jivi®	Bayer HealthCare LLC	Jivi is a recombinant antihemophilic human factor FVIII, PEGylated, indicated for use in previously treated adults and adolescents (12 years of age and older) with hemophilia A (congenital Factor VIII deficiency) for: on demand treatment and control of bleeding episodes; perioperative management of bleeding; and routine prophylaxis to reduce the frequency of bleeding episodes.	Baby Hamster Kidney (BHK) cells
2018	Andexxa	Portola Pharm.	Andexxa is coagulation factor Xa (recombinant), inactivated-zhzo is a recombinant modified human Factor Xa (FXa) protein indicated for patients treated with rivaroxaban and apixaban, when reversal of anticoagulation is needed due to life-threatening or uncontrolled bleeding.	CHO

humans. The inherent conservatism of the industry and regulatory authorities, and the inefficient or non-human glycosylation patterns that make them immunogenic resulting in faster clearance if present in large amounts has so far precluded alternative expression systems from displacing the dominant platforms outside a few niche areas (Foley 2018). Therefore, actually, only mammalian-based production systems are used for the manufacturing of approved biopharmaceuticals, which need the accurate glycosylation. Nevertheless, tremendous efforts are being made both in academic labs and in industry to engineer the glycosylation pathways of mammalian cells into alternative platforms to allow the production of recombinant proteins exhibiting human-like glycosylation (Beck and Reichert 2012). Expanding knowledge of glycosylation pathways and new genetic engineering tools offer a path forward for a wider choice of expression platforms to meet the ever-expanding needs of the biopharmaceutical industry.

4.5 Synthetic Biology

The initiation of recombinant DNA technologies in the 1970s recognised genetic cloning possibilities, escorting the era of biotechnology (Rovner et al. 2015). Over the past decade, the communion of efforts of engineers and biologists has lead to the design and building of novel biomolecular components, networks and pathways, and using them to rewire and reprogram genetically modified organisms (GMOs). These constructs were proposed as common and valued solutions in clinical, industrial and environmental settings fuelling the emergence of synthetic biology (Benner and Sismour 2005, Purnick and Weiss 2009, Khalil and Collins 2010). The designation of "synthetic biology" was first used in 1980 by Barbara Hobom to describe genetically engineered bacteria that had been created via recombinant DNA technology (Hobom 1980). Then, by the year 2000, at the annual meeting of the American Chemical Society in San Francisco, the expression "synthetic biology" was once more presented by the speakers including by the laureate Eric Kool (Award for Achievement in Biomimetic Chemistry, American Chemical Society in 2015 (Kool 2015)). At this time, the term was used to describe the synthesis of unnatural organic molecules that function in living systems and more broadly in this sense, to reference the efforts to "redesign life" (Benner and Sismour 2005). Currently, the scientific community has given further meaning to this concept. This community seeks to extract from living systems interchangeable parts "interoperable parts" that might be tested, validated as basic building units, and reassembled to create functional higher-order networks (Lu et al. 2009, Purnick and Weiss 2009). While the parts come from natural living systems, their assembly is, however, unnatural; therefore such devices might, or not, have analogues in living systems, referencing in a certain way to Mary Shelley's Frankenstein creature (Shelley 1818). Some selected examples of the next-generation of synthetic gene recoded platforms and their potential application in nanomedicine areas are discussed below.

4.5.1 Artificial Recoded Cells

Recoding cellular platforms generate new properties including non-natural amino acid incorporation, virus confrontation, and biocontainment. They can be used to interpret and translate the genetic code, model genetic circuits and metabolic pathways, support portable diagnostics, simplify biomolecular manufacturing on demand and produce MAbs at the commercial scale, among others. The estimated cost of construction that includes DNA synthesis, assembly by recombination, and troubleshooting, is today comparable to the costs of early stage development of drugs or other high-tech products (Kuo et al. 2018). Next, we will discuss some relevant examples of the wide pool of available reports. The recent flow of applications has revitalized interest in cell-free protein synthesis (CFPS) systems, especially in areas where limits imposed by the organism may impede progress. One such area is expanding the genetic code to incorporate non-canonical amino acids (ncAAs) into proteins, where the extent of engineering can be limited by the capability of the organism. Nevertheless, inefficiencies associated with the engineered orthogonal translation (TL) machinery, i.e. TL elements that specifically use ncAA and do not interact with the cell's natural TL apparatus, have limited the ability to incorporate multiple ncAAs into proteins with high purity and yields, limiting applications in both basic and applied science. Recently, Rey W. Martin and co-workers (Martin et al. 2018) have successfully incorporated multiple ncAAs into proteins using crude extract-based cell-free systems avoiding the limiting effect of the release factor 1 competition. To attain their victory, they genomically recoded an *Escherichia coli* bacteria platform that lacked release factor 1. Another point of interest is the study of post-translational phosphorylation route, which is essential to human cellular processes; however, the transient, heterogeneous nature of this modification hinders its study in native systems. Karl W Barber et al. (2018) developed an approach to cross-examine the phosphorylation pathway and its role in protein-protein interactions on a proteome wide scale by an approach that involved genetically encoded phosphoserine in recoded *Escherichia coli* and the generation of a peptide-based heterologous representation of the human serine phosphoproteome. According to the authors, this technology can be used to explore user-defined phosphoproteomes in any organism, tissue, or disease of interest.

As mentioned previously, the glycosylation profile of recombinant proteins, destined for human use, is of critical significance because glycosylation controls the biological activity, function, clearance from blood stream, and antigenicity of recombinant proteins. The glycosylation profile of nonhuman cells is extremely different from humans, especially from those, which are more distant from humans in evolutionary terms like bacteria, yeasts, insects, and plants. However, in the search for the reduction of cost production, the well knowledge of genome and the evolution of recombinant technologies, researchers have been motivated to equip nonhuman cells with human-like glycosylation machinery for the production of therapeutic glycoproteins. By modifying the glycosylation machinery present in the cells of *Pichia pastoris*, N. Sethuraman and co-workers (Sethuraman et al. 2017) patented a route for the attaining of glycoproteins having terminal α-1,3-galactose. This technology is applicable to any number of vaccines that can be developed as recombinant protein-based molecules. A similar methodology was patented by Nien-Yi et al. (Chen et al. 2017) by which genetically engineered host animal cells, Chinese hamster ovary (CHO) cell, rat myeloma cell, baby hamster kidney (BHK) cell, hybridoma cell, Namalwa cell, embryonic stem cell and fertilized egg, may further express an exogenous glycoprotein. Such host animal cells were engineered to be capable of producing glycoproteins such as antibodies having modified glycosylation, including defucosylation and monoglycosylation. Examples include, but are not limited to, antibody, F_c-fusion protein, cytokine, hormone, growth factor, or enzyme production. The authors claimed that changes to the cellular glycosylation machinery in the host animal cells did not result in adverse effects in relation to glycoprotein synthesis and host cell growth.

Biological-based approaches, such as stem cell transplantation, are therefore receiving increasing attention to treat complex diseases that affect multiple pathways and regions (Blurton-Jones et al. 2014); so, they can also be benefited by genetic manipulation. The use of adenoviral vectors to deliver human tissue kallikrein (KLK1) gene or KLK1 protein infusion into injured tissues of animal models has provided particularly encouraging results in attenuating or reversing myocardial, renal and cerebrovascular ischemic phenotype and tissue damage. Such investigations pave the way for the administration of genetically modified mesenchymal stem cells (MSCs) or endothelial progenitor cells (EPCs) with the human tissue kallikrein (KLK1) gene. Collectively, findings from pre-clinical studies raise the possibility that tissue KLK1 may be a novel future therapeutic target in the treatment of a wide range of cardiovascular, cerebrovascular and renal disorders (Devetzi et al. 2018). On the other hand, enhancing endogenous synaptic connectivity in transgenic mice with Alzheimer's disease (AD) was obtained by the use of genetically modified neural stem cell (NSCs). NSCs that were stable to express and secrete a beta-amyloid Aβ-degrading enzyme, neprilysin (sNEP), provides a marked and significant reduction in Aβ pathology and increases synaptic density after implantation on both 3xTg-AD and Thy1-APP transgenic mice. Remarkably, Aβ plaque loads are reduced not only in the hippocampus and in subiculum adjacent to engrafted NSCs, but within the amygdala and medial septum, areas that receive afferent projections from the engrafted region (Blurton-Jones et al. 2014).

Genetic manipulation can also be applied to immunotherapies, which are emerging as highly promising approaches for the treatment of cancer. A key component of many of these approaches is functional tumour-specific T cells, but the existence and activity of sufficient T cells in the immune repertoire is not always the case. Recent methods of generating tumour-specific T cells include the genetic modification of patient lymphocytes with receptors to endow them with tumour specificity and are used to boost immunity against malignant cells (Kershaw et al. 2014, Ahmed et al. 2017, Smith et al. 2017). In addition to a solution against malignant tumour cells, the treatment or cure of HIV infection by cell and gene therapy has been a goal that researchers pursued for decades. Recent advances in both gene editing and chimeric antigen receptor (CAR) technology have created new therapeutic possibilities for a variety of diseases (June et al. 2018). Among them are included the broadly neutralizing monoclonal antibodies (bNMAbs) with specificity for the HIV envelope glycoprotein, which provides a promising approach of targeting HIV-infected cells. M. Hale et al. (2017) presented a study where primary human T cells were engineered to express anti-HIV CARs based on bNMAbs (HIVCAR). Those cells show specific activation and killing of HIV-infected versus uninfected cells in the absence of HIV replication.

4.5.2 Genetically Modified Plants, the Green High-Tech Molecular Farming

Unlike bacteria platforms, plants are effectively living single-use "bioreactors" that can be scaled indefinitely simply by sowing more seeds. In addition to be easily transmuted with a relatively little capital investment, the utilization of plants as "bioreactors" provides a reduction in health risks from human pathogen or endotoxins contamination and high production yields. Thus, they can be considered as a cheap-engineered "biofactories" for the manufacture of pharmaceutical proteins and peptides in large quantities (Giddings et al. 2000). Another benefit of plant-derived recombinant proteins production is that they can be expressed in edible plant organs, allowing them to be administered as unprocessed or partially processed material that could be stored without refrigeration at low cost (Fischer et al. 2004, Takeyama et al. 2015). Biopharmaceutical production and administration is directly related to the selected plant-based host and to the transformation technology involved. We can select our host species from a large pool of available plants. In this way we have:

Tobacco and Other Leafy Crops

Nicotiana tabacum and *Nicotiana benthamiana*, wild relatives of tobacco originally from Australia, are the routinely used plants for transient expression (Giddings et al. 2000, Lomonossoff and D'Aoust 2016). In fact, many of the early plant-derived recombinant proteins were produced in transgenic tobacco plants and were extracted directly from harvested leaves. The continuing popularity of tobacco is based on its fast growth rate, stable genetic system, high density tissue that offers a high protein production, requirements of a simple growth medium, and that it can be kept in a greenhouse. Also, optimized antibody expression can be rapidly verified using transient expression assays (short development time) in the plants before the creation of transgenic suspension cells or stable plant lines (longer development time). Different vector systems, harbouring targeting signals for subcellular compartments, are constructed in parallel and used

for transient expression. Applying this screening approach, high expressing cell lines can rapidly be identified. Currently, tobacco is a well-established expression host for which robust transformation procedures and well-characterized regulatory elements for the control of transgene expression are available. Tobacco has been adopted as a platform system by several biotech companies, including Planet Biotechnology Inc. (founded in 1994, Hayward, CA, U.S., http://www.planetbiotechnology.com/), Large Scale Biology Corp. (founded in 1987, Vacaville, CA, U.S., http://www.lsbc.com/), Meristem Therapeutics (founded in 1997, Clermont-Ferrand, France http://www. meristem-therapeutics.com/), and Phytomedics (founded in 1996, Dayton, NJ, U.S., http://www.phytomedics.com/) (Thiel 2004) . One disadvantage of tobacco is its high content of nicotine and other toxic alkaloids, which must be removed completely during downstream processing steps (Fischer et al. 2004). However, the negative perception of tobacco with respect to its beneficial use might not be an issue, especially when the plant molecular farming proteins are not for human consumption, for example for topical formulations of biopharmaceuticals (Obembe et al. 2011).

Even though low-alkaloid tobacco cultivars are available, attention has turned to other leafy crops for pharmaceutical production (Fischer et al. 2004, Obembe et al. 2011, Fu et al. 2015). These crops include lettuce, which has been used for clinical trials with a hepatitis B virus subunit vaccine (Walmsley and Arntzen 2000), and alfalfa, which is being promoted as a platform system by Medicago Inc. (funded in 1997 Quebec, Canada, http://www.medicago.com/). This Canadian biotech company has isolated novel promoters that allow high-level protein expression in alfalfa leaves, and it has focussed on the early part of the production pipeline by developing alfalfa cell-culture and transient-expression technology. Advantages of alfalfa include its high biomass yield and the fact that it is a perennial plant that fixes its own nitrogen. A strong advantage of alfalfa for pharmaceutical production is the fact that glycoproteins synthesised in alfalfa leaves tend to have homogeneous glycan structures, which is important for batch-to-batch consistency. Additionally, expression in vegetative organs, such as the leaf could affect growth and development of the particular plant; for example, alfalfa is a feed crop and its leaves contain large amounts of oxalic acid, which might interfere with processing. Furthermore, proteins that are expressed in leaves tend to be unstable, which means the harvested material has a limited shelf life and must be desiccated, or frozen or processed immediately after harvest.

Cereals

Even if tobacco is mainly used for basic research, commercial production is more likely to be in food crops. Proteins stored in seeds can be desiccated and remain intact for long periods of time, their purification and extraction is likely to be done by adaptations of current processes for the extraction and/or fractionation. In addition, the seed-based system allows oral delivery of biopharmaceuticals. Thus, several kinds of cereals, including rice, wheat, barley and maize, have been investigated as potential hosts for large-scale commercial production of recombinant proteins (Knäblein 2006). Many of the companies developing transgenic plant expression systems have chosen maize after surveying various crops for potential protein recovery and the economics of production (Giddings et al. 2000). Maize, for example, has been chosen by ProdiGene Inc. (founded in 1995, College Station, U.S.) for the commercial production of the technical proteins avidin and b-glucuronidase (GUS) (Kusnadi et al. 1998). Currently, the heavily glycosylated gastric lipase protein, produced in maize, is in Phase II of clinical trials for the treatment of patients suffering from exocrine pancreatic insufficiency (Gomord et al. 2005). In addition, bovine trypsin produced in maize, TrypZean (Sigma-Aldrich), has been in the market since 2002. TrypZean is particularly useful in animal cell cultures because it has no contaminants of animal origin. Rice, a self-pollinating plant that would have a lower risk of unintended gene flow, has been used by Ventria Bioscience (founded in 1993, Fort Collins, Colorad, U.S., http://www.ventria.com/) to the pioneered manufacture of human lysozyme and lactoferrin (Hennegan et al. 2005). Such proteins have received regulatory approvals, and have already been marketed. Essentially, rice is investigated for the production of human serum albumin (Yang et al. 2018) and soybean, in particular, has been explored for its efficacy to express a humanized antibody against herpes simplex virus, bovine casein, and a human growth hormone. Soybean was also used to express a functional hypotensive peptide, which reduced the systolic blood pressures of model mice (Obembe et al. 2011).

Fruits and Vegetables

Although they are not the most popular, they have also been employed for the attainment of oral plant-based vaccines. Once these vaccines pass through the gastric environment and reach the small intestine, antigens are incorporated into the specialized epithelial cells of the mucosa-associated lymphoid tissues (M cells) for the induction of mucosal and systemic immune responses (Takeyama et al. 2015). The delivery of recombinant vaccines in edible plant organs is an outstanding goal because it would be advantageous to use locally grown plants for vaccination campaigns (Fischer et al. 2004). Several vaccines have been expressed in transgenic potatoes, tomato and banana including the hepatitis B surface antigen and the heat labile enterotoxin B subunit (LTB) of *Escherichia coli*. Vectors derived from tobacco mosaic virus have been used to produce oral antihypertensive peptide (angiotensin-1-converting enzyme inhibitor) in tomato (Giddings et al. 2000) and rabies G protein has been expressed in several species, including tomato, spinach and carrot (Rojas-Anaya et al. 2009).

Oil Crops

Oil crops are useful hosts for protein production because the oil bodies can be exploited to simplify protein isolation (Fischer et al. 2004). As an example, we can mention the oleosin-fusion platform developed by SemBioSys Genetics Inc. (created in 1994 as a spin-off of the University of Calgary, Canada; it terminated its operations in 2012), and the oilseed technology platform proposed by biotech company UniCrop (funded in 1998, Helsinki, Finland, www.sciencepark.helsinki.fi/unicrop_ltd.htm). In the first approach, the target recombinant protein is expressed in oilseed rape or safflower as a fusion with oleosin and in the second isolate recombinant proteins from the rapidly developing sprouts are cultivated in bioreactors.

After the selection of hosts, gene introduction can be performed *in vitro* by different plant transformation strategies: (i) stable nuclear and plastid transformation, (ii) plant cell-suspension cultures, and (ii) transient expression systems like agro-infiltration, virus infection and magnification technology. Readers can find a detailed review of them from Obembe et al. (2011). Comparatively, transient expression of target proteins in plants using modified plant viruses or viral vectors integrated into binary vectors delivered via *Agrobacterium* is often considered a more robust approach when compared to stable transformation, due to its rapid production capabilities and relatively high protein expression (Loh et al. 2017). The promise of this platform has been evidenced in the high number of successful clinical trials attained, which demonstrated the safety and efficacy of plant-made protein therapeutics and biologics. Examples of plant-produced biopharmaceuticals currently in clinical stages of development or on market obtained from different expression hosts transmuted by diverse strategies are summarised in Table 4.2.

4.5.3 Transgenesis in Livestock Animals, Recoded Platforms to Produce Recombinant Human Proteins and Monoclonal Antibodies

The capability to introduce foreign genes into a germ line and the successful expression of them in a specific organism has allowed the genetic manipulation on an unprecedented scale. In this way, Herculean challenges, such as eradicating biological and social diseases, and promoting health and well-being in the population of a more equal world, become closer and possible (Jaenisch 1988). However, in spite of the successful expression of the first recombinant proteins (RPs) in bacteria, yeast and other lower organisms, it became clear that the existing phylogenetic relationship between human beings and such systems guarantees that they could not provide a large number of functional (ensuring the proper amino acid sequence and folding) and efficiently produced complex hominid RPs. In this context, a simultaneous development of two technological models based on mammalian cell cultures and transgenic animals (TAs) was started (Maksimenko et al. 2013). Of the two, the animal platform, centred on the use of animals as bioreactors, became potentially exploitable expression systems for the production of biopharmaceuticals. Taking into consideration that the mammary gland is an organ specializing in protein synthesis, and that milk

Table 4.2. Examples of therapeutic plant-host recombinant proteins. Information extracted from references (Obembe et al. 2011, Lomonossoff and D'Aoust 2016, U.S. Department of Health and Human Services 2018)

| Plant-Host | Biopharmaceuticals | | | | |
| | Description | | Name | Market | |
	Protein	Technology		Inventor	Status/Year	Application/Administration
Potato	E. coli heat-labile enterotoxin B subunit	Stable transformation		University of Maryland, Boyce Thompson Institute for Plant Research, and Tulane University School of Medicine	Phase I (1998)	Oral vaccine against enterotoxin E. coli
Maize	Trypsin	Stable transformation	TrypZean®	Sigma Aldrich	On market (since 2002)	TrypZean® is recombinant bovine trypsin expressed in corn, available either as a lyophilized powder or in sterile solution
Tobacco	Single chain variable fragment (scFv) vaccines for Non-Hodgkin's lymphoma	Transient transformation using replicating viral vector (tobacco mosaic virus)		Large Scale Biology Corporation	Phase I (2005), Phase III failed (2011)	Therapeutic vaccine for non-Hodgkin's lymphoma

	Product	Method	Product name	Company	Status	Application
	Monoclonal antibody. Hybrid secretory IgA-IgG	Stable transformation	CaroRx™	Planet Biotechnology Inc., Hayward, CA, USA	Phase II (2011), marketing approval in the EU as a medical device (2012)	Designed to block adherence to teeth of the bacteria that causes cavities, applied to the treatment of dental caries.
	Monoclonal antibody. Human anti-HIV IgG	Stable transformation		Pharma-Planta Consortium	Phase I (2015)	HIV-neutralizing human monoclonal antibody 2G12. Prevention of HIV transmission
N. benthamiana	H5N1 influenza virus-like particle	Transient expression, agrobacterium infiltration		Medicago	Phase II (2011)	Pandemic influenza vaccine
	Quadrivalent influenza virus-like particle	Transient expression, agrobacterium infiltration		Medicago	Phase I/II (2016)	Seasonal influenza vaccine
	Tumour-derived idiotype IgG	Transient expression, agrobacterium infiltration		Icon Genetics	Phase I (2015)	Therapeutic vaccine for targeted treatment of B-cell follicular lymphoma (FL), non-Hodgkin's lymphoma.

(Contd.)

Table 4.2. (*Contd.*)

		Biopharmaceuticals				
	Description				Market	
Plant-Host	Protein	Technology	Name	Inventor	Status/Year	Application/ Administration
	Chimeric anti-Ebola IgG cocktail	Transient expression, agrobacterium infiltration	ZMapp™	Mapp Biopharma.	Phase I (2015) Ethical experimental treatment declaration by the WHO	Treatment of Ebola virus infection
Carrot (cultured cells)	Taliglucerase alfa	Stable transformation	Elelyso	Protalix and Pfizer Biotherapeutics	FDA approved (2012)	Recombinant glucocerebrosidase used to treat Gaucher's disease

is easily collected in large quantities, as the use of animal platforms goes ahead, the mammary gland of a transgenic animal becomes the most popular protein bioreactor. The expression system of complex recombinant proteins in milk was considered to be the best option available for the production of biopharmaceuticals. It proved to be highly competitive by adding an extra-value via low cost implementation, production and scale up, as well as high efficiency of synthesized proteins (Bertolini et al. 2016). Blood, egg white, seminal plasma, silk gland, and urine are other theoretically possible expression systems (Wang et al. 2013); however, the expression system in the mammary gland is the only one within the animal platform to generate recombinant proteins as legalized trade for therapeutic use (Bertolini et al. 2016).

Several methods are developed for the introduction of genes into animals (Jaenisch 1988, Wheeler et al. 2003): (i) intra-pronuclear zygotic DNA microinjection, MI; (ii) somatic cell nuclear transfer, NT; (iii) site-specific transgenesis technology using embryonic stem cells (ESCs); retrovirus infection, RVI; (iv) mobile genetic vectors, MGV, which are integrated into the genome by transposase; (v) liposome-mediated DNA, LM, transfer into cells and embryos; (vi) electroporation of DNA into sperm, ova or embryos and (vii) biolistic. The majority of the large farm animals generated, have been obtained by NT, while MI is now used mainly to produce transgenic mice, rabbits and pigs (Maksimenko et al. 2013). Today, it remains difficult to compare the efficiency of new and traditional methods for producing TAs; advantages and disadvantages of them are summarized in Table 4.3.

4.5.4 Recombinant Proteins Production in the Milk

In the previous sections, we understood that, according to the bioreactor used to produce a specific recombinant human therapeutic protein, several of them require post-translational modifications for biological activity and stability after their use. The limited capacity of bioreactors has led the biopharmaceutical industry to investigate alternative protein expression systems. The glycosylation pattern of the mammary gland tissue seems to be similar to native human proteins, so it is preferred to obtain recombinant human proteins. In this way, milk became the most mature and proven transgenic system for the production of recombinant pharmaceutical proteins (Bertolini et al. 2016). In 2006, the European Medicine Agency (EMA) approved ATryn®, a recombinant form of human antithrombin accepted for human use in patients with hereditary antithrombin deficiency; developed by GTC Biotherapeutics (currently rEVO Biologics Inc), it was the first commercial biopharmaceutical recombinant protein produced in goat mammary gland. In 2009, the FDA has validated and confirmed the approbation of ATryn® and the transgenic animal manufacturing platform for the last time biopharmaceuticals. Another recombinant protein produced in the mammary gland of transgenic rabbits (Ruconest® a C1-Esterase Inhibitor by Pharming Group NV), used for the treatment of acute episodes in adult

Table 4.3. Production of transgenic animals (TAs) methodologies. Information extracted from references (Pinkert and Murray 1999, Shirahata et al. 2001, Toussaint and Merlin 2002, Wheeler et al. 2003, Huang et al. 2005, Sparks and Jones 2009, Villemejane and Mir 2009, Maksimenko et al. 2013, Wilmut et al. 2015, Bertolini et al. 2016)

	Description	Advantages	Disadvantages
MI	DNA microinjection, also known as pronuclear microinjection, consists of a manually injecting DNA from one organism into the eggs of another by the use of a very fine glass pipette. The best time for injection is early after fertilization when the ova have two pronuclei. When the two fuse to form a single nucleus, the injected DNA may or may not be taken up.	1. Frequency of stable integration of DNA. 2. Effective transformation of primary cells. 3. The DNA injected in this process is subjected to less extensive modifications. 4. Mere precise integration of recombinant gene in limited copy number can be obtained.	1. Costly. 2. Skilled personnel required. The technique is laborious, technically difficult, and limited to the number of cells actually injected. 3. Process causes random integration. 4. Rearrangement or deletion of host DNA adjacent to the site of integration is common.
SCNT	Transfection of somatic cells is carried out and clones characterized by the integration of the transgene into the genome are selected. The nucleus of the somatic cell is then injected into the enucleated oocyte, which is transplanted into female recipients. Fibroblast cells are typically used for NT.	1. Cheaper and easier procedures for producing transgenic livestock. 2. Potential for treating diseases associated with mutations in mitochondrial DNA. Recent studies show SCNT of the nucleus of a body cell afflicted with one of these diseases into a healthy oocyte prevents the inheritance of the mitochondrial disease (Pera and Trounson 2013).	1. Transfected cells in this case are selected using antibiotic resistance marker genes, which complicates the approval of the produced recombinant proteins by the FDA and EMEA. 2. Incomplete reprogramming of the somatic nucleus, resulting in impaired expression of several of the genes required for the proper progression of embryogenesis. 3. Low in utero embryo survival rate and poor health of the newborn animals. 4. The process of obtaining suitable oocytes and their activation requires considerable expenditure of time and financial resources.

ES and/ or EG cells	This method involves the insertion of a transgene into the genome of ES cells, followed by the selection of clones with a proper integration of the required number of copies, before transgenic ES cells are introduced into the cavity of a blastocyst, which is transplanted into a recipient female.	1. The production of transgenes requires a relatively small number of blastocysts and, hence, a small experimental herd. 2. Animal handling can be performed using nonsurgical methods, which are widely used in animal husbandry.	1. This method has only been perfected for mice and rats; ES cell lines for farm animals have yet to be obtained.
DNA-RTVs	Zygotes lacking protective coating are cultured in a medium supplemented with lentiviral particles, followed by transplantation into female recipients.	1. Efficient production of any TAs specie.	1. Inability to use the introns present in the gene construct and the limitation of the transgene length (approximately 8000 bp), which is determined by the size of the viral particle.
VE	It is a promising method for obtaining TAs consisting in the use of vectors based on mobile genetic elements, which are integrated into the genome by transposase. The gene encoding transposase and the transgene flanked by terminal repeats of transposon are coinjected into the zygote. The reaction catalyzed by transposase results in the integration of a single copy of the transgene into one or several sites of the animal's genome.	1. It can be used to produce large farm animals. 2. The efficiency of the integration of the transgene can be as high as 50%.	1. No data regarding the levels of expression of the target gene in the TAs produced using this method has been obtained thus far.
LP-DNA	Liposomes work with the lipid bilayer of cells to allow incorporation of the DNA into the target cells.	1. Efficient transfection and genetic expression *in vitro* and *ex vivo*.	1. Lack of specificity.

(Contd.)

Table 4.3. (*Contd.*)

	Description	Advantages	Disadvantages
EP-DNA	Target cells or embryos are placed in a solution containing the gene of interest. The solution and the cells are exposed to a very short duration (microseconds) of a high voltage electrical pulse, which allows a temporary breakdown of the cell membrane. This technique has been used in attempts to produce transgenic animals by electroporation of DNA into sperm, which then carry the exogenous DNA to the egg at fertilization.	1. Electroporation is effective with nearly all cell and species types. 2. The amount of DNA required is smaller than for other methods. 3. The procedure may be performed *in vivo*, with intact tissue.	1. If the pulses are of the wrong length or intensity, some pores may become too large or fail to close after membrane discharge causing cell damage or rupture. 2. The transport of material into and out of the cell during the time of electro-permeability is relatively nonspecific. This may result in an ion imbalance that could later lead to improper cell function and cell death.
BI	Biolistics or particle bombardment is a physical method that uses accelerated microprojectiles to deliver DNA or other molecules into intact tissues and cells.	1. Highly mechanized or robotic mediated technique. No skilled personnel are required. 2. Fast, simple and safe, and it can transfer genes to a wide variety of tissues. 3. No limits to the size or number of genes that can be delivered. 4. Biolistics do not depend upon the structure and characteristics of target cell membranes	1. Integration is random. 2. Requirement of specific equipment. 3. It can be a challenge to obtain a sufficient number of cells modified by this method to see a biologically significant effect.

| MI-L | Incorporation of exogeneous gene materials (DNA) into cells with a microbeam irradiation from an UV pulsed laser. | 1. Mild treatment, suitable for any cell type.
2. Transfection of selected cells in the presence of morphologically distinguishable cells.
3. Increased rate of DNA transfection successes as well as the efficiency of cell modification by orders of magnitude.
4. Minimal invasive tissue interaction. | 1. Requirement of specific equipment. |

Intra-pronuclear zygotic DNA microinjection (MI); Somatic cell nuclear transfer (SCNT); Site-specific transgenesis technology using stem (ES) cells and/or embryonic germ (EG) cells, DNA transfer by retroviruses (DNA-RTVs); Vectors (VE) based on mobile genetic elements; liposome-mediated DNA (LP-DNA) transfer into cells and embryos; electroporation of DNA (EP-DNA) into sperm, ova or embryos; biolistics (BI), microprojectile/gene gun; Microlaser (MI-L) and adolescent patients with hereditary angioedema, approved by EMA in 2010 and by FDA in 2014, confirmed the robustness and effectiveness of the platform. The approval of the two above-mentioned drugs paved the way for the use of milk as a vehicle for the production of biopharmaceuticals and for future new approvals of drugs produced into transgenic animal platforms. Recently, the expression of a novel recombinant human plasminogen activator (rhPA) in goat mammary glands has been reported by Zhengyi He and co-workers (He et al. 2018). Similarly, Igor de Sá Carneiro et al. (2018) reported the success of a transgenic goat expressing a recombinant human lysozyme (rhLZ) in the milk and its *in vitro* antibacterial activity against pathogens of the gastrointestinal tract. Furthermore, recombinant blood clotting factors, human fibrinogen (rhFib), human bile salt-stimulated lipase (rhBSSL), human lactoferrin (rhLF) and human growth hormone (rhGH) were expressed in the milk of transgenic cows (Salamone et al. 2006, Nelson et al. 2017, Wang et al. 2017a, b, Saavedra et al. 2018) for the application in humans to treat protein deficiencies. Traditional dairy species such as cows, sheep and goats are not only used for recombinant protein expression in the milk, but also rabbits and pigs. Recombinant forms of human protein C (rhPC), coagulation factor VIII (rhFVIII) and erythropoietin (rhEP) were obtained from the milk of transgenic pigs (Velander et al. 1992, Paleyanda et al. 1997, Park et al. 2006). Recently, recombinant human plasma phospholipid transfer protein (rPLTP) from the milk of transgenic rabbits was obtained by Pierre-Jean Ripoll and Laurent Lagrost (Ripoll and Lagrost 2018) and Deckert et al. (Deckert et al. 2017).

The mammary gland is especially appropriate for the expression of monoclonal antibodies (MAbs), both for the fabrication of biosimilars and for the creation of improved molecules (Meade et al. 1998). LFB Group Company has created over a dozen different MAbs expressed in the milk of transgenic goats that have all been shown to have target-binding characteristics equivalent and even enhanced effector functions to the commercial products: Adalimumab, Trastuzumab, and Cetuximab. For example, the equivalent to Cetuximab produced in goat milk was lacking the immunogenic 1, 3 alpha-galactose sugar moiety found in the commercial product, rendering to this protein version a potentially less immune reactive action upon its use (Bertolini et al. 2016). Mammalian glands bioreactors and the alteration of milk composition may have other applications than the production of pharmaceutical proteins. Some of the performed studies aim to increase the overall nutritional values of pig milk with higher levels of β-casein and κ-casein (Brophy et al. 2003). Other modifications of ruminant milk composition may improve quality of the curd or lead to the secretion of antibacterial proteins such as human lactoferrin (Iglesias-Figueroa et al. 2019), lysozyme (Souroullas et al. 2018) or lysostaphin (Kumar et al. 2018) that should simultaneously protect milk, mammary gland and potentially consumers from bacterial infection. The whey protein β-lactoglobulin (BLG) is a major milk allergen which is absent in human milk; cow milk lacking BLG is under study and that might be extremely helpful for such consumers who are intolerant to lactose (Sun et al. 2018). The above-mentioned examples point out that the initially planned methods to achieve biopharmaceutical in milk can also be extended to use milk as a vehicle to provide consumers with molecules beneficial to health, also improving it as a food (Houdebine 2009).

4.6 Recombinant Blood Substitutes

There is a worldwide demand for blood, currently unsatisfied (Slonim et al. 2014); translating these facts into statistics, only in the U.S., approximately 36,000 units of red blood cells, nearly 7,000 units of platelets and about 10,000 units of plasma are needed every day. Almost 21 million blood components are transfused each year with an average of three units of red blood cell and the "O" blood type the most often requested by hospitals. A single car accident victim can demand as many as 50 L of blood. Furthermore, sickle cell patients require blood transfusions throughout their lives. Taking into account that sickle cell disease affects about 90,000 to 100,000 people only in the U.S. and around 1,000 babies are born with this disease each year, the blood supply demand is infinite. According to the American Cancer Society, about 1.7 million people were diagnosed with cancer in 2017; many of them will need blood, sometimes daily, during their chemotherapy treatment (American Red Cross 2018). About 40% of U.S. blood and blood components were provided by The Red Cross (American Red Cross 2018), all from generous volunteer donors. Nevertheless, supply can't always meet demand because only about 10% of eligible people donate blood yearly. A study performed by Akita Tomoyuki and coworkers showed that the number of blood donations increased from 5,020,000 in 2008 to 5,260,000 in 2012, but will decrease to 4,770,000 units by 2025. In particular, the number of donors in their 20s and 30s decreased every year. Moreover, the number of donations required to supply blood products would have been increased from 5,390,000 in 2012 to 5,660,000 units in 2025. Consequently, the projected deficit of blood donations is likely to increase each year from 140,000 in 2012 to 890,000 in 2025 and then more than double to 1,670,000 in 2050 (Akita et al. 2016). Thus, the worldwide blood demands and the shortage of donors has generated a significant request for alternatives to whole blood and packed red blood cells for use in transfusion therapy. One of such alternatives involve the use of acellular recombinant hemoglobin (Hb) as an oxygen carrier (Varnado et al. 2013). The potential benefits of viable recombinant hemoglobin (Hb)-based oxygen carriers (rHBOCs) include universal compatibility without the need for cross matching of donated blood, availability, lack of infection, and long-term storage. The term "blood substitute" is usually used to refer to these type of compounds; however, it has often been inaccurately used to describe these compounds because they do not perform normal blood functions, such as transport of nutrients, immune response, and coagulation. However, as an "oxygen bridge", rHBOCs and their analogous non-rHBOCs, which are constructed from human or animal Hb obtained from whole blood, can be transfused complementing standard blood throughout life-threatening situations, such as trauma, specific surgical settings, and when blood is not an option (Alayash 2014). They can be used in place of packed red blood cells to restore impaired oxygen transport offering the following advantages: (i) universal compatibility; (ii) longer shelf-life; (iii) reduced risk of disease transmission; (iv) enhanced oxygen delivery; (v) improved rheological properties; (vi) improved uniformity in composition; (vii) more reliable availability; and (viii) a choice by persons who cannot receive conventional blood transfusions for clinical, geographical, or religious reasons.

RHBOCs consist of concentrated solutions of purified acellular human Hb, which has been heterologously expressed and purified from transgenic bacteria, mice, swine, yeast, and other organisms (Varnado et al. 2013). The physiological suitability of this material can be enhanced using protein-engineering strategies to address specific efficacy and toxicity issues. Therefore, mutagenesis of Hb can (i) adjust dioxygen affinity over a 100-fold range, (ii) reduce nitric oxide (NO) scavenging over 30-fold without compromising dioxygen binding, (iii) slow the rate of autooxidation, (iv) slow the rate of hemin loss, (v) impede subunit dissociation, and (vi) diminish irreversible subunit denaturation. Recombinant Hb production is potentially unlimited and readily subjected to current good manufacturing practices, but it may be restricted by cost. Acellular Hb-based O_2 carriers have superior shelf-life compared to red blood cells, are universally compatible, and provide an alternative for patients for whom no other alternative blood products are available or acceptable. Notwithstanding, the significant efforts at proceeding HBOCs and rHBOCs have not yet resulted in therapeutic licensure in the United States, although HBOC (Hemopure®; OPK Biotech, Boston, MA) is approved for human use in South Africa and Russia. The lack of approval of these HBOC products in the United States is primarily due to reports of adverse cardiovascular and cerebrovascular events associated with hypertension (Varnado et al. 2013).

In the case of recombinant HBOCs, additional barriers to development include globin denaturation, misfolding, and heme-orientational disorder, issues that arise during the expression of recombinant heme-proteins. There are also significant downstream processing problems, including the removal of large amounts of antigenic *Escherichia coli* lipopolysaccharides (LPS), protein impurities, and modified hemes and free porphyrin, all of which increase costs.

Critical issues remain unsolved including increasing Hb stability, mitigating iron-catalyzed and iron-centered oxidative reactivity, lowering the rate of hemin loss, and lowering the costs of expression and purification. Although many mutations and chemical modifications have been proposed to address these issues, the precise ensemble of mutations has not yet been identified. Future studies are aimed at selecting various combinations of mutations that can reduce NO scavenging, auto-oxidation, oxidative degradation, and denaturation without compromising O_2 delivery, and then investigating their suitability and safety *in vivo* (Varnado et al. 2013, Bracke et al. 2018, Frost et al. 2018, Silkstone et al. 2018).

4.7 Potential of Biotechnology, the *"Promethean Promises"*

In Greek mythology, the Titan Prometheus had a reputation of a clever trickster as he becomes famous for introducing fire to humans (Dougherty 2006). By giving the human race the gift of fire, Prometheus was punished by Zeus, who ensured every day that an eagle ate his liver as he was helplessly chained to a rock. It could be thought that Zeus also penalized humans; it must be noted that, along with all the advantages that fire bestowed upon humanity, a remarkable innovation at the time, it also brought many ills. For example, with the development of flaming forge and the power to work metals, cruel weapons of war emerged, along with all the attendant miseries that followed the disruption of a simple way of life.

Beyond the Promethean promises, currently, it is nearly impossible to imagine our modern biotechnology without the manipulation of DNA sequence. Since the publication of the complete human genome in 2003, the cost of DNA sequencing has dropped dramatically, making it a simple, available and widespread laboratory research tool. As virtually all of the biology focuses around the instructions confined in DNA molecule, biotechnologists who hope to modify the properties of cells, plants, and animals must speak the same molecular language and by tailoring at the atomic scale, the "nanoscale", genetic material from multiple sources created recombinant DNA. Nowadays, recombinant DNA technology can produce new kinds of living organisms or alter the genetic code of existing organisms and, as happens with most technologies, there are great benefits and notable downsides to the use of recombinant DNA knowledge.

By virtue of this technology, crucial proteins required for health problems and dietary purposes can be produced safely, affordably, and in the required amount. Recombinant pharmaceuticals are now being used confidently and are rapidly attaining commercial approvals. Techniques of recombinant DNA technology, gene therapy, and genetic modifications are also widely used for the purpose of bioremediation and treating serious diseases.

This technology has multidisciplinary applications and the potential to deal with important aspects of life; however, the rapid progress of research has raised questions about the consequences of advances of biotechnology.

Most of the downsides of recombinant DNA technology are ethical in nature. Some people feel that recombinant DNA technology goes against the laws of nature, or against their religious beliefs. Because of how much control this technology gives humans over the most basic building blocks of life, biotechnologists are accused of "playing gods".

From a different point of view, some people worry about the biggest technology corporations that can pay scientists to patent, buy and sell genetic material turning it into a luxurious article of trade. It may sound peculiar, but possible. In February 1951, the first human cells were successfully cultivated in a laboratory. These cells were called HeLa cells, named after Henrietta

Lacks, an African-American woman native to the state of Virginia. The cells were used – without her consent – for a scientific discovery that would become a medical miracle and change the scientific landscape for good. Her family did not know about her involuntary donation until after her death, and never received compensation, but others have profited from the use of HeLa cells (Cook 2018).

In addition, some other persons have apprehension about the safety of modifying food and medicines using recombinant DNA technology. Although genetically modified foods seem safe in multiple studies, it is easy to see why such fears exist. Recombinant food allergenic (glycol-) proteins are available in standardised quantity and constant quality. Therefore, they offer new perspectives to overcome current difficulties in the diagnosis, treatment and investigation of food allergies (Lorenz et al. 2001).

Other people worry that humans may start to manipulate their own genetic material too much and create social problems, for example, to live longer, become stronger or handpick certain traits for their offspring. The division between "genetically modified" and "standard" citizens is a concern that scientists and the public will likely continue to consider as humanity moves toward a future where manipulating DNA is easier than ever before.

For the moment, let's take a more positive view, many of the benefits of biotechnology are concrete while many of the risks remain hypothetical, but it is better to be proactive and cognizant of the risks than to wait for something to go wrong first and then attempt to address the damage. I hope that we will not receive a punishment to use biotechnology as it happened to Prometheus' civilisation (Rose and Rose 2014).

References

Ahmed, N., V. Brawley, M. Hegde, K. Bielamowicz, M. Kalra, D. Landi, et al. 2017. Her2-specific chimeric antigen receptor-modified virus-specific T cells for progressive glioblastoma: A phase 1 dose-escalation trial. JAMA Oncol. 3(8): 1094-1101.

Akita, T., J. Tanaka, M. Ohisa, A. Sugiyama, K. Nishida, S. Inoue, et al. 2016. Predicting future blood supply and demand in Japan with a Markov model: Application to the sex- and age-specific probability of blood donation. Transfusion 56: 2750-2759.

Alayash, A. I. 2014. Blood substitutes: Why haven't we been more successful? Trends Biotechnol. 32: 177-185.

Almeida, J. F., N. M. Breyner, M. Mahi, B. Ahmed, B. Benbouziane, P. C. B. V. Boas, et al. 2016. Expression of fibronectin binding protein A (FnBPA) from Staphylococcus aureus at the cell surface of Lactococcus lactis improves its immunomodulatory properties when used as protein delivery vector. Vaccine 34: 1312-1318.

American Red Cross, A. 2018. Blood Needs and Blood Supply.

Ball, P. 2004. Synthetic biology for nanotechnology. Nanotechnology 16, R1.

Barber, K. W., P. Muir, R. V. Szeligowski, S. Rogulina, M. Gerstein, J. R. Sampson, et al. 2018. Encoding human serine phosphopeptides in bacteria for proteome-wide identification of phosphorylation-dependent interactions. Nat. Biotechnol. 36(7): 638-644.

Beck, A. and J. M. Reichert. 2012. Marketing approval of mogamulizumab: A triumph for glyco-engineering. MAbs. 4(4): 419-425.

Benner, S. A. and A. M. Sismour. 2005. Synthetic biology. Nat. Rev. Genet. 6(7): 533-543.

Berg, P. and J. E. Mertz. 2010. Personal reflections on the origins and emergence of recombinant DNA technology. Genetics 184: 9-17.

Bertolini, L., H. Meade, C. Lazzarotto, L. Martins, K. Tavares, M. Bertolini, et al. 2016. The transgenic animal platform for biopharmaceutical production. Transgenic Res. 25(3): 329-343.

Bhatia, S., 2018. History, Scope and Development of Biotechnology. Introduction to Pharmaceutical Biotechnology. IOP Publishing Ltd. Bristol, UK.

Bischof, R.H., J. Ramoni and B. Seiboth. 2016. Cellulases and beyond: The first 70 years of the enzyme producer Trichoderma reesei. Microb. Cell Fact. 15(1): 106.

Blurton-Jones, M., B. Spencer, S. Michael, N.A. Castello, A.A. Agazaryan, J.L. Davis, et al. 2014. Neural stem cells genetically-modified to express neprilysin reduce pathology in Alzheimer transgenic models. Stem Cell Res. Ther. 5(2): 46.

Bracke, A., D. Hoogewijs and S. Dewilde. 2018. Exploring three different expression systems for recombinant expression of globins: Escherichia coli, Pichia pastoris and Spodoptera frugiperda. Anal. Biochem. 543: 62-70.

Brophy, B., G. Smolenski, T. Wheeler, D. Wells, P. L'Huillier and G. Laible. 2003. Cloned transgenic cattle produce milk with higher levels of β-casein and κ-casein. Nature Biotechnol. 21(2): 157-162.

Bud, R. 2002. History of Biotechnology. Encyclopaedia of Life Sciences. Nature Publishing Group, London, UK.

Carneiro, I. S., J. N. R. Menezes, J. A. Maia, A. M. Miranda, V. B. S. Oliveira, J. D. Murray, et al. 2018. Milk from transgenic goat expressing human lysozyme for recovery and treatment of gastrointestinal pathogens. Eur. J Pharm. Sci. 112: 79-86.

Cohen, S. N. and A. C. Chang. 1974. A method for selective cloning of eukaryotic DNA fragments in Escherichia coli by repeated transformation. Mol. Gen. Genet. 134(2): 133-141.

Cohen, S. N., A. C. Chang, H. W. Boyer and R. B. Helling. 1973. Construction of biologically functional bacterial plasmids in vitro. Proc. Natl. Acad. Sci. U.S.A. 70(11): 3240-3244.

Cook, M. 2018. Pros and Cons of Recombinant DNA Technology. sciencing.com, https://sciencing.com/pros-cons-recombinant-dna-technology-8433972.html.

Cox, M. M. J. 2012. Recombinant protein vaccines produced in insect cells. Vaccine 30: 1759-1766.

Chen, N. -Y., C. -H. Wu, H. -C. Chen and W. Town. 2017. Methods for producing recombinant glycoproteins with modified glycosylation. Google Patents.

Choi, B. -K., P. Bobrowicz, R. C. Davidson, S. R. Hamilton, D. H. Kung, H. Li, et al. 2003. Use of combinatorial genetic libraries to humanize N-linked glycosylation in the yeast Pichia pastoris. Proc. Natl. Acad. Sci. U.S.A. 100(9): 5022-5027.

Davies, K. E. 1981. The application of DNA recombinant technology to the analysis of the human genome and genetic disease. Hum. Genet. 58(4): 351-357.

Decker, E. L., J. Parsons and R. Reski. 2014. Glyco-engineering for biopharmaceutical production in moss bioreactors. Front. Plant. Sci. 5: 346.

Decker, E. L. and R. Reski. 2007. Moss bioreactors producing improved biopharmaceuticals. Curr. Op. Biotechnol. 18: 393-398.

Deckert, V., S. Lemaire, P. -J. Ripoll, J. -P. P. de Barros, J. Labbé, C. C. -L. Borgne, et al. 2017. Recombinant human plasma phospholipid transfer protein (PLTP) to prevent bacterial growth and to treat sepsis. Sci. Rep. 7(1): 3053.

Devetzi, M., M. Goulielmaki, N. Khoury, D. A. Spandidos, G. Sotiropoulou, I. Christodoulou, et al. 2018. Genetically-modified stem cells in treatment of human diseases: Tissue kallikrein (KLK1)-based targeted therapy (Review). Int. J. Mol. Med. 41(3): 1177-1186.

Dinnis, D. M. and D. C. James. 2005. Engineering mammalian cell factories for improved recombinant monoclonal antibody production: Lessons from nature? Biotechnol. Bioeng. 91(2): 180-189.

Dougherty, C. 2006. Prometheus. Routledge. London, UK.

Fári, M. G. and U. P. Kralovánszky. 2006. The founding father of biotechnology: Károly (Karl) Ereky. Int. J. Hortic. Sci. Technol. 12: 9-12.

Fischer, R., E. Stoger, S. Schillberg, P. Christou and R. M. Twyman. 2004. Plant-based production of biopharmaceuticals. Curr. Op. Plant Biol. 7: 152-158.

Fleming, A. 1944. The discovery of penicillin. Br. Med. Bull. 2: 4-5.

Foley, S. L. 2018. Plethora of Potential Platforms: A review of current and future trends in expression system choice in the biopharmaceutical industry. Institute of Technology, Sligo.

Frost, A. T., I. H. Jacobsen, A. Worberg and J. L. Martínez Ruiz. 2018. How synthetic biology and metabolic engineering can boost the generation of artificial blood using microbial production hosts. Front. Bioeng. Biotechnol. 6: 186.

Fu, G., V. Grbic, S. Ma and L. Tian. 2015. Evaluation of somatic embryos of alfalfa for recombinant protein expression. Plant Cell Rep. 34: 211-221.

Gasser, B., M. Saloheimo, U. Rinas, M. Dragosits, E. Rodríguez-Carmona, K. Baumann, et al. 2008. Protein folding and conformational stress in microbial cells producing recombinant proteins: A host comparative overview. Microb. Cell Fact. 7: 11.

Geison, G. L. 2014. The Private Science of Louis Pasteur. Princeton University Press. Princeton, New Jersey, USA.

Gerngross, T. U. 2004. Advances in the production of human therapeutic proteins in yeasts and filamentous fungi. Nat. Biotechnol. 22(11): 1409-1414.

Gheysen, D., E. Jacobs, F. de Foresta, C. Thiriart, M. Francotte, D. Thines, et al. 1989. Assembly and release of HIV-1 precursor Pr55gag virus-like particles from recombinant baculovirus-infected insect cells. Cell 59: 103-112.

Giddings, G., G. Allison, D. Brooks and A. Carter. 2000. Transgenic plants as factories for biopharmaceuticals. Nat. Biotechnol. 18(11): 1151-1155.

Gomord, V., P. Chamberlain, R. Jefferis and L. Faye. 2005. Biopharmaceutical production in plants: Problems, solutions and opportunities. Trends Biotechnol. 23(11): 559-565.

Goodsell, D. S. 2004. Bionanotechnology: Lessons from nature. John Wiley & Sons Ltd. Chichester, West Sussex, UK.

Hale, M., T. Mesojednik, G. S. Romano Ibarra, J. Sahni, A. Bernard, K. Sommer, et al. 2017. Engineering HIV-Resistant, Anti-HIV Chimeric Antigen Receptor T Cells. Mol. Ther. 25: 570-579.

He, Z., R. Lu, T. Zhang, L. Jiang, M. Zhou, D. Wu, et al. 2018. A novel recombinant human plasminogen activator: Efficient expression and hereditary stability in transgenic goats and in vitro thrombolytic bioactivity in the milk of transgenic goats. PloS One 13: e0201788.

Hennegan, K., D. Yang, D. Nguyen, L. Wu, J. Goding, J. Huang, et al. 2005. Improvement of human lysozyme expression in transgenic rice grain by combining wheat (Triticum aestivum) puroindoline b and rice (Oryza sativa) Gt1 promoters and signal peptides. Trans. Res. 14: 583-592.

Hobom, B. 1980. Surgery of genes – At the doorstep of synthetic biology. Medizinische Klinik. 75: 14-21.

Houdebine, L. -M. 2009. Production of pharmaceutical proteins by transgenic animals. Comp. Immunol. Microbiol. Infect. Dis. 32(2): 107-121.

Huang, L., M. -C. Hung and E. Wagner. 2005. Nonviral Vectors for Gene Therapy. Elsevier. Oxford, UK.

Iglesias-Figueroa, B. F., E. A. Espinoza-Sánchez, T. S. Siqueiros-Cendón and Q. Rascón-Cruz. 2019. Lactoferrin as a nutraceutical protein from milk, an overview. Int. Dairy J. 89: 37-41.

Jaenisch, R. 1988. Transgenic animals. Science 240(4858): 1468-1474.

June, C. H., R. S. O'Connor, O. U. Kawalekar, S. Ghassemi and M. C. Milone. 2018. CAR T cell immunotherapy for human cancer. Science 359: 1361-1365.

Kay, M. A., J. C. Glorioso and L. Naldini. 2001. Viral vectors for gene therapy: The art of turning infectious agents into vehicles of therapeutics. Nat. Med. 7(1): 33-40.

Kershaw, M. H., J. A. Westwood, C. Y. Slaney and P. K. Darcy. 2014. Clinical application of genetically modified T cells in cancer therapy. Clin. Transl. Immunology. 3(5): e16.

Khalil, A. S. and J. J. Collins. 2010. Synthetic biology: Applications come of age. Nat. Rev. Genet. 11(5): 367-379.

Khan, A. H., H. Bayat, M. Rajabibazl, S. Sabri and A. Rahimpour. 2017. Humanizing glycosylation pathways in eukaryotic expression systems. World J. Microbiol. Biotechnol. 33(1): 4.

Kinch, M. S. 2015. An overview of FDA – Approved biologics medicines. Drug Discov. Today 20(4): 393-398.

Knäblein, J. 2006. Plant-based expression of biopharmaceuticals. *In*: Meyerd, R. A. (Ed.). Encyclopedia of Molecular Cell Biology and Molecular Medicine, 2nd Ed. Vol. 10. Wiley-VCH Verlag GmbH & Co. KGaA, Weinheim, Germany.

Kool, E. 2015. Award Address (Ronald Breslow Award for Achievement in Biomimetic Chemistry sponsored by the Ronald Breslow Award Endowment). Designer DNA bases: Probing molecules and mechanisms in biology. Abstracts of Papers of The American Chemical Society. Amer. Chemical. Soc. 1155 16th st, N. W., Washington, D.C. 20036 U.S.A.

Kumar, M., P. Ratwan and V. Vohra. 2018. Transgenic or Genetically Modified Farm Animals and their Applications: A Review. RRJoVST. 5: 25-34.

Kuo, J., F. Stirling, Y. H. Lau, Y. Shulgina, J. C. Way and P. A. Silver. 2018. Synthetic genome recoding: New genetic codes for new features. Curr. Genet. 64(2): 327-333.

Kusnadi, A. R., E. E. Hood, D. R. Witcher, J. A. Howard and Z. L. Nikolov. 1998. Production and purification of two recombinant proteins from transgenic corn. Biotechnol. Prog. 14(1): 149-155.

Laird, P. W. 2010. Principles and challenges of genome-wide DNA methylation analysis. Nat. Rev. Genet. 11(3): 191-203.

Langer, R. and D. A. Tirrell. 2004. Designing materials for biology and medicine. Nature 428(6982): 487-492.

Li, M. Z. and S. J. Elledge. 2007. Harnessing homologous recombination in vitro to generate recombinant DNA via SLIC. Nat. Methods. 4(3): 251-256.

Lindahl, T. and R. D. Wood. 1999. Quality control by DNA repair. Science 286: 1897-1905.

Loh, H. -S., B. J. Green and V. Yusibov. 2017. Using transgenic plants and modified plant viruses for the development of treatments for human diseases. Curr. Opin. Virol. 26: 81-89.

Lomonossoff, G. P. and M. -A. D'Aoust. 2016. Plant-produced biopharmaceuticals: A case of technical developments driving clinical deployment. Science 353: 1237-1240.

Lone-Star-College. 2018. History of Biotechnology. http://www.lonestar.edu/history-of-biotechnology.htm

Lorenz, A. -R., S. Scheurer, D. Haustein and S. Vieths. 2001. Recombinant food allergens. J. Chromatogr. B Biomed. Sci. Appl. 756(1-2): 255-279.

Lu, T. K., A. S. Khalil and J. J. Collins. 2009. Next-generation synthetic gene networks. Nat. Biotechnol. 27(12): 1139-1150.

Maiorella, B., D. Inlow, A. Shauger and D. Harano. 1988. Large-scale insect cell-culture for recombinant protein production. Bio/Technology 6: 1406-1410.

Maksimenko, O., A. Deykin and P. Georgiev. 2013. Use of transgenic animals in biotechnology: Prospects and problems. Acta Naturae 5(1): 33-46.

Martin, R. W., B. J. Des Soye, Y. -C. Kwon, J. Kay, R. G. Davis, P. M. Thomas, et al. 2018. Cell-free protein synthesis from genomically recoded bacteria enables multisite incorporation of noncanonical amino acids. Nat. Commun. 9(1): 1203.

Meade, H., P. DiTullio and D. Pollock. 1998. Transgenic production of antibodies in milk. Google Patents.

Muramatsu, M., K. Kinoshita, S. Fagarasan, S. Yamada, Y. Shinkai and T. Honjo. 2000. Class switch recombination and hypermutation require activation-induced cytidine deaminase (AID), a potential RNA editing enzyme. Cell 102: 553-563.

Nelson, K. M., M. W. Mosesson, A. Pusateri and E. J. Forsberg. 2017. Use of anticoagulants in the production of recombinant proteins in the milk of transgenic animals. Google Patents.

Obembe, O. O., J. O. Popoola, S. Leelavathi and S. V. Reddy. 2011. Advances in plant molecular farming. Biotechnol. Adv. 29: 210-222.

Olson, M. V. 1993. The human genome project. Proc. Natl. Acad. Sci. U.S.A. 90(10): 4338-4344.

Orel, V. 1996. Gregor Mendel: The First Geneticist. Oxford University Press. U.S.A.

Orlowski, J. and J. M. Bujnicki. 2008. Structural and evolutionary classification of Type II restriction enzymes based on theoretical and experimental analyses. Nucleic Acids Res. 36(11): 3552-3569.

Paleyanda, R. K., W. H. Velander, T. K. Lee, D. H. Scandella, F. C. Gwazdauskas, J. W. Knight, et al. 1997. Transgenic pigs produce functional human factor VIII in milk. Nat. Biotechnol. 15(10): 971-975.

Park, J. -K., Y. -K. Lee, P. Lee, H. -J. Chung, S. Kim, H. -G. Lee, et al. 2006. Recombinant human erythropoietin produced in milk of transgenic pigs. J. Biotechnol. 122(3): 362-371.

Pasteur, L. and J. Lister. 1996. Germ Theory and its Applications to Medicine and on the Antiseptic Principle of the Practice of Surgery. Prometheus Books. Amherst, New York, USA.

Pera, M. and A. Trounson. 2013. Cloning debate: Stem-cell researchers must stay engaged. Nature 498(7453): 159-6.

Pinkert, C. A. and J. D. Murray. 1999. Transgenic farm animals. Transgenic Animals in Agriculture 1-18.

Purnick, P. E. and R. Weiss. 2009. The second wave of synthetic biology: From modules to systems. Nat. Rev. Mol. Cell Biol. 10(6): 410-422.

Kusnadi, A. R., E. E. Hood, D. R. Witcher, J. A. Howard and Z. L. Nikolov. 1988. Production and purification of two recombinant proteins from transgenic corn. Biotechnol. Prog. 14(1): 149-155.

Rahmati, M., A. H. Khan, S. Razavi, M. R. Khorramizadeh, M. J. Rasaee and E. Sadroddiny. 2016. Cloning and expression of human bone morphogenetic protein-2 gene in Leishmania tarentolae. Biocatal. Agric. Biotechnol. 5: 199-203.

Ripoll, P. -J. and L. Lagrost. 2018. Preparation of recombinant human plasma phospholipid transfer protein (PLTP) from the milk of transgenic rabbits. Google Patents.

Roberts, R. J., T. Vincze, J. Posfai and D. Macelis. 2007. REBASE—enzymes and genes for DNA restriction and modification. Nucleic Acids Res. 35: D269-D270.

Rojas-Anaya, E., E. Loza-Rubio, M. T. Olivera-Flores and M. Gomez-Lim. 2009. Expression of rabies virus G protein in carrots (Daucus carota). Transgenic Res. 18(6): 911-919.

Rosano, G. L. and E. A. Ceccarelli. 2014. Recombinant protein expression in Escherichia coli: Advances and challenges. Front. Microbiol. 5: 172.

Rose, H. and S. P. R. Rose. 2014. Genes, cells, and brains: The promethean promises of the new biology. Verso Trade.

Rovner, A. J., A. D. Haimovich, S. R. Katz, Z. Li, M. W. Grome, B. M. Gassaway, et al. 2015. Recoded organisms engineered to depend on synthetic amino acids. Nature. 518(7537): 89-93.

Saavedra, S. L., M. C. Martínez Ceron, S. L. Giudicessi, G. Forno, M. B. Bosco, M. M. Marani, et al. 2018. Single step recombinant human growth hormone (rhGH) purification from milk by peptide affinity chromatography. Biotechnol. Prog. 34(4): 999-1005.

Salamone, D., L. Barañao, C. Santos, L. Bussmann, J. Artuso, C. Werning, et al. 2006. High level expression of bioactive recombinant human growth hormone in the milk of a cloned transgenic cow. J. Biotechnol. 124: 469-472.

Salmond, G. P. C. and P. C. Fineran. 2015. A century of the phage: Past, present and future. Nat. Rev. Microbiol. 13(12): 777-786.

Sanchez-Garcia, L., L. Martín, R. Mangues, N. Ferrer-Miralles, E. Vázquez and A. Villaverde. 2016. Recombinant pharmaceuticals from microbial cells: A 2015 update. Microb. Cell Fact. 15: 33.

Schwartz, M. 2001. The life and works of Louis Pasteur. J. Appl. Microbiol. 91(4): 597-601.

Secore, S. L., J. L. Bodmer, J. H. Heinrichs and J. C. Scholten. 2017. Bacillus megaterium recombinant protein expression system. pp. 1-107. Patent Application No. US15/522,593.

Sethuraman, N., R. C. Davidson, T. A. Stadheim and S. Wildt. 2017. Yeast strain for the production of proteins with terminal alpha-1, 3-linked galactose. Google Patents.

Shelley, M. W. 1818. Frankenstein or the Modern Prometheus. In three volumes. Lackington, Hughes, Harding, Mayor & Jones, Finsbury Square.

Shendure, J. and H. Ji. 2008. Next-generation DNA sequencing. Nat. Biotechnol. 26(10): 1135-1145.

Shirahata, Y., N. Ohkohchi, H. Itagak and S. Satomi. 2001. New technique for gene transfection using laser irradiation. J. Investig. Med. 49(2): 184-190.

Silkstone, G. G. A., M. Simons, B. S. Rajagopal, T. Shaik, B. J. Reeder and C. E. Cooper. 2018. Novel redox active tyrosine mutations enhance the regeneration of functional oxyhemoglobin from methemoglobin: Implications for design of blood substitutes. pp. 221-225. *In*: O. Thews, J. C. LaManna, D. K. Harrison (eds.). Oxygen Transport to Tissue XL. Springer International Publishing. Midtown Manhattan, New York City, USA.

Slonim, R., C. Wang and E. Garbarino. 2014. The market for blood. J. Econ. Perspect. 28(2): 177-196.

Smith, T. T., S. B. Stephan, H. F. Moffett, L. E. McKnight, W. Ji, D. Reiman, et al. 2017. In situ programming of leukaemia-specific T cells using synthetic DNA nanocarriers. Nat. Nanotechnol. 12(8): 813-820.

Souroullas, K., M. Aspri and P. Papademas. 2018. Donkey milk as a supplement in infant formula: Benefits and technological challenges. Food Res. Int. 109: 416-425.

Sparks, C. A. and H. D. Jones. 2009. Biolistics transformation of wheat. Transgenic Wheat, Barley and Oats. Springer, pp. 71-92.

Strasser, R., F. Altmann and H. Steinkellner. 2014. Controlled glycosylation of plant-produced recombinant proteins. Curr. Opin. Biotechnol. 30: 95-100.

Sun, Z., M. Wang, S. Han, S. Ma, Z. Zou, F. Ding, et al. 2018. Production of hypoallergenic milk from DNA-free beta-lactoglobulin (BLG) gene knockout cow using zinc-finger nucleases mRNA. Sci. Rep. 8(1): 15430.

Takeyama, N., Kiyono, H. and Y. Yuki. 2015. Plant-based vaccines for animals and humans: Recent advances in technology and clinical trials. Ther. Adv. Vaccines 3(5-6): 139-154.

Thackray, A. 1998. Private Science: Biotechnology and the Rise of the Molecular Sciences. University of Pennsylvania Press.

Thiel, K. A. 2004. Biomanufacturing, from bust to boom...to bubble? Nat. Biotechnol. 22(11): 1365-1372.

Toussaint, A. and C. Merlin. 2002. Mobile elements as a combination of functional modules. Plasmid 47: 26-35.

Travis, A. S., W. J. Hornix and R. Bud. 1992. The zymotechnic roots of biotechnology. Br. J. Hist. Sci. 25: 127-144.

U.S. Department of Health and Human Services, U.S.D. 2018. U.S. Food and Drug Administration (FDA).

Varnado, C. L., T. L. Mollan, I. Birukou, B. J. Smith, D. P. Henderson and J. S. Olson. 2013. Development of recombinant hemoglobin-based oxygen carriers. A Antioxid. Redox Signal. 18(17): 2314-2328.

Vázquez, E. and A. Villaverde. 2013. Microbial biofabrication for nanomedicine: Biomaterials, nanoparticles and beyond. Nanomedicine 8: 1895-1898.

Velander, W. H., J. L. Johnson, R. L. Page, C. G. Russell, A. Subramanian, T.D. Wilkins, et al. 1992. High-level expression of a heterologous protein in the milk of transgenic swine using the cDNA encoding human protein C. Proc. Natl. Acad. Sci. U.S.A. 89(24): 12003-12007.

Villemejane, J. and L. M. Mir. 2009. Physical methods of nucleic acid transfer: General concepts and applications. Br. J. Pharmacol. 157(2): 207-219.

Walmsley, A. M. and C. J. Arntzen. 2000. Plants for delivery of edible vaccines. Curr. Opin. Biotechnol. 11(2): 126-129.

Wang, M., Z. Sun, T. Yu, F. Ding, L. Li, X. Wang, et al. 2017a. Large-scale production of recombinant human lactoferrin from high-expression, marker-free transgenic cloned cows. Sci. Rep. 7(1): 10733.

Wang, Y., F. Ding, T. Wang, W. Liu, S. Lindquist, O. Hernell, et al. 2017b. Purification and characterization of recombinant human bile salt-stimulated lipase expressed in milk of transgenic cloned cows. PloS One 12: e0176864.

Wang, Y., S. Zhao, L. Bai, J. Fan and E. Liu. 2013. Expression systems and species used for transgenic animal bioreactors. Biomed. Res. Int. 2013: 580463.

Watson, J. D. and F. H. Crick. 1953. The structure of DNA. Cold Spring Harbor symposia on quantitative biology. Cold Spring Harbor Laboratory Press. 123-131.

Wheeler, M., E. Walters and S. Clark. 2003. Transgenic animals in biomedicine and agriculture: Outlook for the future. Anim. Reprod. Sci. 79(3-4): 265-289.

Whitesides, G. M. 2003. The 'right' size in nanobiotechnology. Nat. Biotechnol. 21(10): 1161-1165.

Wickham, T. J., T. Davis, R. R. Granados, M. L. Shuler and H. A. Wood. 1992. Screening of Insect Cell Lines for the Production of Recombinant Proteins and Infectious Virus in the Baculovirus Expression System. Biotechnol. Prog. 8(5): 391-396.

Wilmut, I., Y. Bai and J. Taylor. 2015. Somatic cell nuclear transfer: Origins, the present position and future opportunities. Philos. Trans. R. Soc. Lond. B. Biol. Sci. 370(1680): 20140366.

Yang, D., J. Ou, J. Shi, Z. Guo, B. Shi and N. Abiri. 2018. Transgenic rice for the production of recombinant pharmaceutical proteins: Case study of human serum albumin. pp. 275-307. In: A. R. Kermode and L. Jiang (eds.). Molecular Pharming: Applications, Challenges, and Emerging Areas. John Wiley & Sons. Hoboken, N, USA.

Nanotechnology in Gene Therapy

"...Human beings are ultimately nothing but carriers-passageways- for genes..."

From the dystopian novel "1Q84" written by Haruki Murakami
First published in three volumes in Japan in 2009–2010

5.1 Applying the Commandment Code to Heal

Instructions for the body function is codified into genes- they are like the command code for how the collection of cells, limbs, and organs are supposed to behave and perform. Sometimes, due to multiple factors, like age, sickness, disease or damage, the capacity of the body to respond to the gene instructions can either suffer a breakdown or stop answering back. Gene therapy is a medical practice for treating these bodily malfunctions by administering new DNA codes in the form of genes into the affected area to regulate and normalize the multiplicative pattern and function of the organism. Researchers have shown that they can use nanotechnology in gene therapy as a way of delivering these new genes. Nanotechnology gene therapy is a promising tool for health care. It can be proposed as a modern medical treatment and consequently, it is shown as an alternative for facing new challenges in the medical field. Thus, it would provide a possible cure for multiple complex diseases. Under the word, "cure", the word "expectation" is underlying. The cure is the success that comes after the treatment, and in this sense, gene therapy opens new expectations. Gene therapy is a therapeutic strategy centered on the use of genes, mainly DNA, as a pharmaceutical agent. It is based on the replacement of a malfunctioning gene inside the cell that is adversely affected by a certain condition. This therapy has been proposed and, in few cases, employed successfully for treating a wide range of complex acquired and inherent diseases such as many types of cancers, cystic fibrosis, acquired immunodeficiency syndrome (AIDS), cardiac and metabolic diseases (Ibraheem et al. 2014).

The European Medicines Agency has defined a product based on gene therapy, as a product in which a vector containing a genetic material designed to express a specific protein sequence is used to treat a certain pathology or as a regulatory agent (for repairing) of a genetic sequence. Therefore, this therapy considers a gene as a medicine and consequently, this "genetic medicine" must be carried to the target site inside the body. Therefore, if the gene is considered a drug, a pharmaceutical formulation must be designed to carry the "genetic medicine" and this is one of the mains problems *in vivo* for gene therapy. Due to this,large, fragile and negative charged molecules such as DNA must reach the nucleus of the target cells without degradation. Thus, the success of gene therapy depends on the development of safe and efficient gene vectors. Gene therapy is shown as a method with the capability to restore functional proteins through replacement of missed or distorted genes by healthy genes. This method is displayed as therapeutically simple; however, in practice it implies a more complex development than conventional pharmaceutical formulations. Genetic material must overcome several obstacles due to its intrinsic features and also due to those related to the action which must carry out within the cell. The expression of functional proteins encoded by nucleic acids that belong to gene material is vital for maintaining the homeostasis of the body and for avoiding diseases progression. Therefore, the challenge of this therapy is associated with the correct expression of gene material. Thus, the basis of the treatment implies the correction of protein malfunction. The use of medicines with active compounds based on proteins is used for treating many diseases with a high impact such as diabetes, those based on the replacement of growth factor and others. Due to that fact, the development of DNA recombinant techniques has advanced. However, there are many drawbacks that must be considered for applying this therapy, for example: (i) the low bioavailability of proteins due to their high degradation; (ii) the low time of circulation into blood vessels owing to high hepatic clearance and, finally (iii) their lack of oral availability due to high degradation in stomach, which implies that the injection is the only possible way of administration (Ibraheem et al. 2014). The introduction of genes within cells which could be expressed and therefore generate the protein of interest is the method to replace the introduction of protein. This method is called gene therapy.

In this chapter, a brief history of gene therapy is described. This section covers a description of "transforming principle" in the 1930s, to the approbation of the first protocol used in human beings in the 1990s. In the following section, mechanisms for crossing cell barriers and some considerations about the relation of nucleic acids introduced into the cell and its intracellular machinery are described. Then, a simple classification of the different gene therapy approaches and a description of the vector types used are shown. In this chapter, non-viral vectors are mainly displayed; however, a brief description of viral vectors is shown. Among non-viral vectors, liposomes, liquid crystals, and associations of lipid nanoparticles or polymers nanoparticles with DNA are described. These carriers have been proposed as an alternative to incorporate the specific genes into target cells instead of the virus because of their advantages which are described below. In this context, nanotechnology has taken an overriding place in the development of non-viral formulations to deliver "genetic drugs". Significant advances in this field will soon improve the preparation of nanostructures based on nucleic acids.

5.2 Milestones in Gene Therapy

The current concept of gene therapy emerged between the 1960s and 1970s. However, the basic concepts used in this field to reach the current ones started in 1928. Thus, in that year, Frederick Griffith described the "transforming principle", but this concept was better explained some years later. The concept of this principle shows that a non-virulent type of pneumococcal can become a virulent type (Griffith 1928). In 1931, Dawson and Sia reported the confirmation of this finding (Sia and Dawson 1931) and in 1932, Alloway reported that the addition of the intracellular content of S form of pneumococcus (not the whole bacteria) to an R culture produced a change from R form into S (Alloway 1932).Therefore, Alloway concluded that there is something in the cell-free extracts which is responsible for the change from one to the other form of pneumococcus. Thus, this "something" was named as "transformation principle" but the responsible structure was not identified. In 1941, Avery and McCarty identified and isolated this "something" and then, in 1944, these authors reported that DNA was responsible for that transformation (from one form of bacteria into another) (Avery et al. 1944). Later, Joshua Lederberg described another mechanism of genetic transfer which is produced by the mating of bacteria strains (Zinder and Lederberg 1952). This mechanism, named transduction, was important to understand how wild bacteria strains can show antibiotic resistance when they are previously incubated with resistant ones. As a temporary location, we can mention that a year later of this research, in 1953, Watson and Crick described the double helix of DNA (Watson and Crick 1953). In 1962, Szybalska and Szybalski demonstrated that cells can uptake foreign DNA (Szybalska and Szybalski 1962). Therefore, this concept means that a genetic defect could be rectified by transferring DNA and this material could be provided from a different source. This finding forms the cornerstone of hereditable gene transfection in mammalian cells.

Up to this point, it has been demonstrated that genes could be transferred between cells; however, this transference does not guarantee that we are talking about gene therapy because a vector is needed. The first findings on gene therapy have been laid by Temin in 1961 (Temin 1961). This researcher reported that as a result of a virus infection, mutations on genes of the infected cells had been evidenced. Another study showed that RNA information turns into DNA and the acquisition of a new genetic feature was inherited by means of chromosomal insertion of exogenous genetic material (Sambrook et al. 1968). Therefore, the virus has resulted in a very useful gene delivery system for somatic cells. Consequently, in 1966, the use of the virus as a gene transfection vector to eukaryotic cells had been reported by Tatum (Tatum 1966). Thus, for using a virus as a gene transfer system and in this sense, for using this virus as a certain therapy, it must not necessarily be provided with disease-causing genes. Due to that fact, for the emergence of gene therapy such as we currently know, many years passed.

Some authors consider that gene therapy begins with a specific event in a health centre of the United States in September of 1990 when the FDA approved the first treatment based on gene therapy for human beings. This procedure was applied to patients who had adenosine deaminase (ADA) deficiency. In this fact, they received an intravenous infusion of autologous blood cells (T cells) into which a normal ADA gene had been inserted (Blaese et al. 1995). Therefore, this event was considered as a cultural breakthrough and it is expected that this procedure will be widely available for patients who need it (Anderson 1992, 1998). This fact was the starting point for gene therapy-based treatments in other parts of the world (Bordignon et al. 1995) and it helped in the expansion of this therapy. However, in 1999 was reported the first death of a patient subjected to a gene therapy trial. The patient died by an immunological response to adenovirus high doses and it was established that the causes of his death were directly related to the virus vector used (Stolberg 1999). These described facts have led to the development of safe gene transfer systems. There are several problems associated with the use of viral vectors for gene transfections: their high immunogenicity, carcinogenesis, and limited nucleic acids packaging capacity, among others (Baum et al. 2006, Bessis et al. 2004, Thomas et al. 2003). Thus, non-viral vectors have emerged as an alternative. These systems will be described below; however, it is important to mention that non-viral vectors can produce adverse effects such as other traditional formulations but they are safer than viral vectors.Currently, a number of diseases are being studied for the application of gene therapy, out of which oncologic and infectious diseases are the most investigated (del Pozo-Rodríguez et al. 2016).

5.3 Toward the Target Site: Crossing Barriers

Gene therapy could be considered as a nucleic acid-based medical treatment. Therefore, these therapeutic systems must be able to cross different barriers depending on the type of nucleic acid used in the therapy. RNAs based systems must reach the cytoplasm and those based on DNA must get to the nucleus of the cell. The first step for reaching the cell is spatially defined as the path from the application site of the genetic medicine to the cell membrane. Generally, these formulations are intravenously or intramuscular administrated; hence, this space is defined as an extracellular environment where bare nucleic acids (without the protection given by the formulation; in this case, the non-viral vector) could be easily degraded by nucleases (Lechardeur and Lukacs 2002, Ruponen et al. 2003).Therefore, the genetic material must cross the cell membrane and it should fulfill different steps for reaching the nucleus of the cell: the first step, cell-binding; the second, internalization; then, the material must escape from vesicles inside the cytoplasm (endosomes); and finally, it must get into the nucleus. Therefore, once close to the target cell, the genetic agent must pass through the membrane, but first, a binding process between the surface of the cell membrane and the active molecule must be carried out. This process can be mediated by a specific molecular structure, such as receptors, or by means of unspecific interactions. Depending on the vehicle used to deliver the therapeutic agent, the mechanism for passing through the cell membrane can be different, for example: for non-viral delivery system, such as cationic lipid nanoparticles, the binding is generally promoted by means of electrostatic interactions (unspecific interaction) and therefore, the subsequent pass through the cell membrane is produced by endocytosis. This mechanism can be divided by its intrinsic properties into phagocytosis (delimited to specialized phagocytes such as macrophages, neutrophils, monocytes, and dendritic cells); and pinocytosis (which occurs in all types of cells and has multiple forms). This latter mechanism can be divided into clathrin-dependent endocytosis or clathrin-independent endocytosis and macropinocytosis, among others (del Pozo-Rodríguez et al. 2008). The capability for generating a specific interaction between the active molecule and the cell surface can improve the efficiency of the transfection process. In this context, molecules such as streptavidin on the surface of lipid nanoparticles can be used to bind epidermal growth factor to target cancer cells (Pedersen et al. 2006). By means of the use of transferring onto nanostructured lipid carriers, it is possible to develop an active targeting against lung cancer cells (Shao et al. 2015). Another strategy for passing the cell membrane is based on the use of penetrating peptides onto the surface of the non-viral vectors. Peptides such as TAT-peptide have been anchored to solid lipid nanoparticles surface for improving the transfection efficiency in human bronchoepithelial cells *in vitro* studies (Rudolph et al. 2004). In this context, physicochemical properties of the non-viral carriers, such as particle size and shape, their molecular weight, surface charge, composition, and stability, affect their cellular internalization and trafficking to the nucleus of cells (Perumal et al. 2008).

The intracellular traffic is the next step to be overcome by the active molecule after cell internalization and the system has a tendency to be entrapped by vesicles such as endosomes. The escape of gene material from endosomes environment is one of the major challenges for the vector because of these structures' fusion with lysosomes membrane which provides acidic pH conditions and enzymes for degradation of the internalized structures (Little and Kohane 2008, Wagner 2014). For example, cationic liposomes formulated with fusogenic lipids such as dioleoyl-phosphatidylethanolamine (DOPE) or peptides for escaping from endosomes have been reported (Varkouhi et al. 2011). Another example includes the use of polyethylenimine (PEI) in polymer-based non-viral vectors as a proton scavenger (Cheung et al. 2001). The last step is the nuclear entry; therefore, the DNA must cross the nuclear envelope and the access mechanism is related to the delivery system. Consequently, gene-carrier systems below 25 nm diameter could be actively transported across the nuclear pore system for transferring the gene, but for passive transport, the molecule should be below 9 nm (Allen et al. 2000).

5.4 Classification of Gene Therapy and Its Types of Vectors

The treatment based on gene therapy can be divided into categories depending on the type of target cells involved: Germline gene therapy (Okita et al. 2007) and somatic line gene therapy (Bank 1996). The election of one or another type depends on the type of the disease and on the kind of strategy used for treating the disease. Germline gene therapy is based on the transfection of genes from a vector to reproductive cells such as zygote or sperm. Thus, the integration of these genes within the guest genome will produce heritable phenotypic characteristics (Stribley et al. 2002). Somatic gene therapy is based on the transference of genes to somatic cells and, therefore, they are devoid of reproductive capacity. This concept means that only an individual cell or group of cells can express the phenotype, and this change only affects one-generation cells. In other words, this new phenotypic characteristic is not heritable.

The main therapeutic molecules (nucleic acids) used today for gene therapy treatments are DNA and RNA. One of the first medicine approved for gene therapy was based on DNA. These molecules were the gene material traditionally included into vectors and they are the main type of "nucleic acid medicine" used. Among these nucleic acids, the DNA plasmid (DNAp) was the first type introduced successfully in vectors (Allen and Cullis 2013). A newer strategy for addressing the treatment of complex diseases based on gene therapy was the inclusion of molecules with directly acting capability such as different types of RNA: messenger RNA (mRNA), short interfering RNA (siRNA)(Kanasty et al. 2013, Spagnou et al. 2004), micro RNA (miRNA) (Broderick and Zamore 2011, Rupaimoole and Slack 2017) among others.

There are two main strategies to carry out gene therapy studies: *in vivo* and *ex vivo*. The first scheme involves an application of genes directly to the patient. Genes must be loaded into a formulation and they are administered by injection. Therefore, the transfection is produced inside the patient. Contrarily, in *ex vivo* studies, the cells of the patients are removed from their body, and they are transfected outside the body of the patient and are then returned to the patient (Herzog et al. 2010).

There are general requirements for gene therapy vectors and consequently, many factors must be considered before the application of this therapy to patients (Hug and Sleight 1997). There are many important and necessary steps for reaching the success of gene therapy. Among them area clear understanding of the genesis of disease, the type of target tissue for gene delivery, an effective therapeutic gene, and the animal model that simulates the disease for preclinical testing (Robbins and Ghivizzani 1998). However, in the chapter we will focus on two main approaches, one of them is based on the safety of the gene therapy vectors and the other is based on their manufacturing process and scaling. First, the method and type of treatment should be jointly evaluated with regard to the severity of the disease. Gene therapy should be applied to complex diseases because its application can cause many problems. Therefore, it is expected that, while the treatment provided a cure for the ill, it could develop a new disease. This kind of problems can also emerge for oncologic treatments when traditional medicines are used. It is important to remark that the treatment must not be a danger for other patients. For example, an oncogene, loaded into a virus vector, could be possibly transmitted to a wild adenovirus strain and this wild strain could cause a new disease and its dissemination. Therefore, substantial progress must be carried out in the construction of replication-defective viral vectors for avoiding these problems. There are many types of viral vectors used in gene therapy. Among them are retrovirus (small RNA viruses that replicate through a DNA intermediate) (Guild et al. 1988), adenovirus and herpes simplex. Adenovirus (double-stranded linear DNA viruses) has been used for gene delivery *in vivo* studies and in clinical trials for cystic fibrosis and different types of cancer (Yang and Wilson 1995). Vectors based on herpes simplex virus have the advantage of establishing latency in some cell types (possibility to infect non-dividing cells) and the capacity to be loaded with large regions of exogenous DNA (Fink et al. 1995).

The non-viral vectors have emerged as a tool for addressing issues such as the high immunogenicity of viral vectors. In order to determine how successful is the treatment based on gene therapy, a diminishing of symptoms or cure of disease must be evidenced, such as for traditional treatments. Although retreatments may be acceptable, each of them brings the possibility of mutagenesis generation and/or the trigger of an immune response by the vector used. The second aspect, referenced for the manufacturing process, must include easy preparation, scaling and if possible generate the simplest administration mode. Consequently, these features are more easily reached by non-viral than viral vectors.

The gene transfection efficiency is one of the major challenges of gene delivery. Therefore, it is important to improve the non-viral carrier features for reaching this aim (Liu and Zhang 2011). Thus, in the next section, a description of types and advantages of non-viral carriers is shown.

5.5 Nanotechnology-based Carriers: An Alternative to Virus Vectors

The notion of nanoscience and nanotechnology started with the talk of Richard Feynman in 1959 at Caltech, California. These concepts have been incorporated into the current topics of science and they have answered many questions. For example, they have been used for explaining why nanometric compounds of different nature, such as inorganic-organic or hybrid organic-inorganic, have different properties than those of macroscopic ones (Soler-Illia et al. 2002). Nanoscience also presents paramount advantages to facilitate the integration and miniaturization of devices, and in the biology field, these concepts have elucidated how the bacterial flagellar motor works. They are also used for explaining the self-assembly of colloidal systems used as a template for nanoparticles synthesis (Walz and Caplan 2002). Beyond all mentioned applications, nanotechnology has been also applied to pharmaceutical formulations design and later in gene therapy. At the beginning, viral vectors were used but later, many strategies based on nanotechnology for introducing an exogenous gene into a set of cells of a patient have been developed (Hug and Sleight 1997). In this context, molecular aggregates based on nanotechnologies such as liquid crystals, liposomes, lipoplexes, lipid nanoparticles, nanostructured lipids, polymeric systems, and micelles have gained the space, and they have been proposed as nano-carriers for gene delivery. Therefore, this section is focused on these non-viral vectors, and additionally, their advantages and disadvantages compared to viral systems are addressed.

5.5.1 Liposomes and Liquid Crystals

At the end of the 1990s, the scientific community has focused on the study of interactions between a set of lipids used for drug delivery systems and nucleic acids. The aim of those studies was to elucidate the self-assembly behaviour of cationic lipids for generating liposomes potentially used as DNA-carrier systems (Safinya 2001).

The liposomes have been one of the first nanostructures developed and used in different areas, mainly focused on health care. Within the pharmaceutical field, these systems were oriented towards the improvement of drug delivery systems (Benedini et al. 2017, Gregoiadis 1976). Later, with the advent of gene therapy concept (which considers the DNA molecule as a drug), this material could be included inside these structures (Hug and Sleight 1997, Nicolau and Sene 1982). Discovered in 1965 at Cambridge University, liposomes are vesicles formed by closed single or stacked bilayer which are made from non-toxic phospholipids (Deamer 2010). These lipids contain a hydrophilic head group and hydrophobic tails, and consequently, they have amphiphilic behaviour. In other words, liposomes could be defined as a supramolecular lipid matrix in smectic liquid crystal phase. The number of stacked bilayers can be one or more, and in this last case, they are known as multi-lamellar. These liposomes have been shown as a cloudy solution (suspension) and they have a larger possibility to carry in their membranes' lipophilic drugs. However, they have very limited space for dissolving water-soluble drugs such as DNA or another genetic material (Benedini et al. 2017). Different methods have been applied for changing (e.g. extrusion method) these liposomes into uni-lamellarones (Benedini et al. 2014, Tros de Ilarduya et al. 2010).

Accordingly, these structures could be considered as self-closing laminar mesophases. There are many categories of liposomes and their classification depends on the size, the number of membranes, surface charge, composition of membranes (types of lipids and sterols among others) and methods of preparation (Benedini et al. 2017).

5.5.2 Liposomes and Gene Therapy

The use of lipids for delivery of DNA and nucleic acids is one of the major techniques and one of the first strategies for positioning non-viral vectors as gene carriers (Wasungu and Hoekstra 2006).

One of the first works which show the interaction of DNA with positively charged liposomes has been reported by Radler et al. (Rädler et al. 1997). In this work, the authors have developed mixtures of DOPC (di-oleoyl phosphatidyl choline) (neutral lipid) and DOTAP (di-oleoyl trimethylammonium propane) (univalent cationic lipid). This mixture was complexed with a 48 kD linear lambda-phage DNA. Liposomes formed by DOPC and DOTAP with ~50 nm diameter become spontaneously into dense globules between 100 to 200 nm after interacting with DNA. The size of these aggregates was studied by Dynamic Light Scattering (DLS) and the formation of the membrane of liposomes was shown by means of high-resolution synchrotron Small Angle X-ray Scattering (SAXS). The authors of this work (Rädler et al. 1997) have reported that cationic liposomes decorated by DNA strands and analyzed by cross-polarized microscopy demonstrated birefringence and, therefore, the presence of a mesophase. It is important to remark that water plays an important role in these kinds of systems. Consequently, by increasing the volume fraction of water, the formed liquid crystals may modify their texture (Benedini et al. 2011) and therefore, generate a release of the liposomes content. Additionally, the repulsive forces may increase due to more molecules of water which increase the thickness of the membrane and consequently, their stability. Such results for liposomes loaded with amiodarone were reported by Benedini et al. (Benedini et al. 2014). Lipoplexes are liposomes mainly constituted by cationic lipids. This term is often for referring to all cationic amphiphiles, including cationic cholesterol (Han et al. 2008), bile salt derivatives and other cationic amphiphiles, with capability for developing bilayers (Felgner et al. 1997). Cationic lipids are composed of three main elements: a cationic head group (polar o hydrophilic), a hydrophobic tail and also generally, a linker (Karmali and Chaudhuri 2007). The positive charge of the polar head group of these lipids is commonly ascribed to one or more amines that facilitate the binding with anions such as those found in DNA molecules. Therefore, the resultant liposomes are the product of energetic contributions by Van der Waals forces and electrostatic binding to the DNA which partially direct their shape (Israelachvili 2011). There are many other,well-characterized cationic lipids used for gene therapy such as N-[1-(2,3-dioleyloxy)propyl]-N,N,N-trimethylammonium chloride(DOTMA) (Felgner et al. 1987), 3β[N-(N-,N-dimethylaminoethane)-carbamoyl]cholesterol (DC-Chol) (Gao and Huang 1991), and dioctadecylamidoglycylspermine (DOGS) (Behr et al. 1989). For the development of cationic liposomes, cationic lipids are commonly used in conjunction with neutral ones. Dioleoylphosphatidylethanolamine (DOPE) is a neutral lipid jointly used with cationic ones because DOPE is necessary for stabilizing the membrane of the liposomes at low pH (Felgner et al. 1994), and consequently, these lipids were named as"helpers". A pH decrease is produced into the lysosome environment and therefore, the escape of gene material from liposomes is controlled by the blend formed by the fusion of both membranes (Farhood et al. 1995). The incorporation of helper lipids into the cationic membranes results in an increased conformational disorder of the hydrophobic region of the surfactant and consequently, further dehydration of the linker region is evidenced. These effects decrease with the increase of the lipid/DNA charge ratio (+/−)(Hirsch-Lerner and Barenholz 1999). The condensation effect of DNA is increased when positive single charged lipids are changed by double charged ones (Hyvönen et al. 2000). It is also shown that an increase in the molar ratio of helper lipids into the membranes modify the hydration of the DNA bases (Choosakoonkriang et al. 2001). In contrast to fully charged liposomes, the use of mixtures 1:1 lipid ratio (DOPE:DC-Chol) contains about 50% of the surface positively charged at pH 7.4 (Zuidam and Barenholz 1997). This characteristic leads to greater gene expression because it helps to dissociate the DNA during transport and, consequently, during the transfection process. (Zuidam and Barenholz 1998).

There are other molecules than lipids used for improving liposomes-mediated gene delivery such as Polyethylene glycol (PEG). The incorporation of these molecules to liposomes was one of the main strategies reported to stabilize pristine liposomes through steric hindrance (Allen and Cullis 2013). PEGs are water soluble, non-toxic, biocompatible and non-immunogenic polymers. Liposomal surfaces are modified by including PEGs by physical adsorption onto their surface and/or by covalent attachment onto premade liposomes (Kim et al. 2003). The incorporation of these molecules increases circulation time of liposomes and also lipoplexes; consequently, they are called PEGylated liposomes or stealth liposomes because they avoid their removal by the mononuclear phagocytic system in the liver and spleen (Benedini et al. 2017).

Different techniques have been used for essential lipoplex characterization for understanding their formation, DNA condensation, macroscopic and microscopic features of their structure and eventually, if possible, to predict their behaviour *in vivo*.

The DNA amount frequently used in transfection protocols is 1–2 µg for *in vitro* studies, 20–50 µg/animal for *in vivo* and 50–150 µg/mL by dose for clinical trials. The techniques used in lipoplex characterization should be sensitive to the concentration ranges. However, by using transmission electron microscopy or diffraction techniques for obtaining information about lipoplex structure, a previous sample treatment and higher concentrations of DNA (100–5000 µg/mL) are usually required. Another technique used for characterization of lipoplex macro-structure, with lower DNA amounts (0.01–50 µg/mL), is the atomic force microscopy (Madeira et al. 2011).

Electrophoresis agarose gel is a common and easy technique which allows a first approach for evaluating DNA condensation and for calculating the encapsulation efficiency (Even-Chen and Barenholz 2000, Gershon et al. 1993). However, more rigorous encapsulation efficiencies of these non-viral vectors have been established by fluorescence methods (Madeira et al. 2008, 2011). For measuring lipoplexes size, dynamic light scattering (DLS) is mainly used and, more rarely, microscopy techniques. This parameter is strictly involved not only in transfection studies (*in vitro* and *in vivo*), but also as a measure of colloidal stability (in solution) and the physical stability (after dehydration by lyophilization) during storage (Anchordoquy et al. 1997). The charge of the lipoplexes has been well established by z-potential measurements such as for other liposomal systems. This parameter is important for determining the experimental stability and also for predicting the theoretical stability by means of DLVO (Derjaguin, Landau, Verwey, Overbeek) theory for colloidal systems (Benedini et al. 2014).

For assessing structural changes in DNA upon DNA–lipid contact, several techniques such as circular dichroism (CD) have been used (Simberg et al. 2001). This technique is commonly used for studying the folding of the proteins (Benedini et al. 2018). Nuclear magnetic resonance (NMR) (Scarzello et al. 2005) and Fourier transform infrared spectroscopy (FT-IR) (Choosakoonkriang et al. 2003) along with fluorescence methodologies, allow obtaining structural information on lipid and DNA interaction.

Through fluorescence-based techniques can be carried out the lipoplexes characterization. Thus, the following can be assessed: (a) The kinetic of lipid/DNA interaction for establishing the lipoplex formation; (b) the behaviour of the membranes by modifying the lipid blends; (c) interlamellar distances between lipid/DNA structures; (d) DNA condensation behaviour (Madeira et al. 2011).

The use of liposomes for design and synthesis of optimal carrier systems based on lipids for delivery DNA, short-interfering RNAs, gene silencing applied to complex diseases control is a continuous challenge. The structure-function data obtained from such studies lead to a rational design of these self-assemblies for enhanced nucleic acid delivery applications.

5.5.3 Liquid Crystals

The engineered viruses used for gene delivery and applied to gene therapy are very efficient vectors (Kay et al. 2001, Smith 1995); however, concerns about their safety have increased by severe drawbacks. This fact has further increased the interest in non-viral or synthetic vectors (Huang et al. 2005, Ramamoorth and Narvekar 2015), in which the negatively charged DNA is complexed with cationic lipids (Felgner et al. 1987) or cationic polyelectrolytes (Miller 1998, Tang et al. 1996). Synthetic vectors have several advantages over viral ones: absence of immunogenic protein component, facile and tunable preparation (this feature is improved by a large number of lipids synthesized with this purpose), and the possibility to incorporate an unlimited length of the DNA (Ewert et al. 2006).

There is a very close relationship between liposomes and liquid crystals because liposomes are self-closed lamellar liquid crystals (Benedini et al. 2017). The concentration of the different components (lipids, sterols, etc.) of the systems yields different types of liquid crystal aggregates among liposomes. In this context, Ewet et al. (Ewert et al. 2006) reported the synthesis of a highly charged (16+) multivalent cationic lipid, MVLBG2, and the consequent development of cationic liposomes by using that lipid and another non-charged lipid, DOPC. At high concentration of this neutral lipid, the complex DNA-lipids have shown a lamellar phase. By increasing the multivalent cationic lipid concentration to 25% mol, novel non-lamellar mesophases are shown. Through synchrotron X-Ray diffraction measurements, a hexagonal mesophase was evidenced. The authors propose a model where cylindrical cationic micelles of the lipid mixture, in a hexagonal arrangement, are surrounded by DNA rods negatively charged. This estate, which appears in a narrow range of composition, has shown higher efficient transfection than that of commercially available, optimized DOTAP-based complexes for murine and human cell cultures.

To overcome the environment cellular barriers and to guarantee the delivery of the genetic material into the nucleus of the cell (released endosomal enzymes or the stability of the formulation into serum environment), the improvement of physicochemical properties of the DNA-lipid complex is necessary. In this context, Ahmad et al. (Ahmad et al. 2005) and Lin et al. (Lin et al. 2003) have reported that the membrane charge density, σ_M, of the membranes of cationic lipids and the charge ratio cationic lipid/DNA, ρ_{chg} are two keys parameters which deeply impact transfection efficiency. The transfection efficiency of lamellar complexes DNA/lipid follows a Gaussian bell-shaped curve as a function of membrane charge density. Therefore, an optimal value of this parameter, where a balance is found between efficient endosomal escape (via activated fusion of the cationic membranes of the DNA-lipid complex and the anionic membranes of the endosome) and release of DNA from the highly charged membranes to the cytoplasm, was identified (Ahmad et al. 2005). This work has been carried out using multivalent lipids with head-group valences ranging from 2+ to 5+. However, to investigate higher values of σ_M, a further increase in head-group valence is required. Thus, Zidovska et al. (Zidovska et al. 2009) studied the effect of the charge of structures formed by blends of positively charged lipids and neutral ones on the behaviour of σ_M. For this aim, mixtures between four positive charged multivalent lipids (MVL), MVLG2(4+), MVLG3(8+), MVLBisG1(8+) and MVLBisG2(16+), and the neutral lipid 1,2-dioleoyl-sn-glycerophosphatidyl-choline (DOPC) have been investigated. X-ray diffraction (XRD), optical microscopy and cryo-TEM studies have been performed to understand the transfection efficiency of systems based on cationic/neutral lipids mixtures and DNA.MVLG2(4+)/DOPC–DNA complexes exhibit a lamellar mesophase at all molar fractions of lipids. On the other hand, MVLBisG2 (16+)/DOPC–DNA complexes remain lamellar only for molar fraction ≤ 0.2. For a molar fraction near 0.25 of cationic lipids, a direct hexagonal phase, consisting of a hexagonal lattice of cylindrical micelles and DNA is shown. For molar fractions ≥ 0.5 and under increased ionic strength conditions, this phase coexists with a bundle phase of DNA condensed and surrounded by the cylindrical micelles. At high membrane charge density, the transitions from the lamellar phase to the new hexagonal phase show a deviation of the known transfection efficiency. Accordingly, this behaviour strongly suggests a novel mechanism of transfection for lipids-DNA complex phases (Zidovska et al. 2009).

5.5.4 Nanoparticles

The use of nanoparticles has become a promising tool for developing medical treatments based on gene therapy. They have crucial properties that turn them into highlighted delivery systems for many treatments such as cancer and other complex diseases. The first remarkable property is their size, which is a product of the application of nanotechnology. Thus, their small size (1-100 nm) increases the possibility to overcome different cellular barriers to transfer their gene content into target cells. Additionally, these particles have a large surface-surface to volume ratio and can be developed from different types of materials such as polymers, lipids, surfactants and carbon nanotubes among others (Radenkovic et al. 2016). Through nanotechnology, these materials can be handled to reach the desired objectives, which allow developing drug delivery systems for gene therapy. One of the main aims is to achieve small size particles; however, there are many other properties which are also relevant such as good biocompatibility, biodegradability, lack of immunogenic response and toxicity. Additionally, they must protect the gene material of nucleases activity and they must induce gene expression (Wong et al. 2017).

5.5.4.1 Lipid Nanoparticles and Nanostructured Lipid Carriers

Introduced in 1991, solid lipid nanoparticles (SLN) have formed an interesting drug carrier system as an alternative to traditional colloidal drug delivery systems such as liposomes, micro-emulsions, and micelles among others (Müller et al. 2000, zur Mühlen et al. 1998). SLN and nanostructured lipid carriers (NLC) have demonstrated that they are effective and safe colloidal carriers for different types of active principles. They also provide high flexibilities in the release profiles for those compounds (Han et al. 2016, Olbrich and Müller 1999). They have been recognized as potential pharmaceutical formulations to be used as non-viral vectors for gene transfection and consequently, to potentially treat genetic and non-genetic diseases. Any pharmaceutical formulation must fulfill certain features such as biocompatibility; it must show stability in storage conditions, predictably controlled degradation during its "pathway" to the target site, and it must have large-scale production. Lipid nanoparticles fulfill these requirements and are also easily sterilized. They have the possibility to be lyophilized if necessary (del Pozo-Rodríguez et al. 2016). The use of DNA or RNA as a drug implies the inclusion into a pharmaceutical formulation because they are highly sensitive to the cell medium components and show high interaction capability with other molecules due to their charge. These two factors decrease the availability of these active molecules in the target site. One of the strengths of lipid nanoparticles is their capability to overcome the cell barriers. They keep the chemical stability of molecules against nuclease degradation, generating a cell internalization traffic towards the target and the possibility to interact with a specific cell type by means of selective targeting (Battaglia et al. 2016). SLNs have displayed auspicious effectiveness on DNA and RNA delivery both *in vitro* and *in vivo* conditions (Apaolaza et al. 2015, del Pozo-Rodríguez et al. 2010, Ruiz de Garibay et al. 2015, Torrecilla et al. 2015, 2016). Because lipids are biocompatible and well-tolerated, these non-viral vectors show many advantages such as low toxicity (or absence of it), long-term stability, possibility to be sterilized and low cost (del Pozo-Rodríguez et al. 2016). SLNs are formed by solid lipid core with an external layer of surfactants in an aqueous dispersion. Generally, these particles have a diameter lower than 500 nm, and for producing a good response *in vivo* studies they should reach to 120 nm or less (Rehman et al. 2013). The encapsulation of nucleic acids into SLN of 200 nanometers was first reported by Wheeler et al. (Wheeler et al. 1999). The surfactants used for the development of the external layer of this system have the necessary positive charge (cationic) to bind negatively charged molecules such as DNA. Their zeta potential is surrounding +30 mV or more (del Pozo-Rodríguez et al. 2007). However, the relation diameter/charge is crucial for the stability of the system (Benedini et al. 2014). Improved properties of loading capacity and release profile of SLN have been reached by the development of NLCs. They differ from SLN in that the core of NLCs has a mixture of liquid and solid lipids (Han et al. 2016, Pardeike et al. 2009). SLNs and NSLs have been developed for the potential treatment of many diseases such as ocular, lung, hepatic, cancers and infectious diseases(del Pozo-Rodríguez et al. 2016). Some retinal diseases have been characterized from a genetic approach, and therefore gene therapy has a great potential for their treatment. In this context, Delgado et al. (Delgado et al. 2012) have reported the expression of EGFP in different types of rat cells after intravitreal injection of SLNs bearing the reporter gene pCMS-EGFP. The leading topic for applying SLNs in gene therapy is cancer. Ma et al. have encapsulated into SLN the signal transducer and activator of transcription 3 (STAT3) decoy oligonucleotide (Ma et al. 2015). This system has shown high efficiency for treating human ovarian cancer cells. Another strategy for the delivery of nucleic acids have been reported by Han et al. (Han et al. 2014). In this work is shown the development of a co-delivery system based on a formulation for carrying doxorubicin and the EGFP plasmid by means of SLNs for treating lung cancer cells. SLNs have emerged as an alternative to virus disease treatments. Hepatitis C virus infection is one of the most important infectious diseases because a high percentage of patients have different and persistent hepatic problems, among them hepatocellular carcinoma and ultimately this disease can lead to hepatic failure and death. In this context, Torrecilla et al. (Torrecilla et al. 2015) have reported the use of non-viral vectors based on SLNs for the treatment of this disease. The nanoparticles contained an interference RNA plasmid (shRNA74), which inhibit the internal ribosome entry site (IRES) of the hepatitis C virus, necessary for the replication.

5.5.4.2 Polymer-based Nanoparticles

The polymers are the second main group of synthetic molecules used for the development of non-viral vectors for gene therapy. These vectors formed by a mix of DNA (negatively charged) and self-assembled cationic (polycationic) polymers are called polyplexes (Liu and Zhang 2011). Polyplexes present many advantages over liposome and other lipid-based systems. One of the most important features is their stability. They also show high versatility, small size and narrow size distribution, high protection of gene material against nucleases and easily tunable physical-chemistry properties (affinity for selected solvents and possibility of charge modification) (Al-Dosari and Gao 2009, Fattal and Bochot 2008).

One of the first cationic polymers used for gene transfer in gene therapy is poly-L-lysine. This molecule has amine groups positively charged at physiological pH; therefore, its interaction with DNA molecules negatively charged is favoured. Owing to its biodegradable nature, this polypeptide was quickly employed because it was thought that it would be innocuous *in vivo*. However, low molecular weight polyplexes cannot develop stable complexes and those polyamines with higher molecular weight are suitable for gene delivery but show high toxicity (Morille et al. 2008). Additionally, poly-L-lysine without chemical modifications has poor transfection capability because it shows high plasma protein binding and its rate of clearance from the blood is significantly high. For this reason, numerous modifications have been reported for polyplexes based on poly-L-lysine (Kim 2012). Like it was reported for liposomes, the hydrophilic polymer, polyethylene glycol (PEG) was anchored to the surface of polyplexes for decreasing their interaction with serum proteins and other components. The result of this method yields an increase of circulation time of these particles. The result of this method yields an increase of circulation time of these particles (Alexis et al. 2008, Bazile et al. 1995). The use of PEGylated polyplexes has shown its clinical relevance for treating cystic fibrosis. For this disease, they display good safety and tolerability in clinical trials, as well as some good results for vector gene transfer (Konstan et al. 2004).

Compared with other type of cationic polymers, polyethilenimine (PEI), is one of the most popular used for the development of polyplexes. Due to that,these particles have high efficiency for gene transfer of nucleic acids in a wide variety of cells (*in vitro*) and tissues (*in vivo*) (Liu and Zhang 2011). Used for the first time for nucleotide transfer in 1995 (Boussif et al. 1995), this polymer has been chemically modified for obtaining branched and linear derivatives. This latter derivative is generally preferred for *in vivo* applications owing to its low toxicity. Branched derivatives have a certain ratio of primary, secondary, and tertiary amines (1:2:1) and they have upper values of pKa than physiological pH. Hence, a portion of these groups is non-charged at this pH and consequently, they provide a high buffer capacity. Therefore, polyplexes based on PEI may escape from endosomes because these amines groups can absorb protons. The transfection efficiency and toxicity of polyplexes based on PEI are affected by the kind of PEI (branching degree and molecular weight) (Godbey et al. 1999, Wightman et al. 2001), the size of the nanoparticle, the charge balance of the complex polymer-DNA which plays a crucial role in zeta potential value and the ionic strength of the solution. The degree of binding between the polymer and DNA is produced through primary amine groups; however, this relation contributes to the toxicity during the transfection process. The toxicity is mainly produced by an aggregation process, followed by cell adhesion and finally results in tissue necrosis. Therefore, high charge density improves the transfection process but also increases the toxicity of the nanoparticles. Additionally, the secondary and tertiary amine groups of the polymer provide the buffering capacity. For reducing the interaction between polyplexes and serum proteins, PEGylated polyplexes based on PEI have been developed (Liu and Zhang 2011). Successful outcomes related to the application of intratumorally PEI-DNA polyplexes for gene transfection into mouse lungs cells have been reported (Coll et al. 1999, Goula et al. 1998). In humans, these polyplexes have been used for local treatment of cancer; however, for reducing their toxicity PEGyleated polyplexes have been approved for clinical trials in ovarian and colorectal cancers.

Owing to their effectiveness as a non-viral gene transfer vector, dendrimers have been extensively investigated. They have some advantages compared to conventional polyamines such as easy manufacturing, low toxicity, and high gene transfer ratio. Dendrimers are globular, highly branched molecules with a central core (Gillies and Frechet 2005). Their architecture confers to dendrimers a unique feature: the uniformity (they are monodisperse molecules). Additionally, this structural arrangement gives to these molecules the possibility to encapsulate guest molecules into an inner cavity (Boas and Heegaard 2004, Xu et al. 2010). Polyamidoamine (PAMAM) is one of the most used dendrimers for gene transfer (Tomalia et al. 1985). This compound has a positive charge provided by amine tertiary groups which behaves as proton absorbent. Owing to the low pKa of these groups, dendrimers can buffer the pH inside the endosome environment (Klajnert and Bryszewska 2001, Paleos et al. 2009). For improving gene transfer, the surface of PAMAM can be modified by the addition of hydroxyl groups and L-aginine. The incorporation of this amino acid through ester or amide bond generates non-toxic derivatives with identical transfection properties; however, ester bond shows lower toxicity than the other bond because of their rapid degradation (Nam et al. 2008).

Cyclodextrins (CDs) are other types of polymers used for gene transfection in gene therapy. They are non-hygroscopic, crystalline, and cyclic oligosaccharides derived from starch (Sartuqui et al. 2016). CDs have a non-polar inner cavity, which can form inclusion complexes with hydrophobic substances. Therefore, native CD cannot develop complexes with negatively charged DNA and for this reason, they must be modified by means of cationic polymers such as PAMAM or PEI. Once modified and due to their specific architecture and biocompatibility, CDs have been used in gene therapy (Gonzalez et al. 1999). Svenson and Tomalia (Svenson and Tomalia 2005) have reported the use of CDs functionalized with PAMAM which have shown high gene transfection compared with non-functionalized PAMAM and non-covalent mixtures of CDs and PAMAM. Another example of complexes formed by polymers and CDs have been reported by Burckbuchler et al. (Burckbuchler et al. 2008). In this work, PEI of low molecular weight and CDs have formed a complex for transfection genes into neurons.

The development of supramolecular biomaterials for gene transfer is an emerging area; however, in spite of reached achievements in this field, there are many challenges. We expect higher performances and lower toxicity, for both *in vitro* and *in vivo* studies. Safety of these systems is still the major concern. Accordingly, biocompatibility and biodegradability studies are necessary for supramolecular gene delivery systems.

5.6 Converging Technologies on Duty of Medical Applications

Owing to the importance of gene therapy for the treatment of both genetic and acquired disease, there is a great enthusiasm for working toward the development of safer and targeted viral vectors. However, the scientific community is also focused on carrying out parallel efforts to design more efficient non-viral systems that can successfully achieve the care of a gene therapy for human beings.

Advances in materials science, chemistry, and engineering have contributed to recent breakthroughs in non-viral gene therapy. These results have enabled the synthesis/fabrication of more homogeneous materials with well-defined structures and tunable functionalities. These desirable properties need the evolution of technologies for a more rigorous characterization into the fields of chemistry, physics and biology. For this reason, rapid advances in the materials science field and characterization must be taken together with the progress in our ability to manipulate the genome and different progress at molecular-level to yield clinically relevant breakthroughs in gene therapies based on non-viral vectors. Finally, these tools should provide new and improved treatments for some of our challenging diseases (Werfel and Duvall 2016). The emergence of gene therapy has revolutionized the strategies for facing the treatment of complexes diseases. In the beginning, for carrying out the transfection of genes only viral vectors were used. Relative successful outcomes have been reported by using these types of vectors; however, the death of a patient by an out of control immunological response turned on the alert lights. While the use of viral vectors is being the gold standard for treatment of complex diseases based on gene therapy, the advent of non-viral vectors is the promising tool for their replacement because of their safety. In this chapter, non-viral vector formulations such as liposomes (lipoplexes), solid lipid nanoparticles, nanostructured lipids, and formulations based on polymers (poliplexes) and based on liquid crystals have been described, showing their relevance to this kind of therapy but also their disadvantages. Genetic therapy should be considered an important technological advancement but also as a gateway towards the cure for complex diseases which many times carry with them the suffering of patients and their families. However, safety and ethical considerations for the application of this therapy should be considered above all others, because as the writer Haruki Murakami says, "Human beings are ultimately nothing but carriers for genes".

References

Ahmad, A., H. M. Evans, K. Ewert, C. X. George, C. E. Samuel and C. R. Safinya. 2005. New multivalent cationic lipids reveal bell curve for transfection efficiency versus membrane charge density: Lipid-DNA complexes for gene delivery. J. Gene. Med. 7(6): 739-748.

Ajmani, P. S. and J. A. Hughes. 1999. 3Beta [N-(N',N'-dimethylaminoethane)-carbamoyl] cholesterol (DC-Chol)-mediated gene delivery to primary rat neurons: Characterization and mechanism. Neurochem. Res. 24(5): 699-703.

Al-Dosari, M. S. and X. Gao. 2009. Nonviral gene delivery: Principle, limitations, and recent progress. AAPS J. 11(4): 671-681.

Alexis, F., E. Pridgen, L. K. Molnar and O. C. Farokhzad. 2008. Factors affecting the clearance and biodistribution of polymeric nanoparticles. Mol. Pharm. 5(4): 505-515.

Allen, T. D., J. M. Cronshaw, S. Bagley, E. Kiseleva and M. W. Goldberg. 2000. The nuclear pore complex: Mediator of translocation between nucleus and cytoplasm. J. Cell. Sci. 113(Pt 10): 1651-1659.

Allen, T. M. and P. R. Cullis. 2013. Liposomal drug delivery systems: From concept to clinical applications. Adv. Drug. Deliv. Rev. 65(1): 36-48.

Alloway, J. L. 1932. The transformation in vitro of R pneumococci into S forms of different specific types by the use of filtered pneumococcus extracts. J. Exp. Med. 55(1): 91-99.

Anchordoquy, T. J., J. F. Carpenter and D. J. Kroll. 1997. Maintenance of transfection rates and physical characterization of lipid/DNA complexes after freeze-drying and rehydration. Arch. Biochem. Biophys. 348(1): 199-206.

Anderson, W. F. 1992. Human gene therapy. Science 256(5058): 808-813.

Anderson, W. F. 1998. Human gene therapy. Nature 392(6679): 25-30.

Apaolaza, P. S., A. Del Pozo-Rodríguez, J. Torrecilla, A. Rodríguez-Gascón, J. M. Rodríguez, U. Friedrich, et al. 2015. Solid lipid nanoparticle-based vectors intended for the treatment of X-linked juvenile retinoschisis by gene therapy: In vivo approaches in Rs1h-deficient mouse model. J. Control Release 217: 273-283.

Avery, O. T., C. M. Macleod and M. McCarty. 1944. Studies on the chemical nature of the substance inducing transformation of pneumococcal types: Induction of transformation by a desoxyribonucleic acid fraction isolated from pneumococcus Type III. J. Exp. Med. 79(2): 137-158.

Bank, A. 1996. Human somatic cell gene therapy. BioEssays: News and Reviews in Molecular, Cellular and Developmental Biology. 18(12): 999-1007. Retrieved from http://www.ncbi.nlm.nih.gov/pubmed/8976157

Battaglia, L., L. Serpe, F. Foglietta, E. Muntoni, M. Gallarate, A. Del Pozo Rodriguez, et al. 2016. Application of lipid nanoparticles to ocular drug delivery. Expert Opin. Drug. Deliv. 13(12): 1743-1757.

Baum, C., O. Kustikova, U. Modlich, Z. Li and B. Fehse. 2006. Mutagenesis and oncogenesis by chromosomal insertion of gene transfer vectors. Hum. Gene Ther. 17(3): 253-263.

Bazile, D., C. Prud'homme, M. T. Bassoullet, M. Marlard, G. Spenlehauer and M. Veillard. 1995. Stealth Me.PEG-PLA nanoparticles avoid uptake by the mononuclear phagocytes system. J. Pharm. Sci. 84(4): 493-498.

Behr, J. P., B. Demeneix, J. P. Loeffler and J. Perez-Mutul. 1989. Efficient gene transfer into mammalian primary endocrine cells with lipopolyamine-coated DNA. Proc. Natl. Acad. Sci. U.S.A. 86(18): 6982-6986.

Benedini, L. A., N. L. G. Vidal and M. A. Gonzalez. 2018. Celiac disease: Historical standpoint, new perspectives of treatments and contemporary research techniques. Curr. Protein Pept. Sci. 19(11): 1058-1070.

Benedini, L., N. Andres and M. L. Fanani. 2017. Liposomes: From the pioneers to epigenetic therapy. In: B. Pearson (ed.). Liposomes: Historical, Clinical and Molecular Perspectives (pp. 1-345). Nova Publi, New York.

Benedini, L., S. Antollini, M. L. Fanani, S. Palma, P. Messina and P. Schulz. 2014. Study of the influence of ascorbyl palmitate and amiodarone in the stability of unilamellar liposomes. Mol. Membr. Biol. 31(2-3): 85-94.

Benedini, L., E. P. Schulz, P. V. Messina, S. D. Palma, D. A. Allemandi and P. C. Schulz. 2011. The ascorbyl palmitate-water system: Phase diagram and state of water. Colloids Surf. A. 375(1-3): 178-185.

Bessis, N., F. J. GarciaCozar and M. -C. Boissier. 2004. Immune responses to gene therapy vectors: Influence on vector function and effector mechanisms. Gene Ther. 11(Suppl 1): S10-7.

Blaese, R. M., K. W. Culver, A. D. Miller, C. S. Carter, T. Fleisher, M. Clerici, et al. 1995. T lymphocyte-directed gene therapy for ADA-SCID: Initial trial results after 4 years. Science 270(5235): 475-480.

Boas, U. and P. M. H. Heegaard. 2004. Dendrimers in drug research. Chem. Soc. Rev. 33(1): 43-63.

Bordignon, C., L. D. Notarangelo, N. Nobili, G. Ferrari, G. Casorati, P. Panina, et al. 1995. Gene therapy in peripheral blood lymphocytes and bone marrow for ADA-immunodeficient patients. Science 270(5235): 470-475.

Boussif, O., F. Lezoualc'h, M. A. Zanta, M. D. Mergny, D. Scherman, B. Demeneix, et al. 1995. A versatile vector for gene and oligonucleotide transfer into cells in culture and in vivo: Polyethylenimine. Proc. Natl. Acad. Sci. U.S.A. 92(16): 7297-7301.

Broderick, J. A. and P. D. Zamore. 2011. MicroRNA therapeutics. Gene Ther. 18(12): 1104-1110.

Burckbuchler, V., V. Wintgens, C. Leborgne, S. Lecomte, N. Leygue, D. Scherman, et al. 2008. Development and characterization of new cyclodextrin polymer-based DNA delivery systems. Bioconjug. Chem. 19(12): 2311-2320.

Cheung, C. Y., N. Murthy, P. S. Stayton and A. S. Hoffman. 2001. A pH-sensitive polymer that enhances cationic lipid-mediated gene transfer. Bioconjug. Chem. 12(6): 906-910.

Choosakoonkriang, S., C. M. Wiethoff, T. J. Anchordoquy, G. S. Koe, J. G. Smith and C. R. Middaugh. 2001. Infrared spectroscopic characterization of the interaction of cationic lipids with plasmid DNA. J. Biol. Chem. 276(11): 8037-8043.

Choosakoonkriang, S., C. M. Wiethoff, G. S. Koe, J. G. Koe, T. J. Anchordoquy and C. R. Middaugh. 2003. An infrared spectroscopic study of the effect of hydration on cationic lipid/DNA complexes. J. Pharm. Sci. 92(1): 115-130.

Coll, J. -L., P. Chollet, E. Brambilla, D. Desplanques, J. -P. Behr and M. Favrot. 1999. In vivo delivery to tumors of DNA complexed with linear polyethylenimine. Hum. Gene Ther. 10(10): 1659-1666.

Deamer, D. W. 2010. From "banghasomes" to liposomes: A memoir of Alec Bangham, 1921-2010. FASEB J. 24(5): 1308-1310.

del Pozo-Rodríguez, A., D. Delgado, M. A. Solinís, A. R. Gascón and J. L. Pedraz. 2007. Solid lipid nanoparticles: Formulation factors affecting cell transfection capacity. Int. J. Pharm. 339(1-2): 261-268.

del Pozo-Rodríguez, A., D. Delgado, M. A. Solinís, A. R. Gascón and J. L. Pedraz. 2008. Solid lipid nanoparticles for retinal gene therapy: Transfection and intracellular trafficking in RPE cells. Int. J. Pharm. 360(1-2): 177-183.

del Pozo-Rodríguez, A., D. Delgado, M. A. Solinís, J. L. Pedraz, E. Echevarría, J. M. Rodríguez, et al. 2010. Solid lipid nanoparticles as potential tools for gene therapy: In vivo protein expression after intravenous administration. Int. J. Pharm. 385(1-2): 157-162.

del Pozo-Rodríguez, A., M. Á. Solinís and A. Rodríguez-Gascón. 2016. Applications of lipid nanoparticles in gene therapy. Eur. J. Pharm. Biopharm. 109: 184-193.

Delgado, D., A. del Pozo-Rodríguez, M. Á. Solinís, M. Avilés-Triqueros, B. H. F. Weber, E. Fernández, et al. 2012. Dextran and protamine-based solid lipid nanoparticles as potential vectors for the treatment of X-linked juvenile retinoschisis. Hum. Gene. Ther. 23(4): 345-355.

Even-Chen, S. and Y. Barenholz. 2000. DOTAP cationic liposomes prefer relaxed over supercoiled plasmids. Biochim. Biophys. Acta. 1509(1-2): 176-188.

Ewert, K. K., H. M. Evans, A. Zidovska, N. F. Bouxsein, A. Ahmad and C. R. Safinya. (2006). A columnar phase of dendritic lipid-based cationic liposome-DNA complexes for gene delivery: Hexagonally ordered cylindrical micelles embedded in a DNA honeycomb lattice. J. Am. Chem. Soc. 128(12): 3998-4006.

Farhood, H., N. Serbina and L. Huang. 1995. The role of dioleoyl phosphatidylethanolamine in cationic liposome mediated gene transfer. Biochim. Biophys. Acta. 1235(2): 289-295.

Fattal, E. and A. Bochot. 2008. State of the art and perspectives for the delivery of antisense oligonucleotides and siRNA by polymeric nanocarriers. Int. J. Pharm. 364(2): 237-248.

Felgner, J. H., R. Kumar, C. N. Sridhar, C. J. Wheeler, Y. J. Tsai, R. Border, et al. 1994. Enhanced gene delivery and mechanism studies with a novel series of cationic lipid formulations. J. Biol. Chem. 269(4): 2550-2561.

Felgner, P. L., Y. Barenholz, J. P. Behr, S. H. Cheng, P. Cullis, L. Huang, et al. 1997. Nomenclature for synthetic gene delivery systems. Hum. Gene Ther. 8(5): 511-512.

Felgner, P. L., T. R. Gadek, M. Holm, R. Roman, H. W. Chan, M. Wenz, et al. 1987. Lipofection: A highly efficient, lipid-mediated DNA-transfection procedure. Proc. Natl. Acad. Sci. U.S.A. 84(21): 7413-7417.

Fink, D. J., R. Ramakrishnan, P. Marconi, W. F. Goins, T. C. Holland and J. C. Glorioso. 1995. Advances in the development of herpes simplex virus-based gene transfer vectors for the nervous system. Clin. Neurosci. 3(5): 284-291.

Gao, X. and L. Huang. 1991. A novel cationic liposome reagent for efficient transfection of mammalian cells. Biochem. Biophys. Res. Commun. 179(1): 280-285.

Gershon, H., R. Ghirlando, S. B. Guttman and A. Minsky. 1993. Mode of formation and structural features of DNA-cationic liposome complexes used for transfection. Biochemistry 32(28): 7143-7151.

Gillies, E. and J. Frechet. (2005). Dendrimers and dendritic polymers in drug delivery. Drug Discov. Today 10(1): 35-43.

Godbey, W. T., K. K. Wu and A. G. Mikos. 1999. Size matters: Molecular weight affects the efficiency of poly(ethylenimine) as a gene delivery vehicle. J. Biomed. Mater. Res. 45(3): 268-275.

Gonzalez, H., S. J. Hwang and M. E. Davis. 1999. New class of polymers for the delivery of macromolecular therapeutics. Bioconjug. Chem. 10(6): 1068-1074.

Goula, D., C. Benoist, S. Mantero, G. Merlo, G. Levi and B. A. Demeneix. 1998. Polyethylenimine-based intravenous delivery of transgenes to mouse lung. Gene Ther. 5(9): 1291-1295.

Gregoiadis, G. 1976. The carrier potential of liposomes in biology and medicine. N. Engl. J. Med. 765-770.

Griffith, F. 1928. The significance of pneumococcal types. J. Hyg. 27(2): 113-159.

Guild, B. C., M. H. Finer, D. E. Housman and R. C. Mulligan. 1988. Development of retrovirus vectors useful for expressing genes in cultured murine embryonal cells and hematopoietic cells in vivo. J. Virol. 62(10): 3795-3801.

Han, S., H. Kang, G. Shim, M. Suh, S. Kim, J. Kim, et al. 2008. Novel cationic cholesterol derivative-based liposomes for serum-enhanced delivery of siRNA. Int. J. Pharm. 353(1-2): 260-269.

Han, Y., Y. Li, P. Zhang, J. Sun, X. Li, X. Sun, et al. 2016. Nanostructured lipid carriers as novel drug delivery system for lung cancer gene therapy. Pharm. Dev. Technol. 21(3): 277-281.

Han, Y., P. Zhang, Y. Chen, J. Sun and F. Kong. 2014. Co-delivery of plasmid DNA and doxorubicin by solid lipid nanoparticles for lung cancer therapy. Int. J. Mol. Med. 34(1): 191-196.

Herzog, R. W., O. Cao and A. Srivastava. 2010. Two decades of clinical gene therapy—Success is finally mounting. Discov. Med. 9(45): 105-111.

Hirsch-Lerner, D. and Y. Barenholz. 1999. Hydration of lipoplexes commonly used in gene delivery: Follow-up by laurdan fluorescence changes and quantification by differential scanning calorimetry. Biochim. Biophys. Acta. 1461(1): 47-57.

Huang, L., M. -C. Hung and E. Wagner (Eds.). 2005. Nonviral Vectors for Gene Therapy. Advances in Genetics (2nd edition, Vol. 54). Elsevier, San Diego, California.

Hug, P. and R. G. Sleight. 1997. The advantages of liposome-based gene therapy: A comparison of viral versus liposome-based gene delivery. pp. 345-362. *In*: E. E. Bittar and N. Bittar (Eds.). Principles of Medical Biology. Elsevier B. V. Radarweg 29 Amsterdam 1043 NX, Netherlands.

Hyvönen, Z., A. Plotniece, I. Reine, B. Chekavichus, G. Duburs and A. Urtti. 2000. Novel cationic amphiphilic 1,4-dihydropyridine derivatives for DNA delivery. Biochim. Biophys. Acta. 1509(1-2): 451-466.

Ibraheem, D., A. Elaissari and H. Fessi. 2014. Gene therapy and DNA delivery systems. Int. J. Pharm. 459(1-2): 70-83.

Israelachvili, J. N. 2011. Intermolecular and Surface Forces (3rd edition). Elsevier B. V. Radarweg 29 Amsterdam 1043 NX, Netherlands.

Kanasty, R., J. R. Dorkin, A. Vegas and D. Anderson. 2013. Delivery materials for siRNA therapeutics. Nat. Mater. 12(11): 967-977.

Karmali, P. P. and A. Chaudhuri. 2007. Cationic liposomes as non-viral carriers of gene medicines: Resolved issues, open questions and future promises. Med. Res. Rev. 27(5): 696-722.

Kay, M. A., J. C. Glorioso and L. Naldini. 2001. Viral vectors for gene therapy: The art of turning infectious agents into vehicles of therapeutics. Nat. Med. 7(1): 33-40.

Kim, J.-K., S.-H. Choi, C.-O. Kim, J.-S. Park, W.-S. Ahn and C.-K. Kim. 2003. Enhancement of polyethylene glycol (PEG)-modified cationic liposome-mediated gene deliveries: Effects on serum stability and transfection efficiency. J. Pharm. Pharmacol. 55(4): 453-460.

Kim, S. W. 2012. Polylysine copolymers for gene delivery. Cold. Spring. Harb. Protoc. 2012(4): 433-438.

Klajnert, B. and M. Bryszewska. 2001. Dendrimers: Properties and applications. Acta Biochim. Pol. 48(1): 199-208.

Konstan, M. W., P. B. Davis, J. S. Wagener, K. A. Hilliard, R. C. Stern, L. J. H. Milgram, et al. 2004. Compacted DNA nanoparticles administered to the nasal mucosa of cystic fibrosis subjects are safe and demonstrate partial to complete cystic fibrosis transmembrane regulator reconstitution. Hum. Gene Ther. 15(12): 1255-1269.

Lechardeur, D. and G. L. Lukacs. 2002. Intracellular barriers to non-viral gene transfer. Curr. Gene Ther. 2(2): 183-194.

Lin, A. J., N. L. Slack, A. Ahmad, C. X. George, C. E. Samuel and C. R. Safinya. 2003. Three-dimensional imaging of lipid gene-carriers: Membrane charge density controls universal transfection behavior in lamellar cationic liposome-DNA complexes. Biophys. J. 84(5): 3307-3316.

Little, S. R. and D. S. Kohane. 2008. Polymers for intracellular delivery of nucleic acids. J. Mater. Chem. 18(8): 832-841.

Liu, C. and N. Zhang. 2011. Nanoparticles in gene therapy principles, prospects and challenges. Prog. Mol. Biol. Transl. Sci. 104: 509-562.

Ma, Y., X. Zhang, X. Xu, L. Shen, Y. Yao, Z. Yang, et al. 2015. STAT3 decoy oligodeoxynucleotides-loaded solid lipid nanoparticles induce cell death and inhibit invasion in ovarian cancer cells. PLoS One 10(4): e0124924.

Madeira, C., L. M. S. Loura, M. R. Aires-Barros and M. Prieto. 2011. Fluorescence methods for lipoplex characterization. Biochim. Biophys. Acta 1808(11): 2694-2705.

Madeira, C., L. M. S. Loura, M. Prieto, A. Fedorov and M. R. Aires-Barros. 2008. Effect of ionic strength and presence of serum on lipoplexes structure monitorized by FRET. BMC Biotechnol. 8: 20.

Miller, A. D. 1998. Cationic liposomes for gene therapy. Angew. Chem. Int. Ed. 37: 1768-1785.

Morille, M., C. Passirani, A. Vonarbourg, A. Clavreul and J. -P. Benoit. 2008. Progress in developing cationic vectors for non-viral systemic gene therapy against cancer. Biomaterials 29(24-25): 3477-3496.

Müller, R., K. Mäder and S. Gohla. 2000. Solid lipid nanoparticles (SLN) for controlled drug delivery: A review of the state of the art. Eur. J. Pharm. Biopharm. 50(1): 161-177.

Nam, H. Y., H. J. Hahn, K. Nam, W. -H. Choi, Y. Jeong, D. -E. Kim, et al. (2008). Evaluation of generations 2, 3 and 4 arginine modified PAMAM dendrimers for gene delivery. Int. J. Pharm. 363(1-2): 199-205.

Nicolau, C. and C. Sene. 1982. Liposome-mediated DNA transfer in eukaryotic cells. Dependence of the transfer efficiency upon the type of liposomes used and the host cell cycle stage. Biochim. Biophys. Acta 721(2): 185-190.

Okita, K., T. Ichisaka and S. Yamanaka. 2007. Generation of germline-competent induced pluripotent stem cells. Nature 448(7151): 313-317.

Olbrich, C. and R. H. Müller. 1999. Enzymatic degradation of SLN-effect of surfactant and surfactant mixtures. Int. J. Pharm. 180(1): 31-39.

Paleos, C. M., L. -A. Tziveleka, Z. Sideratou and D. Tsiourvas. (2009). Gene delivery using functional dendritic polymers. Expert. Opin. Drug Deliv. 6(1): 27-38.

Pardeike, J., A. Hommoss and R. H. Müller. 2009. Lipid nanoparticles (SLN, NLC) in cosmetic and pharmaceutical dermal products. Int. J. Pharm. 366(1-2): 170-184.

Pedersen, N., S. Hansen, A. V. Heydenreich, H. G. Kristensen and H. S. Poulsen. 2006. Solid lipid nanoparticles can effectively bind DNA, streptavidin and biotinylated ligands. Eur. J. Pharm. Biopharm. 62(2): 155-162.

Perumal, O. P., R. Inapagolla, S. Kannan and R. M. Kannan. 2008. The effect of surface functionality on cellular trafficking of dendrimers. Biomaterials 29(24-25): 3469-3476.

Radenkovic, D., H. Kobayashi, E. Remsey-Semmelweis and A. M. Seifalian. 2016. Quantum dot nanoparticle for optimization of breast cancer diagnostics and therapy in a clinical setting. Nanomedicine: NBM 12(6): 1581-1592.

Rädler, J. O., I. Koltover, T. Salditt and C. R. Safinya. 1997. Structure of DNA-cationic liposome complexes: DNA intercalation in multilamellar membranes in distinct interhelical packing regimes. Science 275(5301): 810-814.

Ramamoorth, M. and A. Narvekar. 2015. Non viral vectors in gene therapy – An overview. J. Clin. Diagn. Res. 9(1): GE01-6.

Rehman, Z. ur, I. S. Zuhorn and D. Hoekstra. 2013. How cationic lipids transfer nucleic acids into cells and across cellular membranes: Recent advances. J. Control. Release. 166(1): 46-56.

Robbins, P. D. and S. C. Ghivizzani. 1998. Viral vectors for gene therapy. Pharmacol. Ther. 80(1): 35-47.

Rudolph, C., U. Schillinger, A. Ortiz, K. Tabatt, C. Plank, R. H. Müller, et al. 2004. Application of novel solid lipid nanoparticle (SLN)-gene vector formulations based on a dimeric HIV-1 TAT-peptide in vitro and in vivo. Pharm. Res. 21(9): 1662-1669.

Ruiz de Garibay, A. P., M. A. Solinís, A. del Pozo-Rodríguez, P. S. Apaolaza, J. S. Shen and A. Rodríguez-Gascón. 2015. Solid lipid nanoparticles as non-viral vectors for gene transfection in a cell model of fabry disease. J. Biomed. Nanotechnol. 11(3): 500-511.

Rupaimoole, R. and F. J. Slack. 2017. MicroRNA therapeutics: Towards a new era for the management of cancer and other diseases. Nat. Rev. Drug. Discov. 16(3): 203-222.

Ruponen, M., P. Honkakoski, S. Rönkkö, J. Pelkonen, M. Tammi and A. Urtti. 2003. Extracellular and intracellular barriers in non-viral gene delivery. J. Control Release. 93(2): 213-217.

Safinya, C. R. 2001. Structures of lipid-DNA complexes: Supramolecular assembly and gene delivery. Curr. Opin. Struct. Biol. 11(4): 440-448.

Sambrook, J., H. Westphal, P. R. Srinivasan and R. Dulbecco. 1968. The integrated state of viral DNA in SV40-transformed cells. Proc. Natl. Acad. Sci. U.S.A. 60(4): 1288-1295.

Sartuqui, J., N. D'Elía, N. Gravina and L. Benedini. 2016. Application of natural, semi-synthetic, and synthetic biopolymers used in drug delivery systems design. p. 28.

In: J. M. Ruso and P. V. Messina (eds.). Biopolymers for Medical Applications. CRC Press. Taylor and Francis Group.

Scarzello, M., V. Chupin, A. Wagenaar, M. C. A. Stuart, J. B. F. N. Engberts and R. Hulst. 2005. Polymorphism of pyridinium amphiphiles for gene delivery: Influence of ionic strength, helper lipid content, and plasmid DNA complexation. Biophys. J. 88(3): 2104-2113.

Shao, Z., J. Shao, B. Tan, S. Guan, Z. Liu, Z. Zhao, et al. 2015. Targeted lung cancer therapy: Preparation and optimization of transferrin-decorated nanostructured lipid carriers as novel nanomedicine for co-delivery of anticancer drugs and DNA. Int. J. Nanomed. 10: 1223-1233.

Sia, R. H. and M. H. Dawson. 1931. In vitro transformation of pneumococcal types: I. A technique for inducing transformation of pneumococcal types in vitro. J. Exp. Med. 54(5): 701-710.

Simberg, D., D. Danino, Y. Talmon, A. Minsky, M. E. Ferrari, C. J. Wheeler, et al. 2001. Phase behavior, DNA ordering, and size instability of cationic lipoplexes. Relevance to optimal transfection activity. J. Biol. Chem. 276(50): 47453-47459.

Smith, A. E. 1995. Viral vectors in gene therapy. Annu. Rev. Microbiol. 49(1): 807-838.

Soler-Illia, G. J. de A. A., C. Sanchez, B. Lebeau and J. Patarin. 2002. Chemical strategies to design textured materials: From microporous and mesoporous oxides to nanonetworks and hierarchical structures. Chem. Rev. 102(11): 4093-4138.

Spagnou, S., A. D. Miller and M. Keller. 2004. Lipidic carriers of siRNA: Differences in the formulation, cellular uptake, and delivery with plasmid DNA. Biochemistry 43: 13348-13356.

Stolberg, S. G. 1999. The biotech death of Jesse Gelsinger. N Y Times Mag. 136-140, 149-150.

Stribley, J. M., K. S. Rehman, H. Niu and G. M. Christman. 2002. Gene therapy and reproductive medicine. Fertil. Steril. 77(4): 645-657.

Svenson, S. and D. A. Tomalia. 2005. Dendrimers in biomedical applications— Reflections on the field. Adv. Drug. Deliv. Rev. 57(15): 2106-2129.

Szybalska, E. H. and W. Szybalski. 1962. Genetics of human cess line. IV. DNA- mediated heritable transformation of a biochemical trait. Proc. Natl. Acad. Sci. U.S.A. 48: 2026-2034.

Tang, M. X., C. T. Redemann and F. C. Szoka. 1996. In vitro gene delivery by degraded polyamidoamine dendrimers. Bioconjug. Chem. 7(6): 703-714.

Tatum, E. L. 1966. Molecular biology, nucleic acids, and the future of medicine. Perspect. Biol. Med. 10(1): 19-32.

Temin, H. M. 1961. Mixed infection with two types of Rous sarcoma virus. Virology 13: 158-163.

Thomas, C. E., A. Ehrhardt and M. A. Kay. 2003. Progress and problems with the use of viral vectors for gene therapy. Nat. Rev. Genet. 4(5): 346-358.

Tomalia, D. A., H. Baker, J. Dewald, M. Hall, G. Kallos, S. Martin, et al. 1985. A New Class of Polymers: Starburst-Dendritic Macromole. Polym. J. 17(1): 117-132.

Torrecilla, J., A. del Pozo-Rodríguez, P. S. Apaolaza, M. Á. Solinís and A. Rodríguez- Gascón. 2015. Solid lipid nanoparticles as non-viral vector for the treatment of chronic hepatitis C by RNA interference. Int. J. Pharm. 479(1): 181-188.

Torrecilla, J., A. del Pozo-Rodríguez, M. Á. Solinís, P. S. Apaolaza, B. Berzal-Herranz, C. Romero-López, et al. 2016. Silencing of hepatitis C virus replication by a non- viral vector based on solid lipid nanoparticles containing a shRNA targeted to the internal ribosome entry site (IRES). Colloids Surf. B Biointerfaces 146: 808-817.

Tros de Ilarduya, C., Y. Sun and N. Düzgüneş. 2010. Gene delivery by lipoplexes and polyplexes. Eur. J. Pharm. Sci. 40(3): 159-170.

Varkouhi, A. K., M. Scholte, G. Storm and H. J. Haisma. 2011. Endosomal escape pathways for delivery of biologicals. J. Control Release 151(3): 220-228.

Wagner, E. 2014. Polymers for nucleic acid transfer – An overview. Adv. Genet. 88: 231-261.

Walz, D. and S. R. Caplan. 2002. Bacterial flagellar motor and H(+)/ATP synthase: Two proton-driven rotary molecular devices with different functions. Bioelectrochem. 55(1-2): 89-92.

Wasungu, L. and D. Hoekstra. 2006. Cationic lipids, lipoplexes and intracellular delivery of genes. J. Control Release 116(2): 255-264.

Watson, J. D. and F. H. Crick. 1953. Molecular structure of nucleic acids: A structure for deoxyribose nucleic acid. Nature 171(4356): 737-738.

Werfel, T. and C. Duvall. 2016. Polymeric nanoparticles for gene delivery. pp. 147-188. *In*: N. Ravin (ed.). Polymers and Nanomaterials for Gene Therapy. Elsevier B. V. Radarweg 29 Amsterdam 1043 NX, Netherlands.

Wheeler, J. J., L. Palmer, M. Ossanlou, I. MacLachlan, R. W. Graham, Y. P. Zhang, et al. 1999. Stabilized plasmid-lipid particles: Construction and characterization. Gene Ther. 6(2): 271-281.

Wightman, L., R. Kircheis, V. Rössler, S. Carotta, R. Ruzicka, M. Kursa, et al. comment 2001. Different behavior of branched and linear polyethylenimine for gene delivery in vitro and in vivo. J. Gene Med. 3(4): 362-372.

Wong, J. K. L., R. Mohseni, A. A. Hamidieh, R. E. MacLaren, N. Habib and A. M. Seifalian. 2017. Will nanotechnology bring new hope for gene delivery? Trends Biotechnol. 35(5): 434-451.

Xu, Q., C. -H. Wang and D. W. Pack. 2010. Polymeric carriers for gene delivery: Chitosan and poly(amidoamine) dendrimers. Curr. Pharm. Des. 16(21): 2350-2368.

Yang, Y. and J. M. Wilson. 1995. Clearance of adenovirus-infected hepatocytes by MHC class I-restricted CD4+ CTLs in vivo. J. Immunol. 155(5): 2564-2570.

Zidovska, A., H. M. Evans, K. K. Ewert, J. Quispe, B. Carragher, C. S. Potter, et al. 2009. Liquid crystalline phases of dendritic lipid-DNA self-assemblies: Lamellar, hexagonal, and DNA bundles. J. Phys. Chem. B. 113(12): 3694-3703.

Zinder, N. D. and J. Lederberg. 1952. Genetic exchange in Salmonella. J. Bacteriol. 64(5): 679-699.

Zuidam, N. and Y. Barenholz. 1998. Electrostatic and structural properties of complexes involving plasmid DNA and cationic lipids commonly used for gene delivery. Biochim. Biophys. Acta 1368(1): 115-128.

Zuidam, N. J. and Y. Barenholz. 1997. Electrostatic parameters of cationic liposomes commonly used for gene delivery as determined by 4-heptadecyl-7-hydroxycoumarin. Biochim. Biophys. Acta 1329(2): 211-222.

zur Mühlen, A., C. Schwarz and W. Mehnert. 1998. Solid lipid nanoparticles (SLN) for controlled drug delivery – Drug release and release mechanism. Eur. J. Pharm. Biopharm. 45(2): 149-155.

Nano-biosensors: A Custom-built Diagnosis

"In the current climate of streamlined health care with an emphasis on community based care and one stop clinics, the concept of near-patient testing is appealing. Near-patient testing (also known as point-of-care testing) is defined as an investigation taken at the time of the consultation with instant availability of results to make immediate and informed decisions about patient care..."

Kamlesh Khunti, Professor of Primary Care Diabetes and
Vascular Medicine at the University of Leicester (Khunti 2010)

6.1 Forthcoming Diagnosis Platforms: The Necessity of Nano-Sized Biotechnology

Visualize a near future where patients will be able to identify a sign of a pathological change directly through one´s eyes or skin using personalized diagnostic tools, or where precision farming techniques could be used to be cognizant of any potential risk to the environment or to human health. Fortunately, this is not a pure fantasy; the improvement of ultra-sensitive biological and chemical sensors is one of the outstanding scientific, engineering, and educational dares of the 21st century based on the concepts of nanotechnology. Nanotechnology and its influence has bestowed some highly exciting ingredients for the improvement of sensing phenomenon and leads to an efficient nano-biosensor with minuscule assembly compared to the conventional counterparts. Scientific community is attending the challenge by expanding sensor platforms that exploit breakthroughs in nanomaterials, miniaturization, sample collection, preparation strategies, and signal detection. They focus on the development of advanced systems for the rapid, ultra-sensitive detection of pathogens responsible for disease as well as those central to homeland security and the environment. By the use of an integrated approach, that combines biological science, medical science, material science, physics, chemistry, electronics, mechanical engineering, and computers, nano-sensors with immobilized bio-receptor probes that are selective for target analytic molecules are created, the so-called nano-biosensors. Nano-sized dimensions of metal nanoparticles, metal nanoclusters, metal oxide nanoparticles, metal and carbon quantum dots, graphene, carbon nanotubes, and nanocomposites expand the sensitivity by signal amplification and integrate several novel transduction principles such as enhanced electrochemical, optical, Raman effect, enhanced catalytic activity, and superparamagnetic properties into the nano-biosensors. The electrochemical nano-sensors, optical nano-sensors, electronic nose and electronic tongue, nano-barcode technology, and wireless nano-sensors have revolutionized the spatial resolution of new-generation of biosensors down molecular levels, reducing their detection volume to a few cubic micrometres and speeding up their signal response to milliseconds. Thus, the next generation nano-biosensor platforms showed significant improvements in sensitivity and specificity in order to meet the needs of a variety of fields including *in vitro* medical diagnostics, pharmaceutical discovery and pathogen detection, becoming a simple, small and low-cost diagnosis tool. Based on the idea of a personalised nano-biosensor design, extensive work has been done about portable detection system. In this way, because of its handy, well equipped, and classy features such as advanced processors, increased memory, high-resolution camera, high-end security via fingerprinting, and a variety of built-in sensors, smartphones have been currently used to design nano-biosensor prototypes. Notable work using this technology has been done in *in-vitro* and real-time monitoring of clinical condition (Varadan et al. 2011, Zhang and Liu 2016, Hernández-Neuta et al. 2019). The data produced by smartphone based on *in vitro* diagnostics can be added with spatial and temporal information, which can be used for monitoring and management in critical situations. Research on smartphone based nano-biosensor prototypes such as: lateral flow assays (LFA), microscopy, electrochemical sensing, immunoassays, surface plasmon resonance-based biosensing, flow cytometry, and optical detection has been reported. Some of them have even become commercialized: iHealth, AliveCor, GENTAG, Mobile Assay, and CellScope (Vashist et al. 2014).

Together with the medical diagnosis, the detection of nanobiosensors in the food and agriculture sectors expanded with multiplex and real-time detection capabilities for the detection of a wide variety of fertilizers, herbicides, pesticides, insecticides, pathogens, humidity and soil pH. That is extremely important since the food and agriculture sector controls the economic growth of a developing country. Taken together, proper and controlled use of nano-biosensor can support sustainable agriculture for enhancing crop productivity and reducing environmental and human health risks. Food industries have practices of growing crops, raising livestock and seafood, food processing and packaging, regulating production and distribution where the process control and monitoring quality and safety are crucial steps. Despite previous success stories of the remunerative health sector, the approaches are transferred subsequently to food and agriculture sector, with potential application in detection of food contaminants such as preservatives, antibiotics, heavy metal ions, toxins, microbial load, and pathogens along with the rapid monitoring of temperature, traceability, humidity, gas, and aroma of the foodstuff.

To the point, as we have emphasized in chapter 1, the main goal of nanotechnology entails the high performance and efficiency fabrication of devices on a molecular or atomic scale. With the advent of nanotechnology and its impact on developing ultrasensitive devices, it can be stated that it is probably one of the most promising ways to solve some of the problems concerning the increasing need to create highly sensitive, fast and economic system of analysis to apply in medical diagnostics, agriculture, food, drink industries and environment protection. Generating high expectations and promises, such tiny scale technology gained a lot of attention from governments, companies and universities as they see the potential benefit after its integration into real-world systems. This chapter will illustrate the nano-biosensor technology that is envisioned for the detection of chemical and biological threats such as pollutants, microbial contamination, viruses, communicable and genetic diseases, and cancer markers. We review some of the latest applications of nano-biosensors in "point of care" diagnostics field highlighting the fundamental differences in material characteristics realized at nano-size dimensions to detect specific molecular interactions, primarily through the detection of molecular biomarkers, such as proteins and nucleic acids. In addition, we will analyse how the devices progressed to a point where it is possible to determine in which cases their inherent advantages over traditional techniques compensate for the added complexity and cost involved in constructing and assembling them. It will include DNA array technology, which might play an important role in enabling nanofabrication; fast and sensitive drug screening, one of the limiting factors in combinatorial chemistry for drug discovery, and the development of chip-sized biodevices that could revolutionize disease detection and management. Finally, we will summarize the applications of nano-biosensors in the detection of different molecules from industrial emissions whose impact on the environment affects the prosperity of human health.

6.2 Nanotechnology Based Biosensors: Conceptual Idea, Variants and Applications

The term *"biosensor"* was coined by Cammann (Cammann 1977) and its definition was introduced by IUPAC as "...a device that uses specific biochemical reactions mediated by isolated enzymes, immune-systems, tissues, organelles or whole cells to detect chemical compounds usually by electrical, thermal or optical signals..." (McNaught and Wilkinson 1997). Practically and to be easily understood, a biosensor is a device that combines a biological recognition element with a physical or chemical transducer that detects or quantifies a biological analyte, a bioanalyte, such as a particular DNA sequence or a precise configuration of a protein. It is a probe that integrates a biological component with an electronic component to yield a measurable signal (Mehrotra 2016) that essentially consists of three main components (Malik et al. 2013): (i) a bioreceptor that binds to the exact molecular form of the bioanalyte; (ii) an electrochemical interface provided with a transducer where specific biological processes occur giving rise to a biochemical response that is conveniently converted in an electrical signal; and, finally (iii) a signal detector system coupled with an accurate interface for translating the electronic signal into a meaningful physical parameter displaying the results to the operator. Many biosensors are affinity-based, meaning that they use immobilized clamp probe that selectively binds the molecule being sensed, i.e. the target or analyte, thus transferring the conversion of detecting a target in a solution to detecting a change at a localized surface. This difference can then be measured using a battery of techniques, including those that involve light (surface plasmonic resonance [SPR] or fluorescence), mechanical motion (such as, quartz crystal microbalance or resonant cantilever), or magnetic particles (Touhami 2014). Started in the 1960s by the pioneers Clark and Lyons and the first enzyme-based glucose sensor reported by Updike and Hicks in 1967 (Updike and Hicks 1967), various types of affinity-based biosensors are being used; some examples are enzyme-based (Wilson and Hu 2000), tissue-based (Campàs et al. 2008), antibodies- based (North 1985) and DNA-based (Zhai et al. 1997) biosensors. From the palette of techniques available to the user, the electrical probe sensors are widely exploited to design the so-called *"label-free"* biosensors. Such kind of biosensors allows one to investigate the underlying physical and chemical characteristics, and interactions, of target species by relying solely on their intrinsic physicochemical properties and do not require a label or a tag to report the detection of a specific molecule. This has the benefit of reducing sample complexity, preparation time, and analysis cost due to the elimination of potentially confounding molecular labels. Furthermore, because of the relatively minimal sample preparation, label-free sensing approaches are highly amenable to field applications and remote diagnostics where preparation facilities and trained personnel may be limited or unavailable. Electrical *"label-free"* biosensors rely solely on the measurement of currents and/or voltages to detect binding and due to their low cost, low power, and ease of miniaturization, they hold great promise for applications where minimizing size and cost is crucial, such as "point-of-care" disease diagnostics (Prasad 2014). So far, we have summarized the most relevant aspects related to biosensors, the reader can refer to the following references for more information: (Touhami 2014, Narang and Pundir 2017, Prickril and Rasooly 2017, Schöning and Poghossian 2018). Now that we have analysed, understood and exemplified the biosensing concept, we are able to lay the basis for researching and developing the nano-biosensors.

Nano-biosensors are, basically, biosensors, which are made up of nano-sized dimension materials. The nanostructures in nano-biosensors act as an intermediate layer between biological agents and physicochemical detector components or biological agents, and the transducer is combined with nanomaterials to construct the final arrangement (Prasad 2014); examples of nanomaterials used in biosensor construction are summarized in scheme 6.1. Interestingly, their importance does not lie in the fact that they can detect

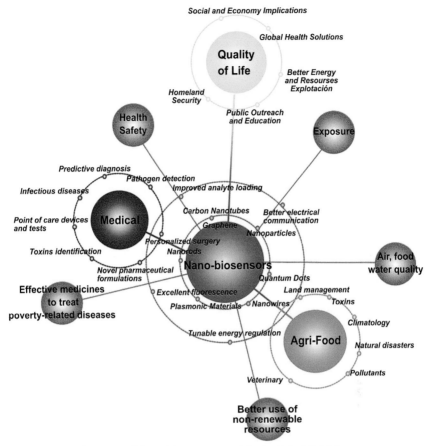

Scheme 6.1. Nano-biosensors and their global field focus.

nanoscale events (Malik et al. 2013). So, what is the need that sustains interest in the use of nano-biosensors? Are they going to drive in any significant difference to the overall technology?

The answer is simply that the size restrictions of these materials makes them unique as they have most of their constituent atoms located at or near their surface to markedly improve the sensitivity and specificity of analyte detection. Along these lines, the integration of nanomaterials with electrical systems gives rise to nanoelectromechanical systems (NEMS), which are very active in their electrical transduction mechanisms, improving electrical biosensors capabilities. In addition, biosensors of nanoscale dimensions may overcome many of the obstacles that prevent wide scale use of affinity-based biosensors. The interest in such devices is great because the vast majority of bioanalytes in disease diagnosis can be best detected using an affinity reaction. Furthermore, the specificity and high affinity of antibodies for their target species potentially allow biosensors to perform analyses at low concentrations of analyte in complex samples. To attain this purpose, there are two main areas where the demand for bio-nanosensors is very significant. The first is to attain the reduction of the detection limit and the second is the ability to detect a number of bioanalytes in the same sample, which is known as multiplexing. Multiplexing is important because reliable diagnosis of disease often requires identification of the levels of a number of molecular markers. The barrier to reducing detection limits in conventional affinity-based biosensors is the thermodynamics of the affinity reaction. Typically, dissociation constants (K_d) for antibody-antigen reactions are about 10^{-8}–10^{-12} M. With conventional transduction techniques, the detection limits are therefore in the range of 10^{-9}–10^{-14} M (Prasad 2014). Detection limits nearly three orders of magnitude below the dissociation constants are achieved by utilizing nanomaterials for the biosensor construction (Di Giusto et al. 2005, Nair and Alam 2008, Duan et al. 2012, Zuo et al. 2013). The drawback of amplification techniques is that they typically require labelling and sample manipulation steps, which restrict these technologies to the hands of experienced operators and preclude real-time analysis. Highly sensitive, label-free transduction of bio recognition reactions has been achieved with cantilevers and field-effect transistor configurations (Ohno et al. 2010). Last, but not the least, monitoring human health for early detection of disease conditions or health disorders is vital for maintaining a healthy life. Good cell viability after testing with nano-probes has been recently demonstrated, which opens the possibility to non-harmful continuous monitoring in real-time (Ewald et al. 2011, Liu et al. 2014, Wiraja et al. 2016, Wang et al. 2018, Wiraja et al. 2018).

6.3 Nano-Biosensors in Medical Practice: Point of Care Diagnosis (POCD) Devices

The international standard ISO 22870: Point-of-care testing (POCT)-Requirements for quality and competence (ISO 22870:2006 2006), defines POCT as: *"... testing that is performed near or at the site of a patient with the result leading to possible change in the care of the patient..."*

Biosensor design is the most critical component of POCD, it is directly responsible for the bioanalytical performance of an assay and must be intended to match the ASSURED criteria: affordable, sensitive, specific, user friendly, robust, equipment free, and deliverable to those in need (Chamorro-Garcia and Merkoçi 2016). The first true POCD device was developed in 1957 to measure urinary protein: the urine dipstick (Gubala et al. 2011). Glucose meters and lateral flow devices are currently the most widely used devices in POCD blood molecular diagnostics, although they are not applicable if quantitative, sensitive, and high-throughput measurements are required. Emerging technologies including label-free biosensors, such as electrochemical, optical, surface plasmon resonance (SPR), white light reflectance spectroscopy (WLRS), among others, have been developed, and are being used for improved POCD (Wang 2006, Ko et al. 2011, Tokel et al.

2014, Yoo and Lee 2016). Complementary technologies, like microfluidics, lab-on-a-chip technologies, system integration, device automation, and signal readout, are providing the desired impetus for continuous improvements in POCD (Nayak et al. 2016, Vashist 2017). Currently, simple, robust, sensitive and cost-effective biosensors joined with nano-sized materials are working as a bridge between advanced detection/diagnostics and daily/routine tests which are paving the way to next-generation of point-of-care tests (POCT) (Syedmoradi et al. 2017). Here we review some of their latest applications.

6.3.1 Label Free Nano-biosensors: Ultrasensitive Immunoassays

Labels structurally and functionally can interfere with an assay, may not be specific and may be difficult to be conjugated: a problem often encountered in single-molecule experiments. Without the need for labels, detection can be done on site and in real time, an important advantage in emerging point-of-care testing applications. Most label-based detection technologies are based on immunoassays, centred on antigen-antibody reactions and the use of enzyme-linked immunosorbent assay (ELISA) experiments, to detect the amount of protein present in the blood (Prasad 2014). ELISA has been the standard diagnostic tool for the last 30 years and it has numerous applications in food quality, environmental, biotechnological, and chemical disciplines among others. In spite of its many advantages, ELISA has certain limitations such as tedious/laborious test procedure and insufficient level of sensitivity in bio-recognition of current attracting biomolecular entities such as microRNAs (Hosseini et al. 2018). These drawbacks make ELISA unsuitable for rapid point of care biosensing applications. Fast and simple point-of-care diagnostics have been developed with the advances in microfabrication and nanotechnology that enable label-free detection of biomolecules. Their introduction dates from the year 2001 (Cui et al. 2001); since then, nanoelectronic transistor sensors based on 1D and 2D nano-sized materials have attained real-time label-free multiplexed detection of key analytes with high sensitivity. In particular, field-effect transistor (FET) biosensors constructed from semiconducting nanowires, single-walled carbon nanotubes, and graphene have been extensively investigated demonstrating great potential to serve as POC detectors. The detection of a varied range of biological species, including protein disease markers, nucleic acids, and viruses, as well as detection of protein–protein interactions and enzymatic activity were reported. Nevertheless, and despite their appeal, electronic nanosensors continue to be a challenge to implement because fundamental limitations render them incapable of sensing molecules in complex, physiological solutions (Stern et al. 2010). Today, important progress has been made in this regard. The direct analysis of high–ionic-strength physiological solutions, usually, limited because of Debye charge screening was overcome by a commodification of the transistor sensor surface with PEG and spacer molecules or PEG and aptamer receptors (Gao et al. 2016). Problems such as biofouling and non-specific binding were also overcome using a microfabricated device that operates upstream of the nanosensors to purify biomarkers of interest (Stern et al. 2010). A microfluidic purification chip (MPC) that captures cancer biomarkers from physiological solutions and, after washing, releases the antigens into a pure buffer suitable for sensing was proposed by E. Stern and co-workers (Stern et al. 2010). They reported a specific and quantitative detection of two model cancer antigens from a 10 ml sample of whole blood in less than 20 min. This study marks the first use of label-free nanosensors with physiological solutions, positioning this technology for rapid translation to clinical settings.

Optical label-free detectors, such as the respected surface plasmon resonance (SPR) sensor, are usually preferred for their capability to attain quantitative data on intermolecular binding. Nevertheless, sensitivity to single binding events was not possible until the recent introduction of resonant microcavities associated with the whispering gallery mode (WGM) resonators; a miniature optical device that recirculates a guided wave to enhance sensitivity down to a single molecule (Vollmer and Arnold 2008, Yang et al. 2015). This scheme was used to quantify several biomarkers of ovarian cancer and multiple myeloma with detection limits comparable to ELISA (Kim 2016) and to monitor DNA–small molecule interactions (Panich et al. 2016). In addition, WGM was integrated with numerous fluidic platforms; Lian et al. inform the implementation of fibre light-coupled optofluidic waveguide (FLOW) immunosensor for the detection of p53 protein, a typical tumour marker. The FLOW consists of a liquid-core capillary and an accompanying optical fibre, which allows evanescent interaction between light and microfluidic sample. Molecular binding at internal surface of the capillary induces a response in wavelength shift of the transmission spectrum in the optical fibre. To enable highly sensitive molecular detection, the evanescent-wave interaction has been strengthened by enlarging shape factor R via fine geometry control. The proposed FLOW immunosensor works with flowing microfluid, which increases the surface molecular coverage and improves the detection limit. As a result, the FLOW immunosensor presents a log–linear response to the tumour protein at concentrations ranging from 10 fg/mL up to 10 ng/mL. In addition, the fluid at an optimal flow rate, which benefits the accuracy of the measurement, can effectively remove the non-specifically adsorbed molecules. Tested in serum samples, the FLOW successfully maintains its sensitivity and specificity on p53 protein, making it suitable for point of care diagnostics applications (Liang et al. 2018). Selective optical biodetection in complex media of nM concentrations of Exotoxin A (limit-of-detection ≈2.45 nM), a virulence factor secreted from *Pseudomonas aeruginosa* in the mucus of Cystic Fibrosis (CF) patients, have been also performed using microtoroids with multifunctional surface functionality (Toren et al. 2018). The same knowhow can provide beneficial biosensing platforms to understand DNA alterations (Toren et al. 2015).

Summarizing, researchers have created high fidelity, miniature and ultrasensitive multiplexed label-free sensors to follow receptor-analyte interactions in real-time. Those can expand the quality and throughput of drug discovery process, making improvements to clinical tools used for diagnosing and drastically increasing the life prognosis of patients, for example to those exhibiting malignant neoplasm pathologies. Their potential application as point-of-care tools brings scientific community's dreams closer to reality.

6.3.2 On-chip Analysis of Tumour-derived Exosomes: Non-Invasive "Liquid Biopsy" Biomarkers for Early Detection

Exosomes are microvesicles found in all living cells formed when vesicles from the endosomal membranes fuse with the cell's plasma membrane instead of fusing with a lysosome, for internal digestion. Resembling their cell of origin, they contain a sample of the cytosolic milieu including an abundance of DNA, RNA, proteins and other analytes (Sheridan 2016). In the last several years, nanoscale exosomes (30 nm-100 nm) and microvesicles (100 nm to 1 µm) have been discovered to contain a wealth of proteomic and genetic information for disease diagnostics as well as the monitoring of cancer progression, metastasis, and drug efficacy. They have their origin in tumour cells and that can be found circulating in the blood, and most bodily fluids. In addition, immunoblotting analysis revealed that Glypican-1 (GPC-1), an exosomal membrane protein, has much higher expression on the cancerous exosomes than on the noncancerous counterparts. This fact highlights its clinical value as an exosomal biomarker for the early diagnosis of pancreatic, breast and colorectal cancer (Liu et al. 2018b). In the light of the obtained results, a great evidence revealed the use of exosomes to the field of cancer diagnostics as a new and exciting platform in the area of "liquid biopsy". However, establishing the clinical utility of exosomes and microvesicles as biomarkers to improve patient care has been limited by fundamental technical challenges that stalk from their small size and the extensive sample preparation required prior to measurement (Ko et al. 2016). These facts have motivated the research of Zheng Zhao and co-workers (Zhao et al. 2016), who developed what they have called *"ExoSearch"*. ExoSearch is a chip-based in a simple microfluidic technology that provides enriched preparation of blood plasma exosomes for *in situ*, multiplexed detection using immunomagnetic beads. It offers a robust, continuous-flow design for quantitative isolation and release of blood plasma exosomes in a wide range of preparation volumes (10 µL to 10 mL). The methodology was successfully employed to detect three exosomal tumour markers of ovarian cancer (CA-125, EpCAM, CD24) showing a significant diagnostic power comparable with the standard Bradford assay (Bradford 1976).

To ensure analytical accuracy, exosomes usually need to be isolated and purified from cell culture supernatant or plasma before analysis. Lewis et al. (Lewis et al. 2018) presented an alternating current electrokinetic (ACE) microarray chip to trap and study exosomes and other cellular nanoparticles (i.e. extracellular vesicles, nucleosomes, cell free DNA, RNA, mitochondria, and necrotic cellular debris) directly from whole blood, plasma, or serum without pre-treatment or dilution of sample, nor need of antibodies or other capture techniques. Following on-chip immunofluorescence analysis, the authors conducted a specific identification and quantification of cancer-related protein and/or nucleic acid biomarkers within as little as 30 min total time. The technology can be performed using less than 25 µL (a drop) of undiluted whole blood that is applied directly to the chip, needing no added reagents. The system was validated to detect glypican-1 and CD63 biomarkers, which were found to reflect the presence of pancreatic ductal adenocarcinoma (PDAC); twenty PDAC patient samples could be distinguished from 11 healthy subjects with 99% sensitivity and 82% specificity.

A droplet-based single-exosome-counting enzyme-linked immunoassay (droplet digital ExoELISA) for cancer diagnostics was developed by Liu et al. (Liu et al. 2018a) using droplet microfluidics. Exosomes were immobilized on magnetic microbeads through sandwich ELISA complexes tagged with an enzymatic reporter that produces a fluorescent signal. The constructed beads were further isolated and encapsulated into a sufficient number of droplets to ensure only a single bead was encapsulated in a droplet. Researchers reported a limit of detection (LOD) down to 10 enzyme-labeled exosome complexes per microliter ($\sim 10^{-17}$ M). The application of ExoELISA approach was validated in plasma samples directly from breast cancer patients.

Despite their diagnostic and therapeutic potential, the clinical use of exosomes as cancer biomarkers is, however, still very limited. For example, a great proportion of non-small cell lung cancer (NSCLC) patients ($\approx 60\%$) developed resistance to targeted epidermal growth factor receptor (EGFR) inhibitor therapy because of EGFR T790M mutation. Patients exhibiting EGFR T790M mutation were treated with third-generation tyrosine kinase inhibitors with a favourable projection; however, obtaining a tissue biopsy to confirm the mutation poses risks and is often not feasible. Castellanos-Rizaldos and co-workers (Castellanos-Rizaldos et al. 2018) developed and validated a novel test (exoNA) that overcomes the limited abundance of the mutation by simultaneously capturing and cross-examining exosomal RNA/DNA and cfDNA in a single step followed by a sensitive allele specific qPCR. The authors reported a 92% sensitivity and 89% of specificity in the detection of the T790M mutation using exoNA compared to tumour biopsy; 88% of sensitivity in patients with intrathoracic disease (M0/M1a) was also attained.

The above-mentioned examples are a small sample of the pool of investigations performed to enhanced cancer diagnosis by the use of exosomes. The speed and simplicity of exosome capture and on-chip biomarker detection, combined with the ability to use a small drop of whole blood, or any body fluid, will enable seamless "liquid biopsy" screening. Expanding these technologies into clinical settings and as "point of care" devices will improve early stage cancer diagnostics.

6.3.3 Beta-amyloid Detection for Alzheimer's Disease Management

Alzheimer's disease (AD) is the leading cause of dementia, affecting > 35 million people worldwide (Forlenza et al. 2015), with a progressive increase of almost double every 20 years, reaching 75 million in 2030 and 131.5 million in 2050 (Song et al. 2018). Histologically, it is related to amyloid-beta (Aβ) deposits and neurofibrillary tangles in the cerebral cortex, hippocampus and sub cortical grey matter, while biochemically there is a reduction in the acetylcholine neurotransmitter. Its cause is actually unknown, clinical diagnosis is still by exclusion and an effective treatment is yet to emerge. The so-called "*AD-signature*" includes a decreased concentration of $A\beta_{1-42}$ peptide and an increased concentration of total tau (T-tau) and hyperphosphorylated tau (P-tau) in the cerebrospinal fluid (CSF) (Forlenza et al. 2015). Amyloid-beta peptides are potential biomarkers for monitoring and detecting AD in its early stage. According to Kaushik et al. (Kaushik et al. 2016), Beta-amyloid (β-A) peptides are very useful biomarkers for AD at the point-of-care (POC) level. Detection of this by affordable nano-enabled electrochemical biosensor device in the body fluids will be the feasible, non-invasive method of the future. Nano-enabled biosensors can detect beta-amyloid at low levels. Such sensor can be developed for AD monitoring at point-of-care at its initial stage and it can be properly and effectively treated.

The degree of tau phosphorylation differs between AD and frontotemporal dementia (FTD). Elucidation of T-tau and P-tau concentrations can be used to early identify AD among other types of degenerative dementia disorders. Chao Song and co-workers (Song et al. 2018) reported the construction of a label-free optical nanosensors through which the multiplexed monitoring of $A\beta_{1-42}$ peptide and T-tau biomarkers in 1 μL of buffer or CSF sample with a very good specificity is possible (7.8 pg/mL of $A\beta_{1-42}$ and 15.6 pg/mL of T-tau).

Other examples include a label-free electrochemiluminescence sensor for highly sensitive detection of $A\beta_{1-42}$ peptide oligomerization (Liu et al. 2018), and a single-step magnetic-beads based immunoassays adjusted to microfluidic droplet operations for the sequential determination of well-established $A\beta_{1-42}$ peptide biomarkers (Mai et al. 2018). Both methodologies were proposed to be adaptable to the construction of low-cost and disposable detection devices. In the first case, the detection platform is based on the bonding of Ru (phenanthroline)$_2$ dipyrido [3,2-a:2',3'-c]phenazine]$^{2+}$, ([Ru (phen)$_2$ dppz]$^{2+}$) to $A\beta_{1-42}$. The formation of aggregates results in an enhanced electrochemiluminescence due to the change in the polarity of the microenvironment of [Ru (phen)$_2$ dppz]$^{2+}$ when it is intercalated into the β-sheets during oligomerization (Liu et al. 2018b). In the second example, a sandwich assay protocol was reported where the captured antibodies are grafted onto magnetic beads and the detection antibodies can simultaneously bind to monomeric β peptides in a single step. The methodology used a series of 4 programmable magnetic tweezers to manipulate a train of nanoscale confined droplets containing magnetic beads, sample, washing and detection solutions providing a sequence of 8 analyses that could be realized within 45 min (Mai et al. 2018). Cited examples highlights the successful integration of nano-enabled device components and miniaturized systems that can become capable of AD biomarkers detection. It is a call to scientists for exploring multi-disciplinary research to fabricate nano-enabled POC biosensors towards diagnosis and monitoring of neurodegenerative diseases.

6.3.4 Detection of Poverty Related Diseases by Microfluidic Lab-On-A-Chip Devices

As we have emphasized in section 2.2.2 of chapter 2, the "Big Three" pathogens that span the spectrum of poverty related diseases (PRD) are: the human immunodeficiency virus (HIV), the *Plasmodia* spp. parasite which causes malaria, and the *Mycobacterium tuberculosis* (TB) bacterium. Together, they are leading the causes of mortality and morbidity in developing countries, accounting for more than half of all infant deaths. Over 95% of deaths linked to PRD are caused by a lack of proper diagnostics and treatments, due to insufficiencies in the health care infrastructure and cost constraints. Standard methods of pathogen detection, including cell culture, nucleic acid amplification, and enzyme-linked immunoassay are impractical because of the overburdened health care providers, the high volume of patients, work-flow and time limitations (Su et al. 2015). Thus, novel and easy to use pathogen detection platforms are emerging as powerful tools that meet the ASSURED criteria. Among them, the most prominent examples are included within microfluidic-based technology (Su et al. 2015). Microfluidic lab-on-a-chip (LOC) devices offer short processing times, reduced sample consumption, complex sample processing and handling of fluids, added to a portability and simplicity assimilated into a miniaturized arrangement (Foudeh et al. 2012). Several literature revisions of recently developed devices along with their respective advantages and limitations are available (Wang et al. 2013, Tay et al. 2016, Nasseri et al. 2018). At this point, we will examine the ways in which the tools at the sub-millimetre scale (microfluidics and nanotechnologies) can be addressed to meet the critical challenges to global public health. Fluorescence microscopy has, for a long time, been the standard method for TB detection in sputum samples, but it has been difficult to implement in the field. An integrated microfluidic system based on a fluorescence immune adsorption reaction for the capture, enrichment, and rapid detection of airborne *Mycobacterium tuberculosis* was proposed by W. Jing and co-workers (Jing et al. 2014). The whole detection consumed less than 50 min comprising 20 min of enrichment and 30 min of immunoreaction analysis. Likewise, Ka-U Ip et al. (Ip et al. 2018) proposed an integrated microfluidic device, including bacteria isolation, on-chip PCR and fluorescence detection, where the full analysis can be completed within 2 hours. The offered system can perform the entire detection of live bacteria from TB samples by the combination of a propidium monoazide (PMA) and real-time polymerase chain reaction (RT-PCR) with a limit of recognition of about 14 colony formation units CFU/reaction.

A dual-targeting strategy integrated into a low-cost, self-driven microfluidic chip, was achieved by Y. Cheng et al. (Cheng et al. 2018) for the rapid labelling and automated quantification of live Bacillus Calmette-Guérin (BCG). The small molecular fluorescence probe (CDG-DNB3) fluoresces upon activation of the β-lactamase BlaC, a hydrolase naturally expressed in *Mycobacterium tuberculosis*, and the fluorescent product is retained through a covalent modification of the decaprenylphosphoryl-β-d-ribose 2'-epimerase (DprE1). Complete analysis proceeds along 1 hour; it discriminates live from dead BCG showing specificity for *Mycobacterium tuberculosis* over other 43 non-tuberculosis mycobacteria species.

Electrical sensing techniques are often simpler and cheaper than optical methods. Xiao Li and Xinyu Liu (Li and Liu 2016) reported a microfluidic paper-based origami nano-biosensor (origami μPAD). The device integrated zinc oxide nanowires (ZnO NWs) and an electrochemical impedance spectroscopy (EIS) biosensing mechanism, for label-free, ultrasensitive immunoassays. The system can detect HIV p24 antigen in human serum with a limit of detection of 300 fg/mL, thus enjoying a significant potential for the diagnosis at early-stage infections. Other approaches may detect mass or mechanical forces as, for example, the work of Majid Ebrahimi Warkiani et al. (Warkiani et al. 2015). The authors constructed a shear-modulated inertial microfluidic system to enrich malaria parasites from blood to facilitate a more reliable and specific PCR-based malaria detection. This technique capitalizes on cell focusing behaviours in high aspect ratio microchannels coupled with pinched flow dynamics to isolate ring-stage malaria parasites from lysed blood containing white blood cells (WBCs). Sensitive detection of parasites ($\sim 2^{-10}$ per μL) has been demonstrated: about 100-fold more sensitive than the gold standard conventional microscopy analysis of thick blood smears. The simplicity and low limit of detection of this device makes it ideal for the application in diagnosis of malaria at the early stage of infection, when the low abundance of parasites in blood makes it a challenge. Likewise, Jeonghun Nam and co-workers (Nam et al. 2016) constructed a microfluidic device for high-throughput particle separation based on size-dependent viscoelasticity-induced lateral migration. The system enabled the rapid detection of extremely rare malaria parasites, *Plasmodium falciparum*, by using PCR analysis.

There is a great need for new diagnostic tools that will allow rapid and effective diagnosis of infectious diseases. Conventional diagnostic tools do not address the needs of the majority of the world's population afflicted with infectious diseases who have, at best, access to poorly resourced health care facilities with almost no supporting clinical laboratory infrastructure. Scientific community is in the quest to develop diagnostic tests to meet the needs of these people and the point-of-care (POC) tests have the potential to be the solution.

6.4 Portable Biosensing Devices: Smartphones in Healthcare Management

Because of their multifunction data transmission capabilities, connectivity, high-resolution cameras, touch-screen displays and high-performance central processing units (CPUs), smartphones has facilitated the development of simple, light-weight and low-cost devices suited for monitoring, sensing, detection and analysis of out-of-the-laboratory requests (Preechaburana et al. 2012, Roda et al. 2014, Liu et al. 2015). The built-in functions of smartphones can be further extended through the addition of accessories and the hundreds of new applications (apps) that are available every day, enabling smartphones to respond to the rising environmental or medical diagnostic needs. Up to now, the smartphone-based sensor systems can be roughly classified into: (i) optical biosensors integrated cameras for high resolution imaging and optical intensity improvements; (ii) surface plasmon resonance (SPR) biosensors developed for bio-interaction quantifications; (iii) electrochemical biosensors that use portable amperometric, potentiometric, and impedimetric detectors; and (iv) wearable near-field communication (NFC) biosensors for non-contacting, and battery-free chemical detections (Zhang and Liu 2016). Several examples reporting the actual opportunity of using smartphone-based platforms to detect biomarkers and analytes of clinical interest in bodily fluids are examined below.

A critical unmet need in the diagnosis of bacterial infections, which remain a major cause of human morbidity and mortality, is the detection of scarce bacterial pathogens in a variety of samples in a rapid and quantitative manner. S. Shrivastava and co-workers (Shrivastava et al. 2018) show a smartphone-based detection of *Staphylococcus aureus* in a rapid and quantitative manner from minimally processed liquid samples. The proposed methodology is based on the preparation of aptamer-functionalized fluorescent magnetic nanoparticles. The tagged bacteria cells were magnetically captured in a detection cartridge, and then imaged using a smartphone camera with a light-emitting diode as excitation source. Likewise, the research team of A. Ozcan (Tseng et al. 2010) combined optofluidic technology with standard cell phones, with the intent of creating an ultra-portable and an ultra-cheap equipment for point-of-care diagnostics and data recording. Design consisted of a simple light-emitting diode (LED) illumination source powered by a battery and filtered through an aperture to project holograms of a sample onto the cell phone provided with a complementary metal oxide semiconductor (CMOS) image sensor. The holograms are then compressed and sent over the network to a computer for holographic reconstruction, and the final processed images are sent back to be viewed on the cell phone. These holographic signatures captured by the cell phone permit the reconstruction of the obtained images through a rapid digital processing. The authors report the implementation of this lens free cell-phone microscope by imaging various sized micro-particles, as well as red blood cells, white blood cells, platelets and a waterborne parasite (*Giardia lamblia*). Initially, fluorescence was utilized as an effective method to label samples to improve the resolution of the microscopic imaging. Far along, the incorporation of fluorescence biosensors on smart phone was developed to satisfy fluorescence labelled detections for DNA, virus, and nanoparticles. As an example, we have mentioned the work of Tong Gou et al. (Gou et al. 2018). In order to enable accurate DNA quantitative analysis, they have developed a smartphone-based mobile digital polymerase chain reaction (dPCR) device integrated with thermal cycling control. All functions of their on-chip dPCR are automatically controlled using a customized Android software. The device is able to quantify accurately down to 10 copies of the human 18 S ribosomal DNA fragment inserted in a plasmid and to detect single molecule of cancer biomarker gene CD147 in a low number of HepG2 cells with a comparable analytical precision to the commercial QuantStudio™ 3D dPCR platform. Colour changes from localized surface plasmon resonance (LSPR) of gold nanoparticles can also be quantified by smart phone for bio- and chemical detections. Jinling Zhang and co-workers (Zhang et al. 2018) reported a smartphone label-free biosensor platform based on grating-coupled surface plasmon resonance (GC-SPR) for lipopolysaccharides (LPS, known as endotoxins) detection. The sensor system relies on the smartphone's built-in flash light source and a camera; a disposable sensor chip with Au diffraction grating and a compact disk (CD) as the spectra dispersive unit. The Au grating sensor chip was modified with a synthetic peptide receptor and employed on the GC-SPR with detection limit of 32.5 ng/mL in water. The authors concluded that the sensor shows feasibility for the detection of LPS in commonly used clinical injectable fluids, such as clinical-grade 0.9% sodium chloride intravenous infusion, compound sodium lactate intravenous infusion and insulin aspart. A similar technology was proposed by Y. Liu et al. (Liu et al. 2015). They used a light-guiding silica capillary that is stripped off its cladding, and coated with 50 nm gold film to fabricate a SPR-sensing element. Utilizing a smart application to extract the light intensity information from the camera images, the light intensities of each channel are recorded every 0.5 s with refractive index (RI) changes. The performance of the smart phone-based SPR platform for accurate and repeatable measurements was evaluated by detecting different concentrations of antibody binding to a functionalized sensing element, and the experiment results were validated through commercial SPR instrument contrast tests.

Smartphone, with many excellent capacities, has become linked with portable apparatus as platforms to control, record, and display electrochemical detections. Thus, Amay J. Bandodkar (Bandodkar et al. 2018) reported the fabrication of a new smartphone-based reusable glucose meter. It included a custom-built smartphone case that houses a permanent bare sensor strip, a stylus that is loaded with enzyme-carbon composite pellets, and sensor instrumentation circuits. In addition, the authors developed a custom-designed Android-based software application to enable the easy and clear display of measured glucose concentration. The regular test implicates, loading the software by the operator, utilization of the stylus to dispense an enzymatic pellet on top of the bare sensor strip affixed to the case, and then introducing the sample. The electronic module then acquires and wirelessly transmits the data to the application software to be displayed on the screen. Finally, the deployed pellet is then discarded to regain the fresh bare sensor surface. Obtained findings reveal that the enzyme loaded in the pellets are stable for up to 8 months at ambient conditions, and generate reproducible sensor signals. Electrochemical smartphone-based sensors can also be proposed to the rapid and precise point-of-measurements of ethanol intoxication in small volumes (40 µL) of whole blood sample (Aymerich et al. 2018). The determination of ethanol intoxication in whole blood samples may open the opportunity for a precise and quick point-of-measurement in the ambit of medical emergency or law enforcement. In contrast with traditional techniques based on breath sampling, direct blood measurements present greater immunity to errors especially in case of unconscious or non-collaborative patients. Wearable near field communication (NFC) biosensing systems integrated biosensors on smart phone can be placed directly on or near the body enabling long-term monitoring (Sun and Hall 2019). Rahim Rahimi and co-workers (Rahimi et al. 2018) proposed a laser-enabled fabrication of flexible and transparent pH sensor with NFC for *in-situ* monitoring of wound infection. Biocompatible sensor has a high sensitivity of -55 mV/pH within the physiologically relevant range of pH 4–10, with optical transparency to enable visual inspection of the underlying tissue. The sensor was demonstrated to successfully monitor pH changes in simulated *in vitro* wound conditions infected with gram-positive *Staphylococcus epidermidis*. In addition to health-care applications, NFC labelling technology integrated to smartphones can also be practical to food status monitoring applications in daily life, storage and supply chains. Ma et al. (Ma et al. 2018) developed a high sensitive nanostructured conductive polymer-based NH_3 gas. The gas sensor plays a critical role as a sensitive switch in the circuit of the NFC tag and enables a smartphone to readout meat spoilage when the concentration of biogenic amines is over a present threshold.

More work should be carried out in sensor fabrication, data communications, and processing algorithm on smartphone to improve the performances while maintaining portability and cost effectiveness. The examples highlighted above demonstrated that smartphones were well-thought-out as the expected evolution of point-of-care (POC) analytical devices (Roda et al. 2014).

6.5 DNA Nano-biosensors

At the hand of Human Genome Project (Collins et al. 1998), massive amounts of genetic information that modernized our comprehension and identification of inherited diseases was shared; pathogens responsible for human and animal diseases, bacteria and viruses became visible via their unique nucleic acid sequences and genetically modified organisms (GMOs) have emerged as one of the mainstays of biomedical research. The doors to an enormous potential for the development of new and specific therapeutic procedures, to new drug research and improvement of gene therapy, to novel food technology and environmental sciences, to mention some, were opened. Therefore, a smart nanomaterial that could both measure physicochemical properties and chemically store information within specific DNA sequences has become increasingly important to meet these needs. The development of a simple, rapid and user-friendly method for readout at a specific point in time would give us unprecedented access to complex systems; for example, along currently inaccessible biological processes within cells or underground water reservoirs (Xu et al. 2009, Brotchie 2016). In order to achieve ultrasensitive methods for sequence-specific DNA detection, researchers have developed many different nano-materials to enhance the response of DNA biosensors. Readers can find many examples in the following literature reviews (Xu et al. 2009, Stougaard and Ho 2014, Abu-Salah et al. 2015, Hou et al. 2018, Kukkar et al. 2018, Kurbanoglu et al. 2018). Here we have emphasised on some selected examples who call our attention.

6.5.1 DNA Monitoring in Real Time

A strong point of biosensing lies in the non-invasive detection and real-time monitoring of binding events, such as antibody–antigen, protein-protein, enzyme-substrate or inhibitor, protein-DNA, receptor-drug, protein-polysaccharide, protein-virus, and living cell-exogenous stimuli. Regarding the biological events where DNA is involved, mutagenesis, carcinogenesis, resistance to genotoxic agents, oxidative DNA damage, abasic lesions, base mismatches and non-covalent interactions with drugs, the development of novel devices to understand and monitoring cellular DNA repair status is highly desirable. For example, Alexander Dobrovic and his team, working at the Olivia Newton-John Cancer Research Institute, Heidelberg, Victoria, Australia evaluated the value of droplet digital PCR (ddPCR) technology to assess the potential of using circulating tumour DNA (ctDNA) as a post therapy monitoring tool in melanoma. ddPCR was shown to be reliable in differentiating mutant from wild type alleles with no false positives (Chang-Hao Tsao et al. 2015). On the other hand, it has been well established that nucleic acids possess their intrinsic electrochemical activity, thus electrochemical techniques were applied to detect DNA damage (Fojta et al. 2016). Jason D. Slinker et al. (Slinker et al. 2015) demonstrated that the repair activity of 8-oxoguanine and uracil glycosylases within DNA monolayers can be sensitive and selective electrochemical sensing with a multiplexed analysis is based on silicon chips and electrospun nanofibres. Their approach involves the contrast of the electrochemical signal of redox probe modified monolayers containing the defect versus the rational control of defect-free monolayers, exhibiting a sequence-specific sensitivity thresholds on the order of femtomoles (fM) of proteins and dynamic ranges of over two orders of magnitude for each target. Electrospun fibres are shown to behave similar to conventional gold-on-silicon devices, showing the potential of these low-cost devices for sensing applications. A pivotal role in DNA replication and DNA repair is played by DNA polymerases, Johannes Hohlbein and his research group proposed an adaptable DNA platform for the DNA polymerization monitoring in real time and at the single-molecule level (Fijen et al. 2017). The assay does not require the labelling of DNA polymerases with fluorophores and the experiments on the ensemble level can be carried out with standard spectrophotometric lab equipment. The whole process involved an acceptor labelled DNA primer reinforced to a DNA template that is labelled on its single stranded, downstream overhung with a donor fluorophore. Upon extension of the primer using a DNA polymerase, the projection of the template altered its conformation from a random coil to the canonical structure of double stranded DNA. This conformational change intensifies the distance between the donor and the acceptor fluorophore and can be detected as a decrease in the Förster resonance energy transfer (FRET) efficiency between both fluorophores. Having micro-well plate imaging instruments, the reaching levels of sensitivity are sufficient to detect fluorescence from samples present in fM quantities. Time dependent imaging of increasing donor (or decreasing acceptor) intensity upon polymerization of DNA allows evaluating large DNA polymerase libraries utilising an instantaneous readout.

Another noteworthy case is the colorimetric multipurpose metallic nano-biosensor to detect p53-DNA binding interactions and screen reactivation compounds for mutant p53 proteins performed by Y.N. Tan et al. (Tan et al. 2016). The p53 protein, known as *"the guardian of the Genome"*, is a tumour suppressor protein that plays a central role in cancer biology. It normalizes the gene expression through binding with specific DNA response elements (RE), thereby enabling many important biological functions such as DNA repair and apoptosis. About 50% of human cancers can be associated with mutated p53 proteins that do not bind with the RE. The nano-biosensors utilize the exclusive light absorption and strong scattering properties of gold nanoparticles (AuNPs). The specific binding of wildtype p53 protein to the RE sequence in the DNA-AuNPs alters the interparticle-distance of RE-AuNPs, resulting in a distinct change in solution colour (red-to-blue) as well as of the UV-vis absorption spectra. The system displays an extraordinary sensitivity to detection (in the order of pM), both for the sequence-specific and/or drug activated protein-DNA binding in complex biological samples.

DNA damage can also cause epigenetic changes (Simmons 2008). A new kind of epigenetic modification, which plays key roles in DNA demethylation, genomic reprogramming, and the gene expression in mammals, is the DNA hydroxymethylation (5-hmC). Shixing Chen et al. (Chen et al. 2016) developed a novel multiplexing electrochemical (MEC) biosensor for 5-hmC detection based on the glycosylation modification of 5-hmC and enzymatic signal amplification. The limit of detection of the MEC biosensor is 20 times lower than that of commercial kits based on optical measurement with high detection specificity. Epigenetic changes, particularly in cancer suppressor genes, are also to be utilized as novel biomarkers for cancer diagnostics and therapeutics. Therefore, tumour DNA-specific mutation and methylation are promising biomarkers for non-invasive cancer assessment. Recently, nano-plasmonics has emerged as a platform for one-step dual detection with high sensitivity and specificity of circulating tumour DNA (ctDNA) behaviour (Nguyen et al. 2015, Nguyen and Sim 2015). Finally, yet importantly, we will mention the determination of binding kinetics and affinity of DNA hybridization and single-base mismatches that plays an essential role in systems biology, personalized and precision medicine. The standard tools are optical-based sensors that are hard to translate to low cost and miniaturize platforms for high-throughput quantification. Xu et al. (2009) constructed an integrated, miniaturized, all-electrical multiplexed, graphene-based DNA array that can reliably and sensitively measure the time- and concentration-dependence of DNA hybridization kinetics and affinity. The authors claimed that the system exhibited a detection limit of 10 pM for DNA and that can distinguish single-base mutations quantitatively in real time suggesting a promising future for cost-effective, high-throughput screening of drug candidates, genetic variations and disease biomarkers.

6.5.2 Detecting the Dynamic Cell Reprogramming

Cancer is an epigenetic disease. In fact, epigenetic changes, particularly DNA methylation, are predisposed to change and are suitable candidates to elucidate how specific environmental aspects may raise the risk of cancer. The delicate organization of methylation and chromatin states that regulates the normal cellular homeostasis of gene expression patterns becomes unrecognizable in the cancer cell. The genome of the transformed cell undergoes, simultaneously, a global genomic hypomethylation and a dense hypermethylation of the CpG islands (high frequency of CpG sites in genome, regions of DNA where a cytosine nucleotide is followed by a guanine nucleotide in the linear sequence of bases along its $5' \rightarrow 3'$ direction) associated with gene regulatory regions. These dramatic changes may lead to chromosomal instability, activation of endogenous parasitic sequences, loss of imprinting, illegitimate expression, aneuploidy, and mutations, and may contribute to the transcriptional silencing of tumour suppressor genes. The aberrant CpG island methylation can also be used as a biomarker of malignant cells and as a predictor of their behaviour (Esteller and Herman 2002). Thus, the attainment of abnormalities' estimation in the living cells gene expression of identifying cancer at the cellular level in an early stage before metastasis holds a great promise for increasing the survival of cancer patients (Laird 2003).

DNA methylation can be visualized using the methyl-CpG-binding domain of the human MBD1 protein. The level and distribution of histone modifications can be monitored by two different methods: (i) fluorescence/Förster resonance energy transfer (FRET)-based sensors and (ii) fluorescent labelled antigen binding fragments of specific antibodies (Kimura et al. 2010). Na Li and co-workers (Li et al. 2012) described a multicolour fluorescence nano-probe that consists of gold nanoparticles (Au NPs) functionalized with a dense shell of recognition sequences (synthetic oligonucleotides) hybridized to three short dye-terminated reporter sequences by gold–thiol bond formation, which simultaneously detects three intracellular tumour-related mRNAs. According to the author's declaration, the nano-probe is capable of detecting changes in gene expression levels in cancer cells. Another example that caught our attention was the device constructed by Wang et al. (Wang et al. 2012); they fabricated a nano-sensor by decorating gold nanoparticles (AuNPs) with Raman reporters and hemi-methylated DNA probes. They proposed a novel concept for enzymatic control of plasmonic coupling as a surface enhanced Raman scattering (SERS) nano-sensor for DNA demethylation. Demethylation of DNA probes initiates a degradation reaction of the probes by methylation-sensitive endonuclease Bsh 1236I and single-strand selective exonuclease I. This destabilizes AuNPs and mediates the aggregation of AuNPs, generating a strong plasmonic coupling SERS signal in response to DNA demethylation. The authors claimed that the nano-sensor possesses a high signal-to-noise ratio, super specificity, and rapid, convenient, and reproducible detection with homogeneous, single-step operation providing a useful platform for detecting DNA demethylation and related molecular diagnostics and drug screening. Likewise, using a protein labelling technique, Yuichiro Hori and co-workers created a hybrid probe containing a DNA-binding fluorogen and a methylated-DNA-binding domain. The hybrid probe enhanced fluorescence intensity upon binding to methylated DNA and successfully monitored methylated DNA during mitosis (Hori et al. 2018).

Photo electrochemical sensors, fabricated on photoactive electrodes, which can convert photo irradiation to electrical signal, can also be adapted to DNA methylation detection. In the work proposed by Shiyun Ai and his research team (Yin et al. 2014), they utilized Bi_2S_3 nanorods as photoelectric conversion material along with a kind of methyl binding domain protein, MBD1, as DNA methylation recognizing unit and an anti-his tag antibody to further inhibit the photocurrent and increase the detection sensitivity. Under optimal experimental conditions, the authors declared that their photoelectrochemical immunosensor showed high detection specificity, distinguishing even among single-base mismatched sequences, and that the photocurrent variation was proportional to the logarithm of methylated target DNA concentration from 10^{-9} to 10^{-13} M with a detection limit of 3.5×10^{-14} M and a signal-to-noise ratio of about 3. Recently, Xi Chen et al. (Chen et al. 2019) proposed an ultrasensitive and specific electrochemical biosensor. The platform for DNA methylation detection is comprised of a stem–loop–tetrahedron composite DNA probe anchored at an Au nanoparticle-coated gold electrode, a restriction enzyme digestion of HpaII, and a signal amplification system. Under optimal conditions, the authors stated that their design exhibited specificity, repeatability, and stability. It showed a broad dynamic range of detection from one attomolar (aM) to one (picomolar) pM concentrations with a detection limit of about 0.93 aM.

This few mentioned examples exposed that detecting live-cell epigenetic phenomena and diseases related to DNA methylation is an intensive field of research with a potential future. Recent advances open the way to a host of innovative diagnostic and therapeutic strategies as powerful future tools in the clinic.

6.6 Nano-biosensors in Environmental Practice: Food and Agriculture

We allocate a small space in this chapter to mention the applications of the bio-nanosensors in agriculture, food and the environment. Although the book is oriented towards medicine and innovations introduced from the development of nanotechnology, the impact on the above-mentioned areas directly affects human health and for this reason, we think it is correct to mention it. Today, the advantage of a new technology that operates at the atomic scale, the nanoscale, is assigned to the agri-food sector to detect contaminants such as preservatives, pesticides, antibiotics, heavy metal ions, toxins, microbial load, and pathogens on postharvest foods, water and arable lands. Furthermore, they provided a rapid monitoring of temperature, traceability, humidity, gas, and the aroma of the foodstuff, among other parameters. The application of metal and metal oxide nanoparticles and nanoclusters, of metal and carbon quantum dots, graphene and carbon nanotube, and even nanocomposites formed by the combination of the aforementioned components, expanded the sensitivity of traditional sensors providing signal amplification and novel transduction principles. In this way, the enhanced electrochemical, optical, Raman, catalytic, and superparamagnetic properties introduced on new electronic noses, tongues and nano-barcode technology revolutionized the sensing in food and agricultural areas through multiplex and real-time sensing capabilities (Dubey and Mailapalli 2016, Fuertes et al. 2016, Srivastava et al. 2018). Wireless equipment to remote sensing has been successfully used to measure crop nutrition, crop disease, water deficiency or surplus, weed infestations, insect damage, plant populations, flood management, and many other field conditions (Dasgupta et al. 2017, Kaushal and Wani 2017, Antonacci et al. 2018, Kumar and Ilango 2018). In addition, biological nanosensors have emerged as a pivotal tool for the detection of a particular biological marker with extreme accuracy. They are developed to detect biological species (pathogen or not) or for the identification of growth regulation immunity mechanisms and to enhance innate biological routes (Rigi et al. 2013, Eleftheriadou et al. 2017, Jafarizadeh-Malmiri et al. 2019). Microfluidics technologies applicable in agriculture and animal sciences enable the construction of point of care devices (POCD) designed for nutrients monitoring and plant cells sorting for improving crop quality and production. Likewise, microfluidic chips are functional to check the effective delivery of biopesticides, to simplify *in vitro* fertilization for animal breeding, animal health monitoring, vaccination and therapeutics (Neethirajan et al. 2011).

From those examples, we can conclude that an intelligent use of nanosensors may help food, agriculture and bio-systems industries to achieve the qualitative security and sustainable environment requirements.

6.7 Nano-biosensors: Historical Perspectives, Current Challenges and Prospective

The clinical treatment and management of diseases are greatly dependent on the availability and promptness of their diagnoses. Currently, the conventional methodologies are extremely powerful; however, most of these are limited by time-consuming protocols, costs, and the necessity of higher volume or difficult to extract test samples. In this context, a new evolving technology called "biosensor" shows an enormous potential for an alternative diagnostic device, which constantly compliments the conventional analyses. Thus, biological detection of serum antigens and many metabolic disorders agents have been detected using biosensors since a long-established time, for example in the detection of diabetes, cancer and many allergic responses. In the recent past, the design of biosensors has adapted to each requirement and perceived noteworthy modifications as novel and potential applications emerged. Nowadays, a diversity of nanomaterials are extensively used to improve not only the performance of biosensing mechanism, but also to obtain robust, cheap, and friendly durable fabrication tools. Nanotechnology evolution led to tremendous advances in point-of-care diagnostics (POCD) and to the real-time sensing which are a result of continuous developments in biosensors, microfluidic, bioanalytical platforms, assay formats, lab-on-a-chip technologies, and complementary technologies. The global point-of-care test (POCT) market is projected to reach USD 38.13 billion by 2022 from USD 23.71 billion in 2017, at a CAGR of 10.0% during the forecast period (Markets and Markets 2018). North America accounts for most of the global POCT market, followed by Europe, while Asia-Pacific's POCT market is expected to be one that grows the most at a CAGR of 14.2%. In the year 2016, the principal players were Roche (Switzerland), Abbott (US), and Siemens (Germany), respectively; other companies involved were Instrumentation Laboratory (US), Nova (US), PTS Diagnostics (US), Sekisui Diagnostics (US), Quidel (US), EKF Diagnostics (UK), Chembio Diagnostics (US), Trinity Biotech (US), and AccuBioTech (China). The high prevalence of infectious diseases in developing countries, the increasing incidence of target diseases, the rising preference for home healthcare across the globe, and the growing government initiatives supporting the adoption of POC diagnostic products are expected to drive the demand for point-of-care diagnostic products to high volumes in the not too distant future.

The advances in developing biosensors, which are sensitive and specific, have opened new opportunities for DNA biosensors. Based on nucleic acid recognition sequence methods, real time detection using DNA biosensors drive global health system towards the assay of rapid, simple and economical testing of epigenetic and infectious diseases. DNA biosensors and gene chips are of major interest due to their great potential for obtaining sequence-specific information in a faster, simpler and cheaper manner compared to the traditional hybridization. Although basic research is still necessary, an increasing interest in DNA based sensors can be expected, together with a commercial production of these devices and their extensive use, in a near future.

In addition to the medical applications, nano-biosensors can also have an impact on precision agriculture, food safety and quality control as well as the efficient consumption and reuse of waste and water sources. However, we cannot deny that for the true and practical implementation of such technology, it is necessary to perform a reliable risk-benefit assessment, as well as a full cost accounting evaluation. We believe that the benefits will outweigh the costs and that we will soon have this new technology accessible to the public market.

References

Abu-Salah, K., M. Zourob, F. Mouffouk, S. Alrokayan, M. Alaamery and A. Ansari. 2015. DNA-based nanobiosensors as an emerging platform for detection of disease. Sensors 15(6): 14539-14568.

Antonacci, A., F. Arduini, D. Moscone, G. Palleschi and V. Scognamiglio. 2018. Nanostructured (Bio)sensors for smart agriculture. Trends Anal. Chem. 98: 95-103.

Aymerich, J., A. Márquez, L. Terés, X. Muñoz-Berbel, C. Jiménez, C. Domínguez, et al. 2018. Cost-effective smartphone-based reconfigurable electrochemical instrument for alcohol determination in whole blood samples. Biosens. Bioelectron. 117: 736-742.

Bandodkar, A. J., S. Imani, R. Nuñez-Flores, R. Kumar, C. Wang, A. M. V. Mohan, et al. 2018. Re-usable electrochemical glucose sensors integrated into a smartphone platform. Biosens. Bioelectron. 101: 181-187.

Bradford, M. M. 1976. A rapid and sensitive method for the quantitation of microgram quantities of protein utilizing the principle of protein-dye binding. Anal. Biochem. 72(1-2): 248-254.

Brotchie, A. 2016. DNA nanosensors: Lab in a particle. Nat. Rev. Mater. 1(3): 16015.

Cammann, K. 1977. Bio-sensors based on ion-selective electrodes (Biosensoren auf der Grundlage von ionenselektiven Elektroden). Fresenius' Zeitschrift für Analytische Chemie 287(1): 1-9.

Campàs, M., R. Carpentier and R. Rouillon. 2008. Plant tissue- and photosynthesis-based biosensors. Biotechnol. Adv. 26(4): 370-378.

Castellanos-Rizaldos, E., D. G. Grimm, V. Tadigotla, J. Hurley, J. Healy, P. L. Neal, et al. 2018. Exosome-based detection of EGFR T790M in plasma from non-small cell lung cancer patients. Clin. Cancer Res. 24(12): 2944-2950.

Collins, F. S., A. Patrinos, E. Jordan, A. Chakravarti, R. Gesteland and L. Walters. 1998. New goals for the US human genome project: 1998-2003. Science 282(5389): 682-689.

Cui, Y., Q. Wei, H. Park and C. M. Lieber. 2001. Nanowire nanosensors for highly sensitive and selective detection of biological and chemical species. Science 293(5533): 1289-1292.

Chamorro-Garcia, A. and A. Merkoçi. 2016. Nanobiosensors in diagnostics. Nanobiomedicine (Rij). 3: 1849543516663574.

Chang-Hao Tsao, S., J. Weiss, C. Hudson, C. Christophi, J. Cebon, A. Behren, et al. 2015. Monitoring response to therapy in melanoma by quantifying circulating tumour DNA with droplet digital PCR for BRAF and NRAS mutations. Sci. Rep. 5: 11198.

Chen, S., Y. Dou, Z. Zhao, F. Li, J. Su, C. Fan, et al. 2016. High-sensitivity and high-efficiency detection of DNA hydroxymethylation in genomic DNA by multiplexing electrochemical biosensing. Anal. Chem. 88(7): 3476-3480.

Chen, X., J. Huang, S. Zhang, F. Mo, S. Su, Y. Li, et al. 2019. Electrochemical biosensor for DNA methylation detection through hybridization chain-amplified reaction coupled with a tetrahedral DNA nanostructure. ACS Appl. Mater. Interfaces. 11(4): 3745-3752.

Cheng, Y., J. Xie, K. -H. Lee, R. L. Gaur, A. Song, T. Dai, et al. 2018. Rapid and specific labeling of single live Mycobacterium tuberculosis with a dual-targeting fluorogenic probe. Sci. Transl. Med. 10(454): eaar4470.

Dasgupta, N., S. Ranjan and C. Ramalingam. 2017. Applications of nanotechnology in agriculture and water quality management. Environ. Chem. Let. 15(4): 591-605.

Di Giusto, D. A., W. A. Wlassoff, J. J. Gooding, B. A. Messerle and G. C. King. 2005. Proximity extension of circular DNA aptamers with real-time protein detection. Nucleic Acids Res. 33(6): e64.

Duan, X., Y. Li, N. K. Rajan, D. A. Routenberg, Y. Modis and M. A. Reed. 2012. Quantification of the affinities and kinetics of protein interactions using silicon nanowire biosensors. Nat. Nanotechnol. 7(6): 401-407.

Dubey, A. and D. R. Mailapalli. 2016. Nanofertilisers, nanopesticides, nanosensors of pest and nanotoxicity in agriculture. Sustainable agriculture reviews. Springer: 307-330.

Eleftheriadou, M., G. Pyrgiotakis and P. Demokritou. 2017. Nanotechnology to the rescue: Using nano-enabled approaches in microbiological food safety and quality. Curr. Opin. Biotechnol. 44: 87-93.

Esteller, M. and J. G. Herman. 2002. Cancer as an epigenetic disease: DNA methylation and chromatin alterations in human tumours. J. Pathol. 196(1): 1-7.

Ewald, J. C., S. Reich, S. Baumann, W. B. Frommer and N. Zamboni. 2011. Engineering genetically encoded nanosensors for real-time in vivo measurements of citrate concentrations. PLoS One 6(12): e28245.

Fijen, C., A. M. Silva, A. Hochkoeppler and J. Hohlbein. 2017. A single-molecule FRET sensor for monitoring DNA synthesis in real time. Phys. Chem. Chem. Phys. 19(6): 4222-4230.

Fojta, M., A. Daňhel, L. Havran and V. Vyskočil. 2016. Recent progress in electrochemical sensors and assays for DNA damage and repair. Trends Anal. Chem. 79: 160-167.

Forlenza, O. V., M. Radanovic, L. L. Talib, I. Aprahamian, B. S. Diniz, H. Zetterberg, et al. 2015. Cerebrospinal fluid biomarkers in Alzheimer's disease: Diagnostic accuracy and prediction of dementia. Alzheimers Dement. 1(4): 455-463.

Foudeh, A. M., T. F. Didar, T. Veres and M. Tabrizian. 2012. Microfluidic designs and techniques using lab-on-a-chip devices for pathogen detection for point-of-care diagnostics. Lab on a Chip 12(18): 3249-3266.

Fuertes, G., I. Soto, R. Carrasco, M. Vargas, J. Sabattin and C. Lagos. 2016. Intelligent packaging systems: Sensors and nanosensors to monitor food quality and safety. J. Sensors 2016: 1-8.

Gao, N., T. Gao, X. Yang, X. Dai, W. Zhou, A. Zhang, et al. 2016. Specific detection of biomolecules in physiological solutions using graphene transistor biosensors. Proc. Natl. Acad. Sci. U.S.A. 113(51): 14633.

Gou, T., J. Hu, J. Wu, X. Ding, S. Zhou, W. Fang, et al. 2018. Smartphone-based mobile digital PCR device for DNA quantitative analysis with high accuracy. Biosens. Bioelectron. 120: 144-152.

Gubala, V., L. F. Harris, A. J. Ricco, M. X. Tan and D. E. Williams. 2011. Point of care diagnostics: Status and future. Anal. Chem. 84(2): 487-515.

Hernández-Neuta, I., F. Neumann, J. Brightmeyer, T. Ba Tis, N. Madaboosi, Q. Wei, et al. 2019. Smartphone-based clinical diagnostics: Towards democratization of evidence-based health care. J. Intern. Med. 285(1): 19-39.

Hori, Y., N. Otomura, A. Nishida, M. Nishiura, M. Umeno, I. Suetake, et al. 2018. Synthetic-molecule/protein hybrid probe with fluorogenic switch for live-cell imaging of DNA methylation. J. Am. Chem. Soc. 140(5): 1686-1690.

Hosseini, S., P. Vázquez-Villegas, M. Rito-Palomares and S. O. Martinez-Chapa. 2018. Advantages, disadvantages and modifications of conventional ELISA. pp. 67-115. *In*: Enzyme-linked Immunosorbent Assay (ELISA)): From A to Z. Springer, Singapore.

Hou, Y., J. Wang, J. Jin and C. Shang. 2018. Research and Application of Nano-Fluorescent Materials based on DNA. IOP Conference Series: Materials Science and Engineering, IOP Publishing.

Ip, K., J. Chang, T. Liu, H. Dou and G. Lee. 2018. An integrated microfluidic system for identification of live mycobacterium tuberculosis by real-time polymerase chain reaction. *In*: 2018 IEEE Micro Electro Mechanical Systems (MEMS). IEEE. 1124-1127.

ISO 22870:2006. 2006. Point-of-care testing (POCT) – Requirements for quality and competence, International Organization for Standardization, Geneva, Switzerland: 11.

Jafarizadeh-Malmiri, H., Z. Sayyar, N. Anarjan and A. Berenjian. 2019. Nano-sensors in food nanobiotechnology. pp. 81-94. *In*: Nanobiotechnology in Food: Concepts, Applications and Perspectives, Springer, Cham, Switzerland.

Jing, W., X. Jiang, W. Zhao, S. Liu, X. Cheng and G. Sui. 2014. Microfluidic platform for direct capture and analysis of airborne mycobacterium tuberculosis. Anal. Chem. 86(12): 5815-5821.

Kaushal, M. and S. P. Wani. 2017. Nanosensors: Frontiers in precision agriculture. pp. 279-291. *In*: Nanotechnology. Springer, Singapore.

Kaushik, A., R. D. Jayant, S. Tiwari, A. Vashist and M. Nair. 2016. Nano-biosensors to detect beta-amyloid for Alzheimer's disease management. Biosens. Bioelectron. 80: 273-287.

Khunti, K. 2010. Near-patient testing in primary care. Br. J. Gen. Pract. 60(572): 157-158.

Kim, D. 2016. Progress in Developing a Clinically Viable Diagnostics Using Whispering Gallery Mode Resonators. Doctoral dissertation. University of Kansas.

Kimura, H., Y. Hayashi-Takanaka and K. Yamagata. 2010. Visualization of DNA methylation and histone modifications in living cells. Curr. Opin. Cell Biol. 22(3): 412-418.

Ko, J., E. Carpenter and D. Issadore. 2016. Detection and isolation of circulating exosomes and microvesicles for cancer monitoring and diagnostics using micro-/nano-based devices. Analyst. 141(2): 450-460.

Ko, P. J., R. Ishikawa, T. Takamura, Y. Morimoto, B. Cho, H. Sohn, et al. 2011. Porous-silicon photonic-crystal platform for the rapid detection of nano-sized superparamagnetic beads for biosensing applications. Nanosci. Nanotechnol. Lett. 3(5): 612-616.

Kukkar, M., G. C. Mohanta, S. K. Tuteja, P. Kumar, A. S. Bhadwal, P. Samaddar, et al. 2018. A comprehensive review on nano-molybdenum disulfide/DNA interfaces as emerging biosensing platforms. Biosens. Bioelectron. 107: 244-258.

Kumar, S. A. and P. Ilango. 2018. The impact of wireless sensor network in the field of precision agriculture: A review. Wirel. Pers. Commun. 98(1): 685-698.

Kurbanoglu, S., S. A. Ozkan and A. Merkoçi. 2018. Electrochemical Nanobiosensors in Pharmaceutical Analysis. Vol 2, pp. 302-353. *In*: Rahman, A., S. A. Ozkan and R. Ahmed (eds). Novel Developments in Pharmaceutical and Biomedical Analysis Betham Books. Sharjah, U.A.E.

Laird, P. W. 2003. The power and the promise of DNA methylation markers. Nat. Rev. Cancer 3(4): 253-266.

Lewis, J. M., A. D. Vyas, Y. Qiu, K. S. Messer, R. White and M. J. Heller. 2018. Integrated analysis of exosomal protein biomarkers on alternating current electrokinetic chips enables rapid detection of pancreatic cancer in patient blood. ACS Nano 12(4): 3311-3320.

Li, N., C. Chang, W. Pan and B. Tang. 2012. A multicolor nanoprobe for detection and imaging of tumor-related mRNAs in living cells. Angew. Chem., Int. Ed. 51(30): 7426-7430.

Li, X. and X. Liu. 2016. A microfluidic paper-based origami nanobiosensor for label-free, ultrasensitive immunoassays. Adv. Healthc. Mater. 5(11): 1326-1335.

Liang, L., L. Jin, Y. Ran, L. -P. Sun and B. -O. Guan. 2018. Fiber Light-Coupled Optofluidic Waveguide (FLOW) immunosensor for highly sensitive detection of p53 protein. Anal. Chem. 90(18): 10851-10857.

Liu, C., X. Xu, B. Li, B. Situ, W. Pan, Y. Hu, et al. 2018. Single-exosome-counting immunoassays for cancer diagnostics. Nano Lett. 18(7): 4226-4232.

Liu, H., X. Zhou, Q. Shen and D. Xing. 2018. Paper-based electrochemiluminescence sensor for highly sensitive detection of amyloid-β oligomerization: Toward potential diagnosis of Alzheimer's disease. Theranostics 8(8): 2289-2299.

Liu, J., J. Bu, W. Bu, S. Zhang, L. Pan, W. Fan, et al. 2014. Real-time in vivo quantitative monitoring of drug release by dual-mode magnetic resonance and upconverted luminescence imaging. Angew. Chem., Int. Ed. 53(18): 4551-4555.

Liu, Y., Q. Liu, S. Chen, F. Cheng, H. Wang and W. Peng. 2015. Surface plasmon resonance biosensor based on smart phone platforms. Sci. Rep. 5: 12864.

Ma, Z., P. Chen, W. Cheng, K. Yan, L. Pan, Y. Shi, et al. 2018. Highly sensitive, printable nanostructured conductive polymer wireless sensor for food spoilage. Detection. Nano Lett. 18(7): 4570-4575.

Mai, T. D., D. Ferraro, N. Aboud, R. Renault, M. Serra, N. T. Tran, et al. 2018. Single-step immunoassays and microfluidic droplet operation: Towards a versatile approach for detection of amyloid-β peptide-based biomarkers of Alzheimer's disease. Sens. Actuators B. 255: 2126-2135.

Malik, P., V. Katyal, V. Malik, A. Asatkar, G. Inwati and T. K. Mukherjee. 2013. Nanobiosensors: Concepts and variations. ISRN Nanomater. 327435 (2013): 92013.

Markets and Markets. 2018. REPORT CODE MD 2702. Point-of-Care/Rapid Diagnostics Market: 226.

McNaught, A. and A. Wilkinson. 1997. Compendium of Chemical Terminology-IUPAC Recommendations (IUPAC Chemical Data), Wiley.

Mehrotra, P. 2016. Biosensors and their applications – A review. J. Oral Biol. Craniofac. Res. 6(2): 153-159.

Nair, P. R. and M. A. Alam. 2008. Screening-limited response of nanobiosensors. Nano Lett. 8(5): 1281-1285.

Nam, J., Y. Shin, J. K. S. Tan, Y. B. Lim, C. T. Lim and S. Kim. 2016. High-throughput malaria parasite separation using a viscoelastic fluid for ultrasensitive PCR detection. Lab on a Chip. 16(11): 2086-2092.

Narang, J. and C. S. Pundir. 2017. Biosensors: An Introductory Textbook. Pan Stanford.

Nasseri, B., N. Soleimani, N. Rabiee, A. Kalbasi, M. Karimi and M. R. Hamblin. 2018. Point-of-care microfluidic devices for pathogen detection. Biosens. Bioelectron. 117: 112-128.

Nayak, S., N. R. Blumenfeld, T. Laksanasopin and S. K. Sia. 2016. Point-of-care diagnostics: Recent developments in a connected age. Anal. Chem. 89(1): 102-123.

Neethirajan, S., I. Kobayashi, M. Nakajima, D. Wu, S. Nandagopal and F. Lin. 2011. Microfluidics for food, agriculture and biosystems industries. Lab on a Chip 11(9): 1574-1586.

Nguyen, A. H., X. Ma and S. J. Sim. 2015. Gold nanostar based biosensor detects epigenetic alterations on promoter of real cells. Biosens. Bioelectron. 66: 497-503.

Nguyen, A. H. and S. J. Sim. 2015. Nanoplasmonic biosensor: Detection and amplification of dual bio-signatures of circulating tumor DNA. Biosens. Bioelectron. 67: 443-449.

North, J. R. 1985. Immunosensors: Antibody-based biosensors. Trends Biotechnol. 3(7): 180-186.

Ohno, Y., K. Maehashi and K. Matsumoto. 2010. Label-free biosensors based on aptamer-modified graphene field-effect transistors. J. Am. Chem. Soc. 132(51): 18012-18013.

Panich, S., M. Haj Sleiman, I. Steer, S. Ladame and J. B. Edel. 2016. Real-time monitoring of ligand binding to G-quadruplex and duplex DNA by whispering gallery mode sensing. ACS Sensors 1(9): 1097-1102.

Prasad, S. 2014. Nanobiosensors: The future for diagnosis of disease? Configurations. 8: 9.

Preechaburana, P., M. C. Gonzalez, A. Suska and D. Filippini. 2012. Surface plasmon resonance chemical sensing on cell phones. Angew. Chem. 124(46): 11753-11756.

Prickril, B. and A. Rasooly. 2017. Biosensors and Biodetection: Methods and Protocols, Volume 2: Electrochemical, Bioelectronic, Piezoelectric, Cellular and Molecular Biosensors. Springer.

Rahimi, R., U. Brener, S. Chittiboyina, T. Soleimani, D. A. Detwiler, S. A. Lelièvre, et al. 2018. Laser-enabled fabrication of flexible and transparent pH sensor with near-field communication for in-situ monitoring of wound infection. Sens. Actuators, B. 267: 198-207.

Rigi, K., S. Sheikhpour and A. Keshtehgar. 2013. Use of bio sensors in agriculture. Intl. J. Farm. & Alli. Sci. 2(23): 1121-1123.

Roda, A., E. Michelini, L. Cevenini, D. Calabria, M. M. Calabretta and P. Simoni. 2014. Integrating biochemiluminescence detection on smartphones: Mobile chemistry platform for point-of-need analysis. Anal. Chem. 86(15): 7299-7304.

Schöning, M. J. and A. Poghossian. 2018. Label-free Biosensing: Advanced Materials, Devices and Applications. Springer.

Sheridan, C. 2016. Exosome cancer diagnostic reaches market. Nat. Biotechnol. 34(4): 359-360.

Shrivastava, S., W. -I. Lee and N. -E. Lee. 2018. Culture-free, highly sensitive, quantitative detection of bacteria from minimally processed samples using fluorescence imaging by smartphone. Biosens. Bioelectron. 109: 90-97.

Simmons, D. 2008. Epigenetic influence and disease. Nature Education 1(1): 6.

Slinker, J. D., M. McWilliams, F. Anka and K. Balkus (2015). Sensitive and selective real-time electrochemical monitoring of DNA repair (Presentation Recording). Organic Field-Effect Transistors XIV; and Organic Sensors and Bioelectronics VIII, International Society for Optics and Photonics.

Song, C., P. Deng and L. Que. 2018. Rapid multiplexed detection of beta-amyloid and total-tau as biomarkers for Alzheimer's disease in cerebrospinal fluid. Nanomedicine. 14(6): 1845-1852.

Srivastava, A. K., A. Dev and S. Karmakar. 2018. Nanosensors and nanobiosensors in food and agriculture. Environ. Chem. Lett. 16(1): 161-182.

Stern, E., A. Vacic, N. K. Rajan, J. M. Criscione, J. Park, B. R. Ilic, et al. 2010. Label-free biomarker detection from whole blood. Nat. Nanotechnol. 5(2): 138-142.

Stougaard, M. and Y. -P. Ho. 2014. DNA-based Nanosensors for Next-generation Clinical Diagnostics via Detection of Enzyme Activity. Taylor & Francis.

Su, W., X. Gao, L. Jiang and J. Qin. 2015. Microfluidic platform towards point-of-care diagnostics in infectious diseases. J. Chromatogr. A. 1377: 13-26.

Sun, A. C. and D. A. Hall. 2019. Point-of-Care smartphone-based electrochemical biosensing. Electroanalysis 31(1): 2-16.

Syedmoradi, L., M. Daneshpour, M. Alvandipour, F. A. Gomez, H. Hajghassem and K. Omidfar. 2017. Point of care testing: The impact of nanotechnology. Biosens. Bioelectron. 87: 373-387.

Tan, Y. N., X. Zheng and E. Assah. 2016. Nanoparticles-based biosensors for cancer diagnostics and drug screening: A study on tumor suppressor protein-DNA Interactions. ECI Symposium Series (2016). http://dc.engconfintl.org/nanotech_med/38.

Tay, A., A. Pavesi, S. R. Yazdi, C. T. Lim and M. E. Warkiani. 2016. Advances in microfluidics in combating infectious diseases. Biotechnol. Adv. 34(4): 404-421.

Tokel, O., F. Inci and U. Demirci. 2014. Advances in plasmonic technologies for point of care applications. Chem. Rev. 114(11): 5728-5752.

Toren, P., E. Ozgur and M. Bayindir. 2015. Real-time and selective detection of single nucleotide DNA mutations using surface engineered microtoroids. Anal. Chem. 87(21): 10920-10926.

Toren, P., E. Ozgur and M. Bayindir. 2018. Label-free optical biodetection of pathogen virulence factors in complex media using microtoroids with multifunctional surface functionality. ACS Sensors 3(2): 352-359.

Touhami, A. 2014. Biosensors and nanobiosensors: Design and applications. Nanomedicine 15: 374-403.

Tseng, D., O. Mudanyali, C. Oztoprak, S. O. Isikman, I. Sencan, O. Yaglidere, et al. 2010. Lensfree microscopy on a cellphone. Lab on a Chip 10(14): 1787-1792.

Updike, S. and G. Hicks. 1967. Reagentless substrate analysis with immobilized enzymes. Science 158(3798): 270-272.

Varadan, V. K., P. S. Kumar, S. Oh, L. Kegley and P. Rai. 2011. e-bra with nanosensors for real time cardiac health monitoring and smartphone communication.   J . Nanotechnol. Eng. Med. 2(2): 021011.

Vashist, S. K. 2017. Point-of-care diagnostics: Recent advances and trends. Biosensors 7(62): 1-4

Vashist, S. K., M. Schneider and J. H.T. Luong. 2014. Commercial smartphone-based devices and smart applications for personalized healthcare monitoring and management. Diagnostics 4(3): 104-128.

Vollmer, F. and S. Arnold. 2008. Whispering-gallery-mode biosensing: Label-free detection down to single molecules. Nat. Methods 5(7): 591-596.

Wang, J. 2006. Electrochemical biosensors: Towards point-of-care cancer diagnostics. Biosens. Bioelectron. 21(10): 1887-1892.

Wang, S., F. Inci, G. De Libero, A. Singhal and U. Demirci. 2013. Point-of-care assays for tuberculosis: Role of nanotechnology/microfluidics. Biotechnol. Adv. 31(4): 438-449.

Wang, S., Y. Xiao, D. D. Zhang and P. K. Wong. 2018. A gapmer aptamer nanobiosensor for real-time monitoring of transcription and translation in single cells. Biomaterials 156: 56-64.

Wang, Y., C. -H. Zhang, L. -J. Tang and J. -H. Jiang. 2012. Enzymatic control of plasmonic coupling and surface enhanced Raman scattering transduction for sensitive detection of DNA demethylation. Anal. Chem. 84(20): 8602-8606.

Warkiani, M. E., A. K. P. Tay, B. L. Khoo, X. Xiaofeng, J. Han and C. T. Lim. 2015. Malaria detection using inertial microfluidics. Lab on a Chip 15(4): 1101-1109.

Wilson, G. S. and Y. Hu (2000). Enzyme-based biosensors for in vivo measurements. Chem. Rev. 100(7): 2693-2704.

Wiraja, C., D. C. Yeo, M. S. K. Chong and C. Xu. 2016. Nanosensors for continuous and noninvasive monitoring of mesenchymal stem cell osteogenic differentiation. Small 12(10): 1342-1350.

Wiraja, C., D. C. Yeo, K. -C. Tham, S. W. T. Chew, X. Lim and C. Xu. 2018. Real-time imaging of dynamic cell reprogramming with nanosensors. Small 14(17): 1703440.

Xu, K., J. Huang, Z. Ye, Y. Ying and Y. Li. 2009. Recent development of nano-materials used in DNA biosensors. Sensors 9(7): 5534-5557.

Yang, S., Y. Wang and H. Sun. 2015. Advances and prospects for whispering gallery mode microcavities. Adv. Opt. Mater. 3(9): 1136-1162.

Yin, H., B. Sun, Y. Zhou, M. Wang, Z. Xu, Z. Fu, et al. 2014. A new strategy for methylated DNA detection based on photoelectrochemical immunosensor using Bi2S3 nanorods, methyl bonding domain protein and anti-his tag antibody. Biosens. Bioelectron. 51: 103-108.

Yoo, S. M. and S. Y. Lee. 2016. Optical biosensors for the detection of pathogenic microorganisms. Trends Biotechnol. 34(1): 7-25.

Zhai, J., H. Cui and R. Yang. 1997. DNA based biosensors. Biotechnol. Adv. 15(1): 43-58.

Zhang, D. and Q. Liu. 2016. Biosensors and bioelectronics on smartphone for portable biochemical detection. Biosens. Bioelectron. 75: 273-284.

Zhang, J., I. Khan, Q. Zhang, X. Liu, J. Dostalek, B. Liedberg, et al. 2018. Lipopolysaccharides detection on a grating-coupled surface plasmon resonance smartphone biosensor. Biosens. Bioelectron. 99: 312-317.

Zhao, Z., Y. Yang, Y. Zeng and M. He. 2016. A microfluidic ExoSearch chip for multiplexed exosome detection towards blood-based ovarian cancer diagnosis. Lab on a Chip 16(3): 489-496.

Zuo, P., X. Li, D. C. Dominguez and B. -C. Ye. 2013. A PDMS/paper/glass hybrid microfluidic biochip integrated with aptamer-functionalized graphene oxide nano-biosensors for one-step multiplexed pathogen detection. Lab on a Chip 13(19): 3921-3928.

The Mightiness of Nanotechnology: Biomolecular Motors

"…The human body is a living, breathing example of the power of nanotechnology. Almost everything happens at the atomic level. Individual molecules are captured and sorted, and individual atoms in these molecules are shuffled from place to place, building entirely new molecules. Individual photons of light are captured and used to direct the motion of individual electrons through electrical circuits. Molecules are packaged and transported expertly over distances of a few nanometres. Tiny molecular machines, orchestrate all of these nanoscale processes of life…"

The Machinery of Life (David S. Goodsell 2009)

7.1 The Efficient Conversion of Chemical Energy

7.1.1 Molecular-level Machines, Motors and Switches. "…swim in molasses and walk in a hurricane…" (Astumian 2007)

We will begin by giving a definition at the molecular level of the expressions "machine", "engine" and "switch". A system in which a stimulus activates the well-ordered motion of a molecular or sub-molecular constituent relative to another, potentially resulting in a net task, is a "molecular machine". In other words, a molecular-level machine can be defined as an assembly of a discrete number of molecular units designed to perform mechanical-like movements (output) because of an appropriate external stimuli (input) (Balzani et al. 2000). Taking as a base the definition of molecular machine, a "molecular switch" is a type of molecular machine in which the change in the relative positions of the components influences a system as a function of the state of the switch. Therefore, when the switch is returned to its original state, any mechanical work performed by the original switching action will be undone. If, on the other hand, the change in a relative position of the elements influences the system as a function of the trajectory of the components, we are in the presence of a "molecular motor". Unlike the molecular switch, when a molecular motor returns to its original state at the end of the motorcycle, the work performed is completed. The distinction between "switch" and "motor" is important because only the molecular motors can be used to progressively drive systems away from equilibrium (Kassem et al. 2017). The challenge of designing molecular motors lies not with producing motion at the molecular level, but in controlling the directionality of movement. Molecular machines operate in conditions where gravity and inertia are irrelevant, and viscous forces and Brownian motion dominate (Astumian 2007, Kassem et al. 2017). As it was affirmed by R. D. Astumian, to move they must *"swim in molasses"* and to do in a specific direction would be as difficult as *"walking in a hurricane"* (Astumian 2007). The main activities are engaged in the design and synthesis of different types of molecular walkers, covalently linked motors, rotaxanes and catenan-based systems, among others. Nevertheless, it became a recent research focus to delve into the improvement of performance and additional control of directional molecular movement. The latent question is simple: how does this movement become useful functionality?

7.1.2 Mechanical Work at the Nanoscale

A conventional biomolecular motor is a protein that uses the Gibbs' free energy from ATP (adenosine tri-phosphate) or from a further nucleotide triphosphate (NTP) hydrolysis as its fuel to produce mechanical force (Bachand et al. 2014). In living organisms, biomolecular motors are involved in processes that range from the cytoskeletal kinesin, dynein, and myosin transport function outside of their native environment to the dissipative self-assembly of biological, biomimetic, and hybrid nanostructures that exhibit non-equilibrium behaviours such as self-healing, see Table 7.1. A widespread common skill exhibited by most, if not of all, is their ability to undergo complex chemo-mechanical conformational changes that, in turn, generates a highly efficient conversion ($\eta > 50\%$) of chemical energy into mechanical work. The knowledge regarding the exact mechanisms of chemo-mechanical conversion and the collective behaviours of biomolecular motors in physiological processes that span from the molecular and nanometre-level to the organismal and mesoscale remains unknown and it is an active area of research. However, the investigation along the past 30 years has established a considerable scientific understanding of biochemistry, biophysics, and physiological structure – function relationships that support biomolecular motors extraordinary properties; those are summarized and reviewed elsewhere (Alberts et al. 2008, Lodish et al. 2008, De et al. 2015).

The comprehension of the biomolecular motors' role *in vivo* stimulates the experimental study of hybrid systems, where bio-inspired functions that include membrane transporters, rotary, nucleic acid and cytoskeletal motors rely on the nanoscale generation of machine-driven commands; this is what this chapter is about. The earliest examples of integrated, biomolecular-powered nano-systems comprise rotary devices powered by ATP synthase and kinesin-powered molecular vehicles (Huang and Juluri 2008). Then, and following the advances in the ability to link biomolecular components with engineered materials and devices, the development of more radical systems such as polymerase-powered nano-devices that move along DNA tracks, optoelectronic devices based on bacteriorhodopsin ion transporters,

Table 7.1. Examples of biomolecular motors (Wikipedia 2018)

Protein	Function
Cytoskeletal motors	
Myosins	They are responsible for muscle contraction, intracellular cargo transport, and producing cellular tension.
Kinesin	They move cargo inside cells away from the nucleus along microtubules.
Dynein	It produces the axonemal beating of cilia and flagella and also transports cargo along microtubules towards the cell nucleus
Polymerisation motors	
Actin	Polymerization generates forces and can be used for propulsion. ATP is used. Microtubule polymerization using guanosine triphosphate (GTP).
Dynamin	It is responsible for the separation of clathrin buds from the plasma membrane. GTP is used instead of ATP.
Rotary motors	
FoF1-ATP synthase	Such family of proteins convert the chemical energy in ATP to the electrochemical potential energy of a proton gradient across a membrane or the other way around. The catalysis of the chemical reaction and the movement of protons are coupled to each other via the mechanical rotation of parts of the complex. This is involved in ATP synthesis in the mitochondria and chloroplasts as well as in pumping of protons across the vacuole membrane.
Bacterial flagellum	It is responsible for the swimming and tumbling of, for example, the *Escherichia coli*. In others, bacteria acts as a rigid propeller that is powered by a rotary motor. This motor is driven by the flow of protons across a membrane, possibly using a similar mechanism to that found in the F_o motor in ATP synthase.
Nucleic acid motors	
RNA polymerase	This machinery transcribes RNA from a DNA template.
DNA polymerase	It turns single-stranded DNA into double-stranded DNA.
Helicases	Fuelled by ATP, they separate double strands of nucleic acids prior to transcription or replication
Topoisomerases	Powered by ATP, they reduce supercoiling of DNA in the cell
RSC and SWI/SNF complexes	They remodel chromatin in eukaryotic cells; ATP is used.

(Contd.)

SMC proteins	They are responsible for chromosome condensation in eukaryotic cells
Viral DNA packaging motors	They inject viral genomic DNA into capsids as part of their replication cycle, packing it very tightly. Several models have been put forward to explain how the protein generates the force required to drive the DNA into the capsid. An alternative proposal is that, in contrast to all other biological motors, the force is not generated directly by the protein, but by the DNA itself. In this model, ATP hydrolysis is used to drive protein conformational changes that alternatively dehydrate and rehydrate the DNA, cyclically driving it from B-DNA to A-DNA and back again. A-DNA is 23% shorter than B-DNA, and the DNA shrink/expand cycle is coupled to a protein-DNA grip/release cycle to generate the forward motion that propels DNA into the capsid.
Enzymatic motors	
Catalase	In nearly all living organisms exposed to oxygen, it catalyzes the decomposition of hydrogen peroxide to water and oxygen. It is a very important enzyme in protecting the cell from oxidative damage by reactive oxygen species (ROS).
Urease	Catalyzes the hydrolysis of urea into carbon dioxide and ammonia. Urease is also found in mammals and humans that is considered very harmful to mammals due to production of the toxic ammonia product in the mammalian cells. Mammalian cells do not produce urease; however, the sources are the various bacteria in the body, specifically in the intestine
Aldolase	It is an enzyme that catalyses the reversible reaction that splits the aldol, fructose 1, 6-bisphosphate, into the triose phosphates dihydroxyacetone phosphate (DHAP) and glyceraldehyde 3-phosphate (G3P).
Hexokinase	It is an enzyme that phosphorylates hexoses (six-carbon sugars), forming hexose phosphate. In most organisms, glucose is the most important substrate of hexokinases, and glucose-6-phosphate is the most important product. Hexokinase possesses the ability to transfer an inorganic phosphate group from ATP to a substrate.
Phosphoglucose isomerase	Encoded by the GPI gene on chromosome 19, this enzyme has been identified as a "moonlighting protein" based on its ability to perform mechanistically distinct functions.
Phosphofructokinase	A kinase enzyme that phosphorylates fructose 6-phosphate in glycolysis.

ATP synthase-powered DNA sensors, and swimming lipid vesicles powered by the rotary flagellar motor of a bacteria were further allowed (Bachand et al. 2014). In addition, the combination of microfluidic flow and cytoskeletal transport was used to process and detect analytes (Bachand et al. 2014). Substituting microfluidics by means of transport motorised by molecular engines allows the inclusion of an ATP energy source into the assay solution. This opens up the opportunity to design highly integrated, miniaturized, independent detection nano-devices where the target analytes can be actively captured, sorted, and transported for autonomous sensing and selective requests. Such devices, in turn, may allow fast and cheap on-site diagnosis of diseases and detection of environmental pathogens and toxins (Korten et al. 2010).

In this chapter, we will give a description of biomolecular motors, from a biological to a synthetical point of view. Then we will move forward to their implementation. We will state a wide range of unique functionalities that are presently limited to living systems and support the development of nanoscale systems for addressing critical engineering challenges. Specifically, we will focus on the two major areas of the *ex-vivo* use of nano-biomolecular motors technology (Bachand et al. 2014): (i) dissipative self-assembly of nanomaterials, and (ii) autonomous, nano-fluidic sensing and diagnostic devices. Finally, we will discuss their application to our current and future nanomedicine. How they can be used as molecule-sized robots that work in molecular factories where small, but complex arrangements are made on tiny assembly lines? How can they be applied to build networks of molecular conductors and transistors for their use as electrical circuits? Biomolecular motors could form the basis of bottom-up approaches for constructing, active structuring and maintenance at the nanometre scale. They can be controlled to continually patrol inside an "adaptive" organism and repair it when necessary. Alberts Hibbs's *"swallowable surgeon"* evolved from *"... not such a bad idea"* according to the Feynman´s concept (Feynman 1993) to take place as a real possibility.

7.2 Nature's Biomolecular Motors

Nature has equipped the living organisms with a variety of molecular machines, predominantly multi-subunit proteins that directly convert the chemical energy stored in ATP into directional movement and thus produce mechanical work: the biomolecular motors. Through them, the cells perform complex automated missions such as division, intracellular transport and sensing functions (Van den Heuvel and Dekker 2007, Bromley et al. 2009). Thinking about the cell as a factory and its biomolecular motors as the machinery enclosed inside, the mightiness of nanotechnology at its maximum expression can be easily appreciated. Through biomolecular motors and operating with similar principles as solar cells or batteries, a cell can create a full copy of itself in less than an hour, can proofread and repair errors in its own DNA, sense its environment and respond to it, change its shape, morphology and transport itself, obtaining energy from photosynthesis or the metabolism circuits (Van den Heuvel and Dekker 2007). Accordingly, many of the emergent and distinguishing phenomena found in living organisms are enabled via the action of this exquisite group of proteins. One fundamental class of biomolecular motors are the cytoskeletal motor proteins: the power of movement at the nanoscale living world. The cytoskeleton of eukaryotic cells possesses three types of proteinaceous polymer filaments: actin, microtubule (MT), and intermediate filaments. They have roles in mechanical stabilization and serve as "pathways" for biomolecular motors enzymes to transport macromolecular cargo and organelles through them (Bachand et al. 2014). These biomolecular motor enzymes are catalytic proteins that contain moving parts and use a source of free energy to direct their motion. There are rotary motors that comprise shafts and bearings, as well as linear motors that move along tracks in a systematic way; examples of both categories have been summarized in reference (Diez et al. 2008). Long-distance transport in cells is powered by linear molecular motors of the kinesin, myosin, and dynein super-families, which move along microtubules or actin filaments. Specimens of each superfamily have been characterized biochemically, structurally, and biophysically in some detail (Goldstein 1991, Kull et al. 1998, DiPaola and Giuliani 2018). Among them, the best-studied motors are myosins moving along actin filaments, *actomyosin systems*, and kinesin moving along microtubules, *kinesin–microtubule systems* (Korten et al. 2013). Members of myosin, kinesin and dynein cytoskeletal motors are involved in organelle and vesicle transport and, in certain types of cell, such motors as conventional kinesin and cytoplasmic dynein can latch onto their cargo via integral membrane proteins. In neurons, for example, the kinesin light chains bind amyloid precursor protein (APP), a transmembrane protein of certain axonally transported vesicles. APP has gained fame as the precursor of a proteolytic fragment that gives rise to amyloid plaques in patients with Alzheimer's disease, this being connexion of potential medical significance. In addition to myosin and kinesin, linear motors also include enzymes that move along DNA and RNA. On the other hand, we have the rotary motors (Kottas et al. 2005) among which we can include the ATP synthase that mediates the proton flow through the cellular membrane driving the synthesis of ATP (Balzani et al. 2000), as well as bacterial flagellum the motor that drives bacteria motility (Diez et al. 2008, Minamino et al. 2008). They generate torque via the rotations of a central core within a larger protein complex. Coordinated protein-directed intracellular transport was found to be responsible for a distributed, motor-based mechanism of gliding motility on bacteria (Sun et al. 2011) and for the ability of organisms, such as certain species of fish, to change colour at the mesoscopic level (Alberts et al. 2008), demonstrating how the adaptive organization of materials can be achieved from active transport at the molecular scale. Among biomolecular motors, we cannot fail to mention the protein machinery involved in copying, transcribing and packaging DNA: the cooperative actions of the DNA polymerase, helicase, and topoisomerase biomolecular motors in order to replicate its genomic DNA prior to cell division (Alberts et al. 2008). RNA polymerases (which synthesize RNA from a DNA template) and helicases (which unwind the double helix to provide single-stranded templates for polymerases) have evolved as motors capable of moving along torsional constrained DNA molecules. DNA-binding proteins can use the polymer's electrostatic potential to cling to DNA while they diffuse along the molecule in search of their target sequences. Topoisomerases break and re-join the DNA to relieve torsional strain that accumulates ahead of the replication fork (Bustamante et al. 2003). Great intricacy biomolecular motors have also evolved to ensure that DNA is replicated with high fidelity, and mismatch repair activities exist to remove the rare misincorporated residues that have escaped proofreading during replication preventing an unacceptably high mutation rate which is a challenge to the cell (Lindahl 1982). Among them we have: DNA Glycosylases (Zhu 2009), AP Endonucleases (Wallace 1988) and UV Endonuclease (Demple and Linn 1980). Since both bacteria and growing mammalian cells contain long stretches of single-stranded DNA, another highly studied protein machinery was comprised by restriction enzymes. These enzymes are unusual molecular motors that bind specifically to DNA and then move the rest of the DNA through this bound complex (Zlatanova and Van Holde 2015); their discovery and manipulation were essential for creating any recombinant DNA molecule and for the birth of modern microbial bio-nanotechnology, please refer to chapter 4.

Crucial to all examples is the ability of biomolecular motors to push the system away from equilibrium dynamics through the dissipation of energy, which in turn enables an array of singularities: like self-healing, that is a precise phenomenon limited to living organisms (Fialkowski et al. 2006). Thereby, if we talk about the ability of biomolecular motors to push the system away from equilibrium dynamics through the dissipation of energy, we cannot forget the mitotic spindle. The mitotic spindle is a highly dynamic molecular machine composed of tubulin, motors, and other molecules that consume energy and resources via irreversible processes to sustain their structure and function. It assembles around the chromosomes and distributes the duplicated genome to the daughter cells during mitosis. Even though the specific biochemical and physical principles that govern the assembly of this machine are still unclear, collected findings show that chromosomes play a key role and its function vary according to the involved species (Karsenti and Vernos 2001).

Therefore, nature provides a vast library of biomolecular motors, optimized by billions of years of evolution. Nevertheless, relatively few of these have been applied as molecular actuators in hybrid and/or integrated nanoscale systems. At the present day, we can only dream with the creation of machines of similar size that hold just a fraction of the functionality of these natural wonders, or to utilize these biological nano-machines in artificial environments outside the cell to perform tasks that we design to our benefit. The development of nanoscale systems for addressing critical engineering challenges will continue to enable a wide range of unique functionalities that are presently limited to living systems, and support the implementation of biomolecular motors. Some of them are highlighted in the following examples.

7.3 Synthetic Control of Biological Processes via Proteins Nano-machines: Can They Provide us with the Inspiration to Mimic Bio-components or Design Artificial Motors on Comparable Scales?

7.3.1 Transport Motor Proteins

High-tech request of biomolecular motors would have been impossible without the pioneering work of biophysicists and cell biologists interested in the design and role of motor proteins (Vale 2003, Van den Heuvel and Dekker 2007, Hirokawa et al. 2009). Upon studying these motors, their resemblance to machines becomes more and more clear. Thus, biomolecular machines that work along artificial tracks at nanoscale dimensions were fabricated and the protein-powered nano-devices based on linear cytoskeletal kinesin and myosin motors have dominated this emerging field. Kinesin and myosin motors are relatively robust, readily available – purified from cells or expressed in recombinant bacterial systems – and harvested in large quantities (van den Heuvel and Dekker 2007). In their most basic geometry, these ATP-utilizing motor systems are employed in a so-called "gliding motility assay", in which they can transport cytoskeletal filaments (microtubules or actin filaments, of about 1 to 20 mm in length that can be commercially purchased) to functionalized surfaces. Initial studies based on the "gliding motility assay" show immobilized motors and filament movement, i.e. microtubules or actin filaments glide in the presence of ATP across surfaces coated with kinesin or myosin motors, respectively. The proposal of a patterned surface that can engender a directed movement led to the idea of exploiting this transport mechanism in nanotechnology. A significant advantage of the employment of such gliding assays is the absence of the need for external driving forces and the requirement of bulky accessory equipment such as pumps and control devices (Hess and Bachand 2005, Lard et al. 2014). Mercy Lard et al. (Lard et al. 2014) demonstrated that the transport of actin filaments propelled by myosin molecular-motors through Al_2O_3 hollow nanowires (HNWs) of approximately 80 nm inner diameter and with a length of several micrometres could be induced. The system showed a motor-driven transport several orders of magnitude faster than passive diffusion, and can be utilized for the construction of lab-on-a-chip advanced devices and as a scaffold to mimic the muscle sarcomere (Lard et al. 2014). Programmable and highly localized patterns could be generated in a highly reproducible manner and proved to be a reliable guide for gliding microtubules, proposing them as an alternative of patterning techniques like electron beam lithography or AFM based nano-shaving. By the way, using a UV-laser-based ablation technique, Cordula Reuther and co-workers (Reuther et al. 2017) created straight and curved motor tracks of functional kinesin-1 on PLL-g-PEG-coated polystyrene surfaces. Lu Chen and co-workers (Chen et al. 2012) also reported that myosin motors can be engineered to reversibly change the course of their movement in response to calcium concentration. Thus, like nanoscale trucks, the microtubules or actin filaments can act as shuttles that transport an attached cargo such as nanoparticles or DNA molecules (Diez et al. 2003, Bachand et al. 2004).

In an alternative transport system, biomolecular motors can move along cytoskeletal filaments. Thus, controlled motion along specific routes, directionality, external control, and steering are highly required. Biomolecular motors can be used as off-the-shelf components to power hybrid nano-systems (Hess and Saper 2018). These hybrid systems combine biological functional modules from different cytoskeletal systems and synthetic toolbox of the nano-engineer to create a new series of possible biomolecular motors to explore novel applications and dynamic design principles. This tactic allows, for example, inferring about the correspondence between a simple moving function and the architecture of a molecule in an inductive manner. Thus, Furuta and co-workers (Furuta et al. 2017) proposed several combinations of a motor core derived from microtubule – based dynein motors and non-motor actin-binding proteins, to drive down movement along an actin filament. Focusing on the mechanism responsible for directional movement, the authors proposed that the direction of the movement along actin strand can be reversed by merely changing the geometric arrangement of motor core building blocks (Furuta et al. 2017).

Added advanced functionalities can be attained by the integration of protein and nucleic acid components in an engineered nucleoprotein motor (Omabegho et al. 2017). Omabegho et al. (2017) designed, prepared and characterized multimeric myosin motors that incorporate RNA lever arms, which walk along actin filaments at speeds of 10–20 nm s^{-1}. The obtained hybrid assemblies displayed amplified conformational changes in the protein motor domain directed by nucleic acid structures. The RNA lever arm geometry regulated the speed and direction of motor transport that can be dynamically controlled using programmed transitions in the lever arm structure. The proposed designs comprise nucleoprotein motors that reversibly change their direction in response to the strand displacement of RNA sequence-specific variants to oligonucleotide signals. Applications of hybrid motors may include artificial molecular transport systems in programmable molecular robots (Muscat et al. 2011).

7.3.2 Microtubule Arrays Motor Proteins Coupled to Microfluidic Channels Systems

Numerous essential processes in the cell, including transport of subcellular structures like nuclei, chromosomes, organelles and, the self-assembly and positioning of the mitotic spindle, are mediated by active motor proteins moving along rigid microtubule arrangements (Shelley 2016). These reduced systems are also the building blocks of new biosynthetic active schemes that are ideal for driving *in vitro* nano-transport, which permit the controlled movement of a selected cargo along predetermined paths (Clemmens et al. 2004). Hence, motor proteins have been designed to utilize their mechanical functions to manipulate cargo molecules as done by liquid solution in micro total analysis systems (Fujimoto et al. 2013). The active, chemically powered transport of biomolecules by molecular motors is an attractive alternative to the use of micro- and nano-fluidics for a wide range of applications. Microtubules driven by kinesin motors have been utilised as "molecular shuttles" in microfluidic environments with potential applications in autonomous nanoscale manipulations such as capturing, separating, and/or concentrating biomolecules (Fujimoto et al. 2015). Fujimoto and co-workers (Fujimoto et al. 2013), for example, reported a reconstructed cellular kinesin/ dynein-microtubule system as nano-tracks to control massively parallel chemical reactions. Microtubules were integrated into a total molecular analysis system to perform parallel biochemical reactions at the molecular level, where the responsibility for the placement of quantum dots (Q-dots) is the binding of glutathione S-transferase (GST) to glutathione (GSH) and streptavidin binding to biotin. The same author (Fujimoto et al. 2015) described the creation of a microtubule-based transport technique based on a single-micrometre-scale channel array that forms dynamically via pneumatic actuation of a polydimethylsiloxane membrane. The proposed method demonstrated, for the first time, the elimination of contamination caused by the diffusion of target molecules and proved the efficiency and unidirectional transport capacity of kinesin microtubule incorporation on a chip.

Microtubule systems may enhance the detection efficiency of an analytical system or facilitate the controlled assembly of sophisticated nanostructures if transport can be coordinated through complex track networks (Clemmens et al. 2004). Improvement of miniaturized devices for the rapid and sensitive detection of analyte is crucial for various applications across healthcare, pharmaceutical, environmental, and other industries. Particularly, multiple myeloma cells are heterogenous in terms of the expression of CD45 and previous studies have suggested important clinical and biological implications of the CD45 expression. It was demonstrated that the CD45+ population has a higher apoptotic rate compared to CD45 cells and that CD45+ cells represented myeloma cells earlier (Kimlimger et al. 2006). S Chaudhuri and co-workers (Chaudhuri et al. 2017) reported the detection of CD45+ microvesicles derived from leukemia cells and label-free detection of multivalent proteins at sub-nanomolar concentrations by using fluorescently-labelled, antibody conjugated microtubules in a kinesin-1 gliding motility assay. According to the authors, the formation of fluorescent supramolecular assemblies of microtubule bundles and spools in the presence of multivalent analytes can perform a rapid, label-free detection that could find a general application to identify any analyte, including clinically relevant microvesicles and proteins.

7.3.3 Artificial Molecular Motors: Nanoscale Controls Movement Hoping to Mimic Nature

The topic of artificial molecular motors takes inspiration from nature's motor proteins to direct the movement at molecular level. Although these tiny, but powerful, synthetic molecular machines and the directional motion on the nanoscale performed by them is a relatively new development, significant advances have been made (Kassem et al. 2017). Some selected and, in our opinion, meaningful designs of artificial molecular motors and their modes of operation are summarized below.

An outstanding aspect of living systems is their facility to generate motility by magnification of collective molecular motion from the nanoscale up to macroscopic dimensions. Some of nature's protein motors, such as myosin in muscle tissue, consist of a hierarchical supramolecular assembly of very large proteins, in which mechanical stress induces a coordinated movement. Jiawen Chen and co-workers (Chen et al. 2018b) reported a supramolecular system (containing 95% of water) formed by the hierarchical self-assembly of a photo-responsive amphiphilic molecular motor that can recreate the macroscopic contractile muscle-like motion. The molecular motor first assembles into nano-fibres, which further assemble into aligned bundles that make up centimetre-long strings. Irradiation induces rotary motion of the molecular motor. The propagation and accumulation of this motion lead to contraction of the fibres towards the light source. This system supports large-amplitude motion, fast response, precise control over shape, as well as weight-lifting experiments in water and air. This system addresses the fundamental question of whether cooperative non-covalent interactions in water can allow amplification of a photo-induced response along many length scales in order to sustain a macroscopic mechanical motion. Similarly, Quan Li et al. (Li et al. 2015) integrated light-driven unidirectional molecular rotors as reticulating units in a polymer gel, enantiopure polymer–motor conjugates. Through this system, it was possible to amplify their individual motions to achieve macroscopic contraction of the global material, utilizing the incoming light to operate under far-from-equilibrium conditions. Researchers stated that their design could be an initial step towards the integration of molecular motors in metastable materials to store and, eventually, to convert energy. Inspired by the hierarchical organization of myosin and actin filaments found in biological systems that exhibit contraction and expansion conducts to produce work and force, Iwaso and co-workers (Iwaso et al. 2016) reported an improved photo responsive and reversible expansion–contraction wet-type molecular motor built from rotaxane-based units. The authors argued that molecular actuator could be fabricated with an alternating layered structure similar to muscle fibrils and that it could be applied in stents to selectively release drugs or in medical devices capable of embolizing blood vessels adjacent to target cancer cells and malignant myoma.

Writing in *Nature Nanotechnology*, Chuyang Cheng et al. (Cheng et al. 2015) proposed a completely artificial compound that acts on small molecules to create a gradient in their local concentration. The proposed system mimics the function of native carrier proteins, which utilise energy in order to pump ions or molecules across cell membranes, creating concentration gradients. The artificial molecular motor proposed by Chuyang Cheng et al. consumes redox energy and induces organized and precise noncovalent bonding interactions to pump positively charged rings from solution and entangle them around an oligo methylene chain, as part of a kinetically trapped entanglement. The authors claimed that the artificial molecular pump realises an efficient work after two repetitive cycles of operation and drives rings away from equilibrium toward a higher local concentration.

Although biological molecular motors are powered by chemical gradients or the hydrolysis of adenosine triphosphate (ATP), it is difficult to obtain synthetic small-molecule motors that can operate autonomously using chemical energy. David Leigh and co-workers (Wilson et al. 2016) from the University of Manchester, UK, reported a system in which a small molecular ring (macrocycle) is continuously transported directionally around a cyclic molecular track when powered by irreversible reactions of a chemical fuel, 9-fluorenylmethoxycarbonyl chloride. In the same way, Erbas-Cakmak and co-workers (Erbas-Cakmak et al. 2017) described the construction of a chemically-driven artificial rotary and linear molecular motors, where the directional rotation and the transport of substrates away from equilibrium by a linear molecular pump are induced by acid-base oscillations. The acid-base oscillations were induced by aliquots of trichloroacetic acid (Cl_3CCO_2H), a chemical fuel that undergoes base-promoted decarboxylation, generating chloroform ($CHCl_3$) and carbon dioxide (CO2) as the only waste products of motor operation. A single fuel pulse generates 360° unidirectional rotation of up to 87% of crown ethers in a [2] catenane rotary motor. Mimicking natural biomolecular motors that catalyse the hydrolysis of chemical fuels, such as ATP, and use the energy released to direct motion through information ratchet mechanisms, the switch simultaneously controls the binding site affinities and the labilities of barriers on the track, creating an energy ratchet. Those autonomous chemically fuelled molecular motors will find application as engines in molecular nanotechnology.

Certain molecular structures that have controlled rolling properties on surfaces have been termed as nanocars (Sasaki et al. 2008). While not technically motors because they are not self-motorized, they are also illustrative of recent efforts towards synthetic nanoscale machines. Towards the realization of motor-driven nanocars, light-driven (Chiang et al. 2011) and chemically powered (Godoy et al. 2011) molecules have been synthetized and shown to operate in solution phase (García-López et al. 2015, 2016). These facts demonstrate the potential use of such systems to transport cargos from one place to another in several biomedical applications.

7.4 Dissipative Self-assembly of Nanomaterials to Boost Biomolecular Motors

Generally, synthetic materials are in equilibrium. Colloidal chemists have conquered the art of assembling small molecules into complex nanostructures using non-covalent interactions (Israelachvili 2011, della Sala et al. 2017), where the forward and backward rates of assembly and bond formation are balanced. The driving force for self-assembly is thermodynamics: the self-assembled structure is more stable than the separate components. In those conditions, at equilibrium, we understand many of the processes involved because we understand that we are able to control the properties of existing materials or even create new materials with new functions. However, structures and materials may also exist out-of-equilibrium in which there is a net exchange of matter and energy with their environment. In fact, life and the structures it comprises are thermodynamically unstable, are often energetically uphill, and can therefore not exist in equilibrium. To occur, biological self-assembly processes require the consumption of chemical energy (van Rossum et al. 2017). Furthermore, the definition of self-organization as a characteristic of living systems implied the existence of an interplay between organization and function. That is to say, self-organization arises when elements mutually and dynamically co-operate to produce a complete arrangement that acquires emergent properties that cannot be directly predicted from the individual properties of their fundamentals units; in nature, this only happens through the dissipation of ATP or GTP (guanosine triphosphate) (Karsenti 2008). Synthetic chemical systems that operate in the same way are essential for creating the next generation of intelligent, adaptive materials, nano-machines and delivery systems. As an example, we can mention the work of Leonard J. Prins and co-workers (Maiti et al. 2016). They reported a novel strategy for the dissipative self-assembly of functional supramolecular vesicles with high structural complexity, which in turn acted as nano-reactors, driven by the noncovalent interactions with adenosine triphosphate (ATP), which functions as the chemical fuel. The authors demonstrated that the lifetime of the vesicles could be regulated by controlling the hydrolysis rate of ATP and that their function as nano-reactor was possible only as long as chemical fuel is present to keep the system in the out-of-equilibrium state. Another example that we found important to highlight is the self-assembling system designed by Alberto Credi et al. (Ragazzon et al. 2015). They have reported a relative unidirectional transit of a non-symmetric molecular axle through a macrocycle powered solely by light. The molecular machine rectifies Brownian fluctuations by energy and information ratchet mechanisms and can repeat its working cycle under photo-stationary conditions (Ragazzon et al. 2015). Likewise and also using radiation as an energy source, Sadamu Takeda and his collaborators (Ikegami et al. 2016) showed a noncovalent assembly of oleic acid and an azobenzene derivative that, as they said, was the first example of a macroscopic square-wave, self-oscillatory motion system. The assembly flips repeatedly in an autonomous manner under continuous blue-light irradiation.

A general route to dissipative adapting nano-systems exhibiting life-like behaviour was showed by Evan Spruijt and Wilhelm T. S. Huck (te Brinke et al. 2018). Along with several collaborators, they designed and constructed a functional nano-system based on the dissipative self-assembly of a bacterial homologue of tubulin, the FtsZ protein, within coacervate droplets. They have demonstrated that barrier-free compartments govern the local availability of the energy-rich building block guanosine triphosphate, yielding highly dynamic fibrils. Likewise, Qiang Yan and co-workers (Hao et al. 2017) created a life-like polymer micellar system that displayed a periodic and self-adaptive pulsating motion fuelled by ATP. The micelles catching ATPs will deviate away from the thermodynamic equilibrium state, driving a continuous micellar expansion that temporarily breaks the amphiphilic balance, until a competing ATP hydrolysis consumes energy to result in an opposing micellar contraction. As long as ATP energy is supplied to keep the system in out-of-equilibrium, this reciprocating process can be sustained, and the ATP level can orchestrate the rhythm and amplitude of nano-particulate pulsation. This synthetic assembly provides a potential periodic nano-carrier for programmed drug delivery, micro-reactors, and cell membrane mimetics. Last, but not the least, DNA based systems have paved the way for the development of autonomous molecular walkers and nano-motors mediated by fuel and catalytic units (Che et al. 2018). DNA is a perfect component of programming chemically fuelled dissipative self assembly structures. Advantages of the DNA-based systems were exposed by the work of Prof. Leonard J. Prins and Francesco Ricci (Del Grosso et al. 2018). They demonstrated that through DNA-based systems, a perfect control over the activation site for the chemical fuel in terms of selectivity and affinity, highly selective fuel consumption that occurs exclusively in the activated complex, and a high tolerance for the presence of waste products could be attained.

7.5 Molecular Motor-driven Diagnosis

There is a great potential in the way of the application of biomolecular motors in detection and diagnostic devices. This opens up the opportunity to design highly integrated, miniaturized, autonomous detection procedures that, in turn, may allow fast and cheap on-site diagnosis of diseases and recognition of environmental pathogens and contaminants. In chapter 6, many examples were listed around microfluidics machinery and how this technology has greatly improved the efficiency of immunoassays by providing a means for active transport of the analyte through a small detection device. Nevertheless, with the continuous decrease of the detection devices dimensions, it becomes impractical to use microfluidics for active transport because of the need for external pumps and high pressure. The integration of cytoskeletal motor proteins that replace fluidic flow for analyte transport and an energy source

(ATP) into the assay solution that make the device completely independent of external power sources might lead to point of care test (POCT) that should be easy to handle as a urine test strip or a home-use pregnancy proof (Korten et al. 2010).

Writing in *Nature Nanotechnology*, Thorsten Fischer, Ashutosh Agarwal and Henry Hess from the University of Florida reported the construction of a "smart dust" biosensor based on a hybrid micro-device powered by ATP and that depended on antibody-functionalized microtubules and kinesin motors to transport the target analyte into a detection region (Fischer et al. 2009). They demonstrated that the micro-device replaces the "capture–wash–tag–wash–detect" sequence in traditional double antibody sandwich (DAS) assays with a "capture–transport–tag–transport–detect" sequence relying on molecular shuttle. The authors argued that the "capture–transport–tag–transport–detect" sequence can serve as the basis of a smart dust micro-device that can be fabricated in large numbers, stored in an inactivated state, be reconstituted, distributed in an aqueous solution, activated by light, and finally read out by stand-off fluorescence detection (Fischer et al. 2009). The potential of "smart dust" to collect information about any environment in incredible detail could impact plenty of things the the healthcare industry, for example, from diagnostic procedures without surgery to monitoring devices that help people with disabilities to interact with tools that allow them to live independently.

In general, nano-devices based on cytoskeletal molecular motor possess a great aptitude in this regard due to unprecedented level of miniaturization and the absence of the requirement of external propulsion. In this regard, Chih-Ting Lin and co-workers (Lin et al. 2008) created a molecular classifier that works without external power or control by incorporating a kinesin microtubule-based biological motor into a microfluidic channel network to sort, transport, and concentrate molecules. In their devices, the authors presented the use of functionalized microtubules that capture analyte molecules and then are directed along kinesin-coated microchannel tracks toward a collector structure, which concentrates and confines them. The whole device and the assays were extremely robust, presenting a femtomolar (fM) sensitivity by the use of a total volume of detection within the pL range matching the measurement scales to those of eukaryotic cells. More examples can be found in the literature revisions written by Hess and Saper (Hess and Saper 2018) and Jia and Li (Jia and Li 2019). Despite its potential and applicability, the motor function has been found to be limited and subjected to essential constraints imposed by the body fluids conditions (Korten et al. 2013). In the eagerness to solve this problem, Saroj Kumar et al. (Kumar et al. 2013) used commercially available ferromagnetic metal nanoparticles (MP) coated with a thin layer of carboxy (COOH) functionalized carbon graphite in a pre-separation step to avoid the recently discovered deleterious effects of complex body fluid environments on the actin filaments and actomyosin motor function. The low flexural rigidity of the actomyosin motor system can also be used to overcome speed limitation of previously reported microtubule-kinesin based devices that are slow compared to several existing biosensor systems. Using a device designed through optimization by Monte Carlo simulations, Mercy Lard and co-workers (Lard et al. 2013a) demonstrated a rapid nano-separation with unprecedented miniaturization (device area: 0.01 mm²). The passage of a given filament at certain positions along a track is required in some of the applications mentioned throughout this chapter, for example in molecular sorting to enable steering into a certain direction at a junction or to detect changes in velocity during drug delivery or diagnosis. For this purpose, fluorescence interference contrast (FLIC) is used to locally enhance the signal strength in the detector regions (Lard et al. 2013b).

7.6 Self-propelling Motors

Self-propelling motors are a nanotechnology dream, thus researchers are beginning to explore novel macroscopic methodologies for the exploitation of bacteria as mechanical power sources and to harness microorganisms as propelling units for micro-devices (Di Leonardo et al. 2010). Coupled into microfluidic devices, flagellar motors could provide a unique biological means to supply either mechanical or electrical power to these systems with high-power conversion efficiency without sophisticated control electronics since the motors are biologically self-sustained. Additionally, flagellar motors are relatively cost-effective. They can be harvested easily from cell growth using established biological protocols and can live on small amounts of simple nutrients, such as sugars, so no external power source would be required. Since bacteria are equipped with extraordinarily sensitive detectors of physical or chemical stimuli that modulate the direction of rotation of their flagellar motors, it may even be possible to design devices to respond to specific stimuli, such as light or chemical substances (Spetzler et al. 2007). Nevertheless, incorporating flagellar motors into artificial devices is extremely difficult. One of the major obstacles is maintaining the motility of the tethered motors in a micro-fabricated environment. To overcome this, research work has been focused on optimizing the chemo-mechanical behaviour of the motors through genetic engineering and the development of an effective integration scheme for selective motor tethering at designated locations in a microfluidic device (Tung and Kim 2006).

Di Leonardo and co-workers (Di Leonardo et al. 2010) proposed schemes and demonstrated that it was possible, through a careful task of aligning, to bind motile *Escherichia coli* cells along synthetic surfaces in order to have them work cooperatively. The gliding bacterium *Mycoplasma mobile* activates the motor in which a cogwheel-like 20-μm-diameter silicon dioxide rotor rotates on a silicon track in a predefined direction. The micro-rotary motor is fuelled by glucose and displays some of the properties normally attributed to living systems (Hiratsuka et al. 2006).

The magnetic actuation of self-assembled bacterial flagellar nano-robotic swimmers in response to changes in the fluidic environment was reported by Jamel Ali and co-workers (Ali et al. 2017). The design of the nano-swimmers, which can serve as nanoscale probes, consists of one or more bacterial flagella attached to a superparamagnetic nanoparticle (40–400 nm in diameter) via biotin-streptavidin complexes. The key to the fabrication of these nano-machines is the ability of flagella to depolymerize into the flagellin homopolymer and re-polymerize into long flagellar filaments through a 'seeded' polymerization mechanism. Thus, bacterial flagella was induced to change their helical form in response to environmental stimuli, leading to a difference in propulsion before and after the change in flagellar form. In addition, the authors demonstrated the aptitude to steer these devices and induce flagellar bundling in multi-flagellated nano-swimmers. Zihan Huang et al. (Huang et al. 2018), meanwhile, presented a theoretical demonstration of the transport of a bacteria-activated Janus particle driven by chemotaxis. The authors claimed that upon increasing the stimuli intensity, the transport of the Janus particle undergoes an intriguing second-order state transition: from a composite random walk, combining power-law-distributed truncated Lévy flights with Brownian jiggling, to an enhanced directional transport with size-dependent reversal of locomotion. Their method could increase the knowledge of molecular motors, contribute to enhance the approaches to achieve efficient targeted delivery to a specific lesion based on the stimulus-response techniques and become promising as bio-actuator and biosensors. By capturing motile *Escherichia coli* inside electropolymerized microtubes, Morgan M. Stanton and co-workers (Stanton et al. 2017) exploited the motility and sensing capability of biological cells generating a functional micro-machine. The main body of the conical microtube was fabricated with electropolymerized polypyrrole (Ppy) deposited on an Au-coated polycarbonate membrane. After Ppy deposition, microtubes were modified internally with an adsorbed two nano-sized layer of polydopamine (PDA). An additional three layers of Ni nanoparticles for magnetic guidance or urease to act as a chemical trigger could also be added to increase the functionality of the system. The micro-swimmers demonstrated cell guided adhesion capacity, magnetic steerability, and a chemically activated termination switch.

7.7 Enzyme Powered Nano-Motors. Enzymatic Reactions to Active Self-Propulsion of Nano- / Micro-architectures

Enzymes – the workhorse proteins of cells – harness and convert chemically free energy into kinetic forces in order to achieve a catalytic activity. Fundamental studies on enzymes revealed that, in addition to their capacity of energy transformation, the catalytic activity which is carried out through them enhances diffusion at the single-molecule level, which thus gives further evidence that enzymes have been well-considered as nano-machines (Ma et al. 2016). Thus, enlightened by nature's design, scientists explore new avenues around the construction and control of self-propelled catalytic micro- and nano-motors; their findings have brought potential applications closer to biomedical requests such as the active transport and delivery of specific drugs to the site of interest (Ma et al. 2016, Tu et al. 2017). Ana C. Hortelão and co-workers (Hortelão et al. 2018) proposed urease-powered mesoporous silica-based core–shell nano-bots that are able to self-propel in ionic media. Such nano-motors can be efficiently loaded with the anticancer drug doxorubicin (Dox), and in specific operational conditions, they are capable to release and deliver the active principle to HeLa cells. This research group gives a step forward, attaching both PEG and anti-FGFR3 antibody on their urease-powered nano-motors' outer surface to target bladder cancer cells (Hortelão et al. 2018). The autonomous motion is promoted by urea, which acts as fuel and is inherently present at high concentrations in the bladder.

The movement of self-propelled catalytic micro- and nano-motors is generally fully dependent on the concentration of accessible fuel, with propulsive movement only ceasing when the fuel consumption is complete; thus the construction of an effective system with an effective speed-regulation mechanism continues to be a challenge. In this sense, researchers from the Institute for Molecules and Materials at the Radboud University, the Netherlands reported evidence that sustains the control over the movement of self-assembled stomatocyte nano-motors via a molecularly built, stimulus-responsive regulatory mechanism (Tu et al. 2017). The authors chemically attached a temperature-sensitive polymer brush onto the nano-motor surface, whereby the opening of the stomatocytes is enlarged or narrowed according to the temperature variation, controlling, consequently, the access of hydrogen peroxide fuel and, in turn, regulating their movement. Responsiveness of nano-motors to the presence of sarin stimulant, diethyl chlorophosphate (DCP), a nerve agent, was constructed and reported by Virendra V Singh and co-workers (Singh et al. 2016). The enzyme-powered system, poly(3,4-ethylenedioxythiophene)/Au tubular micro-engines functionalized with the catalase, was used for the remote detection of chemical vapour threats changes in the surrounding atmosphere. The proposed vapour detection approach relies on the biocatalytic micromotor's response to plumes of nerve-agent vapour and offers a direct real-time visualization of such chemical warfare agents' emanations by monitoring changes in their propulsion behaviour, associated with the partition of the sarin simulant. Artificial responsive nano-systems could have potential applications in controllable cargo transportation and as detector, but they can also be used as antibacterial systems. Researchers from the Department of Nano-engineering, working at the University of California, San Diego, United States, demonstrated that lysozyme-modified fuel-free nano-motors could be adapted to perform an effective and ′rapid bacterial killing nanotechnology approach (Kiristi et al. 2015). This system combines the effective cleavage of glycosidic bonds of peptidoglycans present in the bacterial cell wall from the lysozyme and ultrasound (US)-propelled porous gold nanowire (p-AuNW) motors.

The preceding examples suggest that, despite the debate on their motion mechanism, nano-machines' self-propelled by enzymes have been highly investigated. At present, enzyme-powered micro/nanomachines have been proven to be useful tools in various proof-of-concept applications, presenting possible solutions for many engineering problems from different fields, such as environmental protection and biosensing; they have a promising future in nanomedicine (Ma et al. 2016).

7.8 Man-Made Nano-machines vs. Nature Bio-motors

"…There is a certain irony when 'nanotechnologists' espouse the molecular world of synthetic chemistry (see Eric Drexler's Engines of Creation) and the possibility of making machines on a nanometre scale when Nature has been doing this very successfully for 3.5 billion years!…" Extract from Royal Irish Academy Medal Lecture provided at the Queen's University of Belfast, on 6 September 2000 by T. J. Mantle from Department of Biochemistry, Trinity College, Dublin, Ireland (Mantle 2001).

Biomolecular motors are the precision sensing and actuation systems engineered by nature over millions of years. Displaying unquestionable advantages over other chemo-mechanical systems, they exhibit, to name a few, no wear characteristics, self-regulating and self-healing capabilities, and very high efficiency of energy conversion. In addition, natural motors are autonomous, i.e. they keep operating, in a constant environment, as long as the energy source is available. By contrast, apart from a few examples, the fuel-powered artificial motors described so far are not autonomous because, after the mechanical movement induced by a chemical input, they need another, opposite chemical input to reset, which also implies generation of waste products and potential risks. Nature biomolecular motors offered exciting examples to scientist and, with the look on them and following the understanding of the action of such biological molecules, the design of bio-hybrid engineered systems were improved to allow real applications ranging from micro/nano fluidics, to biosensing or to medical diagnoses. On the micro- and nano-scales, biological organisms have always used their machinery to convert ATP into mechanical energy with a concrete function: ranging from intracellular transport to the large-scale actuation performed by muscles. Therefore, the concrete evidence that this idea could be rational has existed since many years ago; the question is "how to assemble engineering devices power-driven by ATP?" The design of artificial molecular motors and machines is one of the major challenges in contemporary molecular sciences and bottom-up molecular nanotechnology. Whereas the protein-based molecular motors found in the living cell are amongst the most fascinating and complex structures found in nature and crucial to nearly every key biological process, the field of synthetic, linear and rotary motors is still in its infancy. In a broader context, moving molecular sciences from the current situation with a focus on static structures and operation under thermodynamic control to dynamic chemistries with systems under kinetic control will represent a major step beyond current frontiers of chemical sciences. Furthermore, a shift from control of structure to dynamic control of function and from molecules to molecular systems, where several components act in concert often at different hierarchical levels, makes the discovery of fascinating and unique properties possible; their comprehension is a pre-requisite for rational advances in health sciences and related fields. As a solution, scientists have investigated bioinspired and biomimetic approaches and engineered mechanisms for nano-robot locomotion by mimicking the bacteria-motile structure or by using biological motors which can convert chemical energy to linear or rotation motion. However, many molecular engineers agree that the science of molecular machines is not ready for market; this domain is still a fantasy due to the lack of a well-established technology for manufacturing systems specifically for biomedical applications; however, in the longer term, it is possible that all these limitations will be overcome and thereby nano-robots may become ubiquitous in medicine.

References

Alberts, B., A. Johnson, P. Walter, J. Lewis, M. Raff and K. Roberts. 2008. Molecular Cell Biology. New York: Garland Science.

Ali, J., U. K. Cheang, J. D. Martindale, M. Jabbarzadeh, H. C. Fu and M. Jun Kim. 2017. Bacteria-inspired nanorobots with flagellar polymorphic transformations and bundling. Sci. Rep. 7(1): 14098.

Astumian, R. D. 2007. Design principles for Brownian molecular machines: How to swim in molasses and walk in a hurricane. Phys. Chem. Chem. Phys. 9(37): 5067-5083.

Bachand, G. D., N. F. Bouxsein, V. Van Delinder and M. Bachand. 2014. Biomolecular motors in nanoscale materials, devices, and systems. Wiley Interdiscip. Rev. Nanomed. Nanobiotechnol. 6(2): 163-177.

Bachand, G. D., S. B. Rivera, A. K. Boal, J. Gaudioso, J. Liu and B. C. Bunker. 2004. Assembly and transport of nanocrystal CdSe quantum dot nanocomposites using microtubules and kinesin motor proteins. Nano Lett. 4(5): 817-821.

Balzani, V., A. Credi, F. M. Raymo and J. F. Stoddart. 2000. Artificial molecular machines. Angew. Chem., Int. Ed. 39(19): 3348-3391.

Bromley, E. H., N. J. Kuwada, M. J. Zuckermann, R. Donadini, L. Samii, G. A. Blab, et al. 2009. The tumbleweed: Towards a synthetic protein motor. HFSP J. 3(3): 204-212.

Bustamante, C., Z. Bryant and S. B. Smith. 2003. Ten years of tension: Single-molecule DNA mechanics. Nature 421(6921): 423-427.

Clemmens, J., H. Hess, R. Doot, C. M. Matzke, G. D. Bachand and V. Vogel. 2004. Motor-protein "roundabouts": Microtubules moving on kinesin-coated tracks through engineered networks. Lab on a Chip 4(2): 83-86.

Chaudhuri, S., T. Korten, S. Korten, G. Milani, T. Lana, G. te Kronnie and S. Diez. 2017. Label-free detection of microvesicles and proteins by the bundling of gliding microtubules. Nano Lett. 18(1): 117-123.

Che, H., S. Cao and J. C. M. van Hest. 2018a. Feedback-induced temporal control of "breathing" polymersomes to create self-adaptive nanoreactors. J. Am. Chem. Soc. 140(16): 5356-5359.

Chen, J., F. K. -C. Leung, M. C. Stuart, T. Kajitani, T. Fukushima, E. van der Giessen, et al. 2018b. Artificial muscle-like function from hierarchical supramolecular assembly of photoresponsive molecular motors. Nat. Chem. 10(2): 132-138.

Chen, L., M. Nakamura, T. D. Schindler, D. Parker and Z. Bryant. 2012. Engineering controllable bidirectional molecular motors based on myosin. Nat. Nanotechnol. 7(4): 252-256.

Cheng, C., P. R. McGonigal, S. T. Schneebeli, H. Li, N. A. Vermeulen, C. Ke, et al. 2015. An artificial molecular pump. Nat. Nanotechnol. 10(6): 547-553.

Chiang, P. -T., J. Mielke, J. Godoy, J. M. Guerrero, L. B. Alemany, C. J. Villagomez, et al. 2011. Toward a light-driven motorized nanocar: Synthesis and initial imaging of single molecules. ACS Nano 6(1): 592-597.

De, S., W. Hwang and E. Kuhl. 2015. Multiscale modeling in biomechanics and mechanobiology. Springer.

Del Grosso, E., A. Amodio, G. Ragazzon, L. J. Prins and F. Ricci. 2018. Dissipative synthetic DNA-based receptors for the transient loading and release of molecular Cargo. Angew. Chem., Int. Ed. 57(33): 10489-10493.

della Sala, F., S. Neri, S. Maiti, J. L. Y. Chen and L. J. Prins. 2017. Transient self-assembly of molecular nanostructures driven by chemical fuels. Curr. Opin. Biotechnol. 46: 27-33.

Demple, B. and S. Linn. 1980. DNA N-glycosylases and UV repair. Nature 287(5779): 203-208.

Di Leonardo, R., L. Angelani, D. Dell'Arciprete, G. Ruocco, V. Iebba, S. Schippa, et al. 2010. Bacterial ratchet motors. Proc. Natl. Acad. Sci. U.S.A. 107(21): 9541-9545.

Diez, S., J. H. Helenius and J. Howard. 2008. Biomolecular motors operating in engineered environments. Protein Science Encyclopedia: online 185-199.

Diez, S., C. Reuther, C. Dinu, R. Seidel, M. Mertig, W. Pompe, et al. 2003. Stretching and transporting DNA molecules using motor proteins. Nano Lett. 3(9): 1251-1254.

DiPaola, L. and A. Giuliani. 2018. Multiscale synthetic biology: From molecules to ecosystems. ?? *In*: V. Piemonte, A. Basile, T. Ito and L. Marrelli (eds.). Biomedical Engineering Challenges: A Chemical Engineering Insight. John Wiley & Sons, Inc., Hoboken, NY, USA.

Erbas-Cakmak, S., S. D. Fielden, U. Karaca, D. A. Leigh, C. T. McTernan, D. J. Tetlow and M. R. Wilson. 2017. Rotary and linear molecular motors driven by pulses of a chemical fuel. Science 358(6361): 340-343.

Feynman, R. (1993). Infinitesimal machinery. J. Microelectromech. Syst. 2(1): 4-14.

Fialkowski, M., K. J. Bishop, R. Klajn, S. K. Smoukov, C. J. Campbell and B. A. Grzybowski. 2006. Principles and implementations of dissipative (dynamic) self-assembly, ACS Publications.

Fischer, T., A. Agarwal and H. Hess. 2009. A smart dust biosensor powered by kinesin motors. Nat. Nanotechnol. 4(3): 162-166.

Fujimoto, K., M. Kitamura, M. Yokokawa, I. Kanno, H. Kotera and R. Yokokawa. 2013. Colocalization of quantum dots by reactive molecules carried by motor proteins on polarized microtubule arrays. ACS Nano 7(1): 447-455.

Fujimoto, K., M. Nagai, H. Shintaku, H. Kotera and R. Yokokawa. 2015. Dynamic formation of a microchannel array enabling kinesin-driven microtubule transport between separate compartments on a chip. Lab on a Chip 15(9): 2055-2063.

Furuta, A., M. Amino, M. Yoshio, K. Oiwa, H. Kojima and K. Furuta. 2017. Creating biomolecular motors based on dynein and actin-binding proteins. Nat. Nanotechnol. 12(3): 233-237.

García-López, V., P.-T. Chiang, F. Chen, G. Ruan, A. A. Martí, A. B. Kolomeisky, et al. 2015. Unimolecular submersible nanomachines. Synthesis, actuation, and monitoring. Nano Lett. 15(12): 8229-8239.

García-López, V., J. Jeffet, S. Kuwahara, A. A. Martí, Y. Ebenstein and J. M. Tour. 2016. Synthesis and photostability of unimolecular submersible nanomachines: Toward single-molecule tracking in solution. Org. Lett. 18(10): 2343-2346.

Godoy, J., G. Vives and J. M. Tour. 2011. Toward chemical propulsion: Synthesis of ROMP-propelled nanocars. ACS Nano 5(1): 85-90.

Goldstein, L. S. 1991. The kinesin superfamily: Tails of functional redundancy. Trends Cell Biol. 1(4): 93-98.

Goodsell, D. S. 2009. The Machinery of Life. Springer Science & Business Media. Berlin Germany.

Hao, X., W. Sang, J. Hu and Q. Yan. 2017. Pulsating polymer micelles via ATP-fueled dissipative self-assembly. ACS Macro Lett. 6(10): 1151-1155.

Hess, H. and G. D. Bachand. 2005. Biomolecular motors. Mater. Today 8(12): 22-29.

Hess, H. and G. Saper. 2018. Engineering with biomolecular motors. Acc. Chem. Res. 51(12): 3015-3022.

Hiratsuka, Y., M. Miyata, T. Tada and T. Q. P. Uyeda. 2006. A microrotary motor powered by bacteria. Proc. Natl. Acad. Sci. U.S.A. 103(37): 13618-13623.

Hirokawa, N., Y. Noda, Y. Tanaka and S. Niwa. 2009. Kinesin superfamily motor proteins and intracellular transport. Nat. Rev. Mol. Cell Biol. 10(10): 682-696.

Hortelão, A. C., R. Carrascosa, N. Murillo-Cremaes, T. Patiño and S. Sánchez. 2018. Targeting 3D bladder cancer spheroids with urease-powered nanomotors. ACS Nano 13(1): 429-439.

Hortelão, A. C., T. Patiño, A. Perez-Jiménez, À. Blanco and S. Sánchez. 2018. Enzyme-powered nanobots enhance anticancer drug delivery. Adv. Funct. Mater. 28(25): 1705086.

Huang, T. J. and B. K. Juluri. 2008. Biological and biomimetic molecular machines. Nanomedicine 3(1): 107-124.

Huang, Z., P. Chen, G. Zhu, Y. Yang, Z. Xu and L.-T. Yan. 2018. Bacteria-activated janus particles driven by chemotaxis. ACS Nano 12(7): 6725-6733.

Ikegami, T., Y. Kageyama, K. Obara and S. Takeda. 2016. Dissipative and autonomous square-wave self-oscillation of a macroscopic hybrid self-assembly under continuous light irradiation. Angew. Chem., Int. Ed. 55(29): 8239-8243.

Israelachvili, J. N. 2011. Intermolecular and Surface Forces. Academic Press. London, UK.

Iwaso, K., Y. Takashima and A. Harada. 2016. Fast response dry-type artificial molecular muscles with [c2] daisy chains. Nat. Chem. 8(6): 625-632.

Jia, Y. and J. Li. 2019. Molecular assembly of rotary and linear motor proteins. Acc. Chem. Res. 9b00015.

Karsenti, E. 2008. Self-organization in cell biology: A brief history. Nat. Rev. Mol. Cell Biol. 9(3): 255-262.

Karsenti, E. and I. Vernos. 2001. The mitotic spindle: A self-made machine. Science 294(5542): 543-547.

Kassem, S., T. van Leeuwen, A. S. Lubbe, M. R. Wilson, B. L. Feringa and D. A. Leigh. 2017. Artificial molecular motors. Chem. Soc. Rev. 46(9): 2592-2621.

Kimlimger, T. K., M. M. Timm, S. V. Rajkumar, J. L. Haug, M. P. Kline, T. E. Witzig, et al. 2006. Phenotypic characterization of the CD45+ and CD45– plasma cell compartments in monoclonal gammopathies. Blood 108(11): 3505.

Kiristi, M., V. V. Singh, B. Esteban-Fernández de Ávila, M. Uygun, F. Soto, D. Aktas-Uygun, et al. 2015. Lysozyme-based antibacterial nanomotors. ACS Nano. 9(9): 9252-9259.

Korten, S., N. Albet-Torres, F. Paderi, L. ten Siethoff, S. Diez, T. Korten, et al. 2013. Sample solution constraints on motor-driven diagnostic nanodevices. Lab on a Chip 13(5): 866-876.

Korten, T., A. Månsson and S. Diez. 2010. Towards the application of cytoskeletal motor proteins in molecular detection and diagnostic devices. Curr. Opin. Biotechnol. 21(4): 477-488.

Kottas, G. S., L. I. Clarke, D. Horinek and J. Michl. 2005. Artificial molecular rotors. Chem. Rev. 105(4): 1281-1376.

Kull, F. J., R. D. Vale and R. J. Fletterick. 1998. The case for a common ancestor: Kinesin and myosin motor proteins and G proteins. J. Muscle Res. Cell Motil. 19(8): 877-886.

Kumar, S., L. ten Siethoff, M. Persson, N. Albet-Torres and A. Månsson. 2013. Magnetic capture from blood rescues molecular motor function in diagnostic nanodevices. J. Nanobiotechnol. 11(1): 14.

Lard, M., L. ten Siethoff, J. Generosi, A. Månsson and H. Linke. 2014. Molecular motor transport through hollow nanowires. Nano Lett. 14(6): 3041-3046.

Lard, M., L. ten Siethoff, S. Kumar, M. Persson, G. te Kronnie, H. Linke, et al. 2013a. Ultrafast molecular motor driven nanoseparation and biosensing. Biosens. Bioelectron. 48: 145-152.

Lard, M., L. ten Siethoff, A. Månsson and H. Linke. 2013b. Tracking actomyosin at fluorescence check points. Sci. Rep. 3: 1092.

Li, Q., G. Fuks, E. Moulin, M. Maaloum, M. Rawiso, I. Kulic, et al. 2015. Macroscopic contraction of a gel induced by the integrated motion of light-driven molecular motors. Nat. Nanotechnol. 10(2): 161-165.

Lin, C.-T., M.-T. Kao, K. Kurabayashi and E. Meyhofer. 2008. Self-contained, biomolecular motor-driven protein sorting and concentrating in an ultrasensitive microfluidic chip. Nano Lett. 8(4): 1041-1046.

Lindahl, T. 1982. DNA repair enzymes. Annu. Rev. Biochem. 51(1): 61-87.

Lodish, H., J. E. Darnell, A. Berk, C. A. Kaiser, M. Krieger, M. P. Scott, et al. 2008. Molecular Cell Biology. Macmillan.

Ma, X., A. C. Hortelão, T. Patino and S. Sanchez. 2016. Enzyme catalysis to power micro/nanomachines. ACS Nano 10(10): 9111-9122.

Maiti, S., I. Fortunati, C. Ferrante, P. Scrimin and L. J. Prins. 2016. Dissipative self-assembly of vesicular nanoreactors. Nat. Chem. 8(7): 725-731.

Mantle, T. 2001. Enzymes: Nature's Nanomachines. Portland Press Limited. 331-336.

Minamino, T., K. Imada and K. Namba. 2008. Molecular motors of the bacterial flagella. Curr. Opin. Struct. Biol. 18(6): 693-701.

Muscat, R. A., J. Bath and A. J. Turberfield. 2011. A programmable molecular robot. Nano Lett. 11(3): 982-987.

Omabegho, T., P. S. Gurel, C. Y. Cheng, L. Y. Kim, P. V. Ruijgrok, R. Das, et al. 2017. Controllable molecular motors engineered from myosin and RNA. Biophys. J. 112(3): 5a.

Ragazzon, G., M. Baroncini, S. Silvi, M. Venturi and A. Credi. 2015. Light-powered autonomous and directional molecular motion of a dissipative self-assembling system. Nat. Nanotechnol. 10(1): 70-75.

Reuther, C., M. Mittasch, S. R. Naganathan, S. W. Grill and S. Diez. 2017. Highly-efficient guiding of motile microtubules on non-topographical motor patterns. Nano Lett. 17(9): 5699-5705.

Sasaki, T., J. M. Guerrero and J. M. Tour. 2008. The assembly line: Self-assembling nanocars. Tetrahedron 64(36): 8522-8529.

Shelley, M. J. 2016. The dynamics of microtubule/motor-protein assemblies in biology and physics. Annu. Rev. Fluid Mech. 48(1): 487-506.

Singh, V. V., K. Kaufmann, B. E.-F. de Ávila, M. Uygun and J. Wang. 2016. Nanomotors responsive to nerve-agent vapor plumes. Chem. Commun. 52(16): 3360-3363.

Spetzler, D., J. York, C. Dobbin, J. Martin, R. Ishmukhametov, L. Day, et al. 2007. Recent developments of bio-molecular motors as on-chip devices using single molecule techniques. Lab on a Chip 7(12): 1633-1643.

Stanton, M. M., B.-W. Park, A. Miguel-López, X. Ma, M. Sitti and S. Sánchez. 2017. Biohybrid microtube swimmers driven by single captured bacteria. Small 13(19): 1603679.

Sun, M., M. Wartel, E. Cascales, J. W. Shaevitz and T. Mignot. 2011. Motor-driven intracellular transport powers bacterial gliding motility. Proc. Natl. Acad. Sci. U.S.A. 108(18): 7559-7564.

te Brinke, E., J. Groen, A. Herrmann, H. A. Heus, G. Rivas, E. Spruijt, et al. 2018. Dissipative adaptation in driven self-assembly leading to self-dividing fibrils. Nat. Nanotechnol. 13(9): 849-855.

Tu, Y., F. Peng, X. Sui, Y. Men, P. B. White, J. C. van Hest and D. A. Wilson. 2017. Self-propelled supramolecular nanomotors with temperature-responsive speed regulation. Nat. Chem. 9(5): 480-486.

Tung, S. and J. W. Kim. 2006. Microscale hybrid devices powered by biological flagellar motors. IEEE T. Autom. Sci. Eng. 3(3): 260-263.

Vale, R. D. 2003. The molecular motor toolbox for intracellular transport. Cell 112(4): 467-480.

van den Heuvel, M. G. L. and C. Dekker. 2007. Motor proteins at work for nanotechnology. Science 317(5836): 333-336.

van Rossum, S. A., M. Tena-Solsona, J. H. van Esch, R. Eelkema and J. Boekhoven. 2017. Dissipative out-of-equilibrium assembly of man-made supramolecular materials. Chem. Soc. Rev. 46(18): 5519-5535.

Wallace, S. S. 1988. AP endonucleases and DNA glycosylases that recognize oxidative DNA damage. Environ. Mutagen. 12(4): 431-477.

Wikipedia, t. f. e. 2018. Molecular motor. from https://en.wikipedia.org/wiki/Molecular_motor

Wilson, M. R., J. Solà, A. Carlone, S. M. Goldup, N. Lebrasseur and D. A. Leigh. 2016. An autonomous chemically fuelled small-molecule motor. Nature 534(7606): 235-240.

Zhu, J.-K. 2009. Active DNA demethylation mediated by DNA glycosylases. Annu. Rev. Genet. 43: 143-166.

Zlatanova, J. and K. E. Van Holde. 2015. Molecular Biology: Structure and Dynamics of Genomes and Proteomes. Garland Science. New York City, USA.

Nano-targeted Smart Drug Delivery Systems and Nanorobotics

"…You, on the other hand, will be in a ship, a kind of modified submarine. It, too, has been miniaturized and deminiaturized without harm…" "…We will not be crossing an ocean or penetrating the vacuum of space. We will instead enter a microscopic ocean and penetrate the human brain. Can you be a scientist – a neurophysicist – and resist that? …"

From "Fantastic Voyage II – Destination Brain", written by Isaac Asimov

8.1 Advanced Nano-therapeutics: Control of Pharmacokinetic Parameters Using Nano-scale Size Devices

One of the major challenges of nanotechnology has focused on the development of smart drug nano-carriers for improving the effectiveness and accuracy of medical treatments. Through the controlled release of their content and decrease of toxicity, these formulations have offered outstanding advantages over conventional therapies' platforms (Baeza et al. 2017). Therefore, formulations based on this technology have been conceived for improving the biodistribution and for achieving target site accumulation of systemically administered (chemo) therapeutic agents. Many different types of systems have been evaluated over the years, including for instance liposomes, micelles or polymer devices, and a significant amount of evidence has been obtained showing that these sub micrometer-sized carrier materials are able to improve the balance between the efficacy and toxicity of therapeutic interventions. Hence, in this chapter, we will provide a rational classification of the different smart drug delivery systems based on nanotechnology. We will analyze their origin, how they were improved and finally, we will evaluate their positive impact on medical treatments. Among these systems, micelles, liposomes, strategies based on polymers, and those based on lipid nanostructures are described. Additionally, highly porous nanomaterials are other important devices, which have become into ideal candidates as controlled drug delivery systems but, contrary to liposomes or micelles, they are formed by rigid structures. Nevertheless, the importance of highly porous materials in this field lay on their high drugs loading-capability. In the last section of the chapter, we have described one of the newest nano-device systems based on not only smart or environmentally responsive systems but also designed and developed using robotic concepts: the nanorobots. These systems, which emerged as "sci-fiction based devices", have shown essential conditions for acting as nano smart drug delivery systems and for carrying other therapeutic agents such as oxygen. Additionally, in this proposal, some strategies for the development of combined devices which can enhance the outcomes for certain individual treatments are described.

8.2 Nano-targeted Smart Drug Delivery Systems

Literature is brimming with examples of where a reduction of the particle size of an active pharmaceutical ingredient (API) results in an increase in its bioavailability. However, drugs are rarely administered to patients as a pure substance, due to which they have, under this condition, a lack of predictable therapeutic response and can exert irritation in the applied body region or toxicity (Müller 1991). Therefore, pharmaceutical formulations have been developed for giving predictability during the storage, administration, release, and distribution of the API, reaching the target site, and their large-scale manufacturability (Aulton 2004). Formulations avoid the premature degradation of drugs and increase their bioavailability (Torchilin 2006). As a consequence of that, pharmaceutical formulations modify pharmacokinetic profiles of drugs, and hence, they change their adsorption and metabolism (degradation and excretion). Thus, the scientific community has focused on the development of different formulations using a wide variety of excipients, active molecules, and preparation methods (Aulton 2004). Some formulations are based simply on solutions, which are the simplest dosage forms still used for health care. Others have been developed through nanotechnology and show a high complexity and active targeting. Among these formulations, we can mention those used for the treatment of some types of cancer such as nanoparticles based on polymers.

Accordingly, some objectives were raised: decrease of the size particle, generate a specific release and transform the drugs into water-soluble substances. Thus, the researchers have focused their work on the design of drug carriers systems for reaching these features. The development of a medicine or a pharmaceutical formulation involves several steps: (i) obtaining an active principle of proven safety, (ii) finding reliable excipients and (iii) developing the necessary technology to carry out these dosage systems. It is therefore apparent that before an API can be included in a dosage form, we have to consider some points: (i) biopharmaceutical considerations, which affect the absorption of the drug; (ii) the physical and chemical properties of the drug, and (iii) therapeutic considerations (Aulton 2004). This last point will not be widely considered in this text.

Kawabata et al. (Kawabata et al. 2011) have reported that around 70% of the new drug candidates show poor water solubility and approximately 40% of the drugs formulated in immediate release dosage forms are practically insoluble (<100 µg/ml) in water (Takagi et al. 2006). Therefore, drug delivery in the human body of many therapeutic groups is engaged. Active agents have been ordered by means of the biopharmaceutical classification, which correlates the *in vitro* dissolution of agents with their *in vivo* bioavailability. This classification considers two main parameters: intestinal permeability and drug solubility (Yu et al. 2002). The classification has four categories which were extensively described in many reports (Daousani and Macheras 2016, Khadka et al. 2014, Yu et al. 2002). Active pharmaceutical agents which belong to the groups II and IV of this classification show low bioavailability because they have low water-solubility; therefore, in these cases, the formulation plays a very important role (Pouton 2006). In addition, drugs belonging to group IV show low intestinal penetration (Kawabata et al. 2011). For these reasons, it is suitable to propose that drugs belonging to these two groups need a device for improving their delivery and specific release in the target site and, in this context, a nano-smart drug delivery system could be used.

In this section, we address the drug delivery system design based on four different strategies. First sub-section describes drug delivery systems based on surfactants, specifically those systems developed by using micelles and liposomes. The second sub-section addresses those delivery systems based on polymers and within this group, cyclodextrins, poly-lactic acid (and derivatives) and alginate, are shown. In the third sub-section, systems based on lipid nanoparticles and nanostructured lipids are described. In the last sub-section, highly porous materials based systems are shown.

8.2.1 Drug Delivery Systems Based on Surfactants

Surfactants are organic molecules with double affinity by polar and non-polar solvents. This feature is due to the fact that their structure is formed by two parts clearly different: the polar region (commonly known as head) and the non-polar region (also known as tail) (Langevin 1992). These molecules are able to adopt different types of spatial arrangements which generate complex structures successfully applied in pharmaceutical formulations' design (Quirolo et al. 2014). The application in human medicine of these supramolecular associations with colloidal size (between 1 to 1000 nm) has been growing as well as their specificity. These kinds of systems have been reported as very useful devices for delivery of poor water-solubility drugs and for preserving their features against external conditions such as an oxidative atmosphere and extreme pH environments among others (Torchilin 2006). Additionally, these systems can be designed as environmental responsive devices and with capacity for active-targeting of drugs (Liu et al. 2016).

8.2.1.1 *Micelles*

When surfactants are dissolved in water, they are absorbed in the free surface (or into the interface if the surfactants are dissolved into a mix of polar and non-polar solvents) decreasing the surface tension (Casper 1977). While the concentration of surfactants is increased, a reduction of surface tension is evidenced, but there is a concentration in which the surface tension starts to keep constant. This concentration (really, it is a range of concentrations) depends on the type of surfactant (ionic or non-ionic) and also depends on temperature. This point, known as Krafft point, is defined by a concentration, the critical micelle concentration (CMC), and by temperature, known as critical micellization temperature (Benedini et al. 2010). Up to this point, the micelles adopt a spherical form with a certain aggregation number (monomers of surfactant for forming micelles), but by means of an increase of concentration, their size is enlarged and they adopt rod-like structures or lamellar ones (Israelachvili 2011). Different shapes of micelles are not defined only by the concentration of surfactants or temperature; the critical packing parameter, which depends on the relationship between polar and non-polar portions of the surfactant, is another parameter to predict the arrangements adopted by amphiphiles which define the shape of micelles (Israelachvili et al. 1976). Within an aqueous environment, the removal of any non-polar substance (for example, a poorly-soluble drug) re-establishes the hydrogen-bond network among water molecules. Consequently, it produces a decrease of free energy of the system which drives the formation of micelles. In addition to the decrease of free energy, Van der Waals bonds between hydrophobic portions of the surfactants molecules inside the core of the micelles contribute to their stabilization. These facts lead to micelles to be produced spontaneously, under certain conditions. Hence, the core will be formed by the hydrophobic fragments of the surfactant molecules, while the hydrophilic portion will form their shell (Lasic 1992).

Briefly, micelles are structures formed by surfactants arranged in a defined form dispersed in a solution (generally aqueous). They have an oily liquid nucleus (direct micelles) (Israelachvili 2011, Langevin 1992) and consequently, they have the potential to dissolve non-polar drugs into their core (Liu 2018, Torchilin 2006).

Increased water solubility of poorly soluble drugs improves their bioavailability, reduces their toxicity and decreases the occurrence of adverse effects, enhance their permeability for crossing the physiological barriers, and modify favourably the profile of drug biodistribution. These are some of the advantages of using micelle-forming surfactants for carrying and delivering drugs (Jones and Leroux 1999, Kumar and Rajeshwarrao 2011, Torchilin 2001). Owing to their small size (from 5 to 100 nm, although less than 80 nm is preferred), they have demonstrated a spontaneous penetration into weakly vasculature by means of the enhanced permeability and retention effect (EPR). This is a selective mechanism of penetration also known as passive-targeting (Maeda et al. 2000, Stapleton et al. 2013).

Polyoxyethylated non-ionic surfactants such as polysorbates, Brijs, Cremophor EL, polyoxyl stearate, and others are generally chosen for drug delivery systems' development as they are less toxic, less hemolytic and less irritant of cell surfaces than ionic ones and additionally, improve the therapeutic index of the drug. Their CMC for carrying a weight percentage of hydrophobic drugs between 5 to 25% should be 1.10^{-4} M or less (Torchilin 2006).

Jiao et al. (Jiao 2008) describe how these surfactants are widely used in topical delivery of ophthalmic drugs for treating inflammation, allergy, ocular hypertension and glaucoma. Among the drugs carried by these surfactants are tropicamide, pilocarpine, atenolol, timolol, indomethacin, cyclosporine A, and others. Other authors have reported the uses of polysorbates for improving the delivery of zidovudine and paclitaxel by other administration routes (Bayindir and Yuksel 2010, Kumar and Rajeshwarrao 2011, Ruckmani et al. 2010).

Polymeric micelles are another type of micelles formed by polymeric surfactants. These structures have a CMC 10 to 1000-fold lower than micelles formed of low-molecular-weight surfactants ($\sim 10^{-6}$ M). Their size lies within the range 10–100 nm and they can undergo several structural modifications and due to that fact, they further augment cell uptake which ensure that they can selectively leave the circulation into the areas with the compromised vasculature and, therefore, they could reach for example to tumour cells (Mohamed et al. 2014). Like the low-molecular-weight surfactants, polymeric ones consist of a hydrophobic block and a hydrophilic block. Micelles formed by these molecules possess high stability both *in vitro* and *in vivo* and good biocompatibility. They can solubilize a broad variety of poorly-water soluble drugs. Among the compounds used for polymeric micelles development, we can mention polyethylene glycol/phosphatidyl ethanolamine conjugate (PEG/PE), poloxamers (block co-polymers of polyethylene glycol and polypropylene glycol), poly (oxyethylene) poly(oxypropylene), poly(ethylene oxide)-poly(aspartic acid), poly(ethylene oxide)-b-poly(ε-caprolactone), poly(N-vinylpyrrolidone)-block-poly (D,L-lactide) and others (Torchilin 2006). Many systems based on these polymeric surfactants have been used for carrying different active agents, not only drugs but also contrast agents. Among drugs, we find paclitaxel, docetaxel, teniposide, and etoposide (Le Garrec et al. 2004, Shuai et al. 2004); risperidone (Mathot et al. 2006); cyclosporin A (Aliabadi et al. 2005); doxorubicin (Kwon et al. 1997); cisplatin (Xu et al. 2006); and others.

For the reasons mentioned above, drug delivery systems based on micelles are one of the most popular and versatile devices for carrying poorly-soluble and labile drugs which can be administered by different routes.

8.2.1.2 Liposomes

Liposomes are one of the first nano-devices focused on the development of drug carrier systems. Discovered in 1965 by A. D. Bangham, these structures were conceived as closed single vesicles and are formed by a stacked bilayer of non-toxic phospholipids and, if necessary, other molecules. Additionally, their structure can also be explained as a self-closing lamellar liquid crystal matrix or in other words, as a supramolecular lipid matrix in a smectic mesophase (Bangham 1972, Benedini et al. 2017a, Deamer and Bangham 1976, Deamer 2010).

The classification of liposomes depends on their size (giants, large and smalls) from 20 nm to 1000 nm, the number of membranes (multi or unilamellar), surface charge (anion, cation or neutral) and also on the method of preparation (Akbarzadeh et al. 2013, Bozzuto and Molinari 2015, Zylberberg and Matosevic 2016). There are different methods for carrying out liposomes and critical parameters which must be considered such as size, surface charge, number of layers, the disturbance of layers among others, but they are not addressed in this section. Detailed works have been reported by many authors (Akbarzadeh et al. 2013, Benedini et al. 2017a, Khadke et al. 2018, Patil and Jadhav 2014). Liposomes have been designed and applied for different purposes. Among them, they have been used for the development of artificial membranes, have been applied to food and farming industries, as well as formulations for delivering active agents (Akbarzadeh et al. 2013). In this sub-section, we will focus on liposomes applied to the development of pharmaceutical formulations.

Owing to their structural features that were rapidly thought, designed and developed as drug delivery systems (Gregoriadis 1976, 1978), the first drug included in these vehicles was doxorubicin (anti-cancer drug) (Benedini et al. 2017a). This is due to the fact that most of the chemotherapeutic agents have poor pharmacokinetic profiles, cause side effects, and high toxicity because of their non-specific distribution in the body tissues and organs (Olusanya et al. 2018). After that, other drugs have been incorporated such as amphotericin B (Hann and Prentice 2001, Stone et al. 2016), propofol (Kastner et al. 2015), morphine (Alam and Hartrick 2005, Yaksh et al. 2000), paclitaxel (Liu et al. 2018, Xu et al. 2013), amiodarone (Benedini et al. 2010), and others. With the advent of gene therapy, other type of therapeutic molecules, mainly hydrophilic, have been entrapped in liposomes; among them are DNA (Rasoulianboroujeni et al. 2017, Yang et al. 2016), interference RNA or silencing RNA (siRNA) (Hattori et al. 2017, Xia et al. 2016) and plasmids (Dimitriadis 1979, Levine et al. 2013).

Owing to the own features of the lipid membrane and those of the aqueous core, both hydrophobic and hydrophilic drugs can be carried by these vesicles. Therefore, hydrophobic drugs are in contact with the non-polar portion of the membranes (between hydrophobic tails of phospholipids) and hydrophilic drugs are related to polar groups within the aqueous core of liposomes (Gregoriadis and Florence 1993). Consequently, hydrophilic drugs are released first and then, when the membrane is disassembled, the hydrophobic drugs are released (Fahr et al. 2005, Juliano et al. 1987).

One of the major challenges that liposomes have been exposed to are related to their stability in storage containers, and after their administration in the body. Therefore, many strategies have been carried out for increasing their time of permanence in blood circulation to reach a decreased elimination by the cells of the mononuclear phagocyte system in the liver and spleen. The first strategies were focused on the reduction of their size and then modified their membrane fluidity by using cholesterol and sphingomyelins but the results were not improving as expected. As a consequence of that, another approach has been used for reaching success. Molecules of polyethylene glycol were anchored to the surface of liposomes, and thus, PEGylated liposomes were developed. These systems are also named "stealth liposomes" (Immordino et al. 2006). By using this strategy, liposomes have increased their circulating time. Additionally, for improving their properties, small-molecule ligands, peptides, and monoclonal antibodies can be coupled to passive liposomes for becoming active-targeting liposomes (Torchilin 2010, Zhu et al. 2012). Actively targeted liposomes have been used in cancer therapy. For instance, molecules of transferrin and folic acid have been anchored to the surface of liposomes for guiding liposomes loaded with an oncologic drug owing to the receptors of folate and transferrin are overexpressed on the surface of many cancer cells. Therefore, the release of the anticancer drug is mediated by these previous specific interactions (Zylberberg and Matosevic 2016).

The launch of liposomes as an alternative for carrying drugs with poor pharmacokinetic profiles, sensitive to the body environment or with high toxicity started in the middle of 1990s (Barenholz 2012). AmBisome® was one of the first pharmaceutical formulations based on conventional liposomes for carrying amphotericin B, an antifungal drug used for systemic infections. This is a hydrophobic drug and the size of the liposomes was ranged from 45 to 80 nm. Similar sizes were reached by other two liposomal formulations of amphotericin B, AmBil® and Amphocil® with 130 and 110 nm, respectively (Bozzuto and Molinari 2015). Another drug, daunorubicin, was also formulated in conventional liposomes of 45 nm (Chang and Yeh 2012). Doxil® was one of the first pharmaceutical formulations based on PEGylated liposomes of 100 nm for carrying the hydrophilic drug, doxorubicin hydrochloride (Allen and Cullis 2013). Paclitaxel, another anti-cancer drug (but hydrophobic), has improved its pharmacokinetic and pharmacodynamic profiles. Due to that, once formulated in cationic liposomes of 200 nm, two excipients used for dissolving paclitaxel, cremophor EL and ethanol, are taken off of the formulation and consequently, the adverse effects produced by these co-solvents are avoided (Strieth et al. 2014).

Sometimes, lipid membranes can be used as a coat of an inorganic matrix for giving suitable properties to a drug delivery system. Consequently, in this case, phospholipids are not used as liposomes-forming surfactants in the final dosage form, but liposomes are used as intermediate structures for covering the inorganic nanoparticles in the first steps of the development. For instance, Placente et al. (Placente et al. 2018) have developed a system based on hydroxyapatite nano-rods covered by a lipid membrane. In this work, the authors carried out a local delivery system from an osteoconductive biomaterial for simultaneously avoiding peri-implant traumas and inducing osseous tissue regeneration. Placente et al. (2018) have designed and developed a multi-drug delivery formulation based on nano-rods of hydroxyapatite (as osteoconductive material) coated by a lipid membrane mimetic. Ciprofloxacin and Ibuprofen (antibiotic and anti-inflammatory, respectively) were simultaneously loaded and then they were released efficiently. Additionally, the device has shown a pH-responsive release profile and good biocompatibility.

Therefore, liposomes have been and still are one of the most chosen nanodevices with active-targeting capabilities and are environmental responsive for drugs delivery.

8.2.2 Drug Delivery Systems Based on Polymers

Drug delivery systems are designed for improving the efficacy of therapeutic agents and consequently for achieving an effective treatment against diseases. In this context, the use of different types of polymers (synthetic, semi-synthetic, or natural) as a platform for developing drug delivery systems is a good choice. For carrying out this aim, polymers should meet a number of features. The most relevant is the lack of toxicity; they must be biocompatible before and after their degradation (products of their degradation must be innocuous). As any dosage form, containing polymers formulations must carry and release the drug in a predictable form in the target site, therefore, polymers must show a specific (beneficial) interaction with the drug (Sartuqui et al. 2017). In this sub-section, we will focus on three polymers widely used for pharmaceutical formulations design with nanometric size: cyclodextrins, poly-lactic acid (and its derivatives) and alginate.

8.2.2.1 *Cyclodextrins*

Cyclodextrins are the polysaccharides (cyclic oligosaccharides) more widely used as drug carrier systems for increasing the solubility of poorly water-soluble drugs (Puoci 2015). They are crystalline and non-hygroscopic molecules derived from starch. Their cyclic chemical structure contains at least six D-(+)-glucopyranose units attached by α $(1 \rightarrow 4)$ glucoside bonds. Cyclodextrins are divided in: α, β, γ, and δ. They differ in the size of their ring, solubility, and their content of glucose units, and they have 6, 7, 8 and 9 units, respectively (Challa et al. 2005, Rowe et al. 2009). The hydrophobic interior cavity of cyclodextrins is lined with carbons and ethereal oxygen of the glucose residues and the exterior surface, hydrophilic, is lined with hydroxyl groups which are functionalized to obtain more hydrophilic, hydrophobic, or ionizable derivatives (Loftsson et al. 2005, Zafar et al. 2014). Cyclodextrins have an extraordinary advantage compared with other polymers because they have the ability to form inclusion complexes with a variety of organic and inorganic lipophilic molecules. This feature is promising for application on nanotechnology developments such as drug delivery systems, cancer therapy (by increasing the loading capacity and encapsulation efficiency of tumour-targeting nanoparticles), gene delivery, and biosensing devices. A drug forms an inclusion complex with cyclodextrins through weak bonds which allow being temporarily lodged yielding in improved solubility and bioavailability (Arima et al. 2015). β-cyclodextrins are the cheapest and, consequently, the most commonly used, even though they are the least soluble. Therefore, to overcome this drawback, numerous derivatives were obtained for improving their physicochemical properties, such as hydroxyethyl-β-CD, hydroxypropyl-β-CD, sulfobutylether-β-CD, methyl-β-CD among others (Sartuqui et al. 2017). Cyclodextrins have been individually used for improving the solubility of drugs; however, one of their newest applications is to raise the solubility of the nanoparticles. With this association, cyclodextrins improve water solubility and drug loading while nanoparticles provide the targeting-effect drug delivery. As a consequence of that, a wide variety of cyclodextrins-nanoparticles association has been reported, for instance, with lipids, polymers, magnetic and gold nanoparticles among others (Shelley and Babu 2018).

These cyclic polysaccharides have been used for improving the solubility of the drugs in oral formulations. Among these drugs, we find naproxen, tropicamide, bromazepam, zolpidem, theophylline (Tiwari et al. 2010), ketorolac (Nagarsenker et al. 2008) and others. Additionally, other formulations containing plasmid DNA (pDNA) and small interfering RNA (siRNA) have been conceived by means of forming inclusion complexes with cyclodextrins (Bartlett and Davis 2008). In the case of plasmids, this hydrophilic active agent was incorporated into a solution with cyclodextrins and then, this loaded solution was incorporated into a chitosan solution (with a crosslinking agent). Finally, both solutions were mixed together for obtaining the final formulation (Nahaei et al. 2013).

Cyclodextrins have been also used as a coating agent of nanoparticles. For instance, Anirudhan et al. (Anirudhan et al. 2015) have developed silane magnetic nanoparticles coated by cyclodextrins for the delivery of 5-fluorouracil. This drug is used for treating different types of cancers (breast, stomach, colon, and lung) but it has prejudicial effects on normal cells if it is not targeted specifically for the tumour cells. Therefore, by means of application of an external magnetic field, these magnetic nanoparticles can be guided to the target site avoiding their effect on the healthy cells.

Polymer nanoparticles can be also coated by cyclodextrins such as those reported by Zhang et al. (Zhang et al. 2010). In this work, Zhang et al. have developed acid-resistant nanoparticles based on chitosan for carrying insulin. This protein is degraded by the acid media of the stomach, and for this reason, the use of β-cyclodextrins for covering the polymeric nanoparticles loaded with insulin have been proposed for oral administration.

Cyclodextrins can be used as structures for solubilizing poorly-water soluble drugs and also hydrophilic ones. Additionally, they can form complexes with nanoparticles based on polymers or silane and can positively contribute to formulations when combined with polymer solutions. Therefore, these polysaccharides form nanometric inclusion complexes with suitable properties for acting as systems for carrying drugs.

8.2.2.2 *Poly-Lactic Acid and Derivatives*

In the 1960s, the use of poly-lactic acid (PLA), and their derivatives, such as poly-co-glycolic lactic acid (PLGA) and poly-glycolic acid (PGA) in tissue repair (bone plates, abdominal mesh, and others) started. After that, they were rapidly applied to drug delivery systems (Lee et al. 2016) because they are non-toxic, biocompatible, renewable, and biodegradable substances. There are three types of PLA: D, L and racemic (Sha et al. 2016) and they are soluble in organic solvents as dioxane and chloroform. Therefore, the elimination of these solvents from the formulation is a highlight point to consider because of their toxicity (Farah et al. 2016).

Biodegradable PLA nanoparticles applied on drug delivery are mainly developed by solvent displacement, solvent diffusion, solvent evaporation, and salting-out method (Lee et al. 2016). However, PLA-based particles have challenges to overcome, such as low drug-loading capacity, low encapsulation efficiency, and terminal sterilization; for these reasons, PGLA has been chosen as improved copolymer for drug delivery systems development (Makadia and Siegel 2011). PLGA is a copolymer formed by different ratios of PLA and PGA; hence, its mechanical and physicochemical properties are deeply related to the ratio of these polymers within the composition of the copolymer. Owing to that, crystallinity and mean molecular weight are changed with proportions of PLA and PGA in this copolymer; drug loading and size of formed nanoparticles are also changed. For example, PLGA 50:50 (PLA:PGA) suffers faster degradation than PLGA 65:35 because there is a preferential degradation of glycolic acid proportion due to its hydrosolubility (Park 1995).

First bared PLGA nanoparticles proposed as drug delivery systems were rapidly removed by cells of the mononuclear phagocyte system (liver and spleen) as well as other bared nano-systems used with the same purpose such as liposomes or lipid nanoparticles. In this context, Betancourt et al. (Betancourt et al. 2009) have reported some strategies for transforming bare PLGA nanoparticles in "stealthy" nanoparticles by using polyethylene glycol (PEGylated nanoparticles). Other authors also used this strategy. Thus, Rafiei and Haddadi (Rafiei and Haddadi 2017) have analyzed the pharmacokinetic profile and the biodistribution of docetaxel loaded on PLGA nanoparticles and PLGA nanoparticles with polyethyleneglycol anchored onto their surface for application by the intravenous route. In animal models, both these types of nanoparticles have shown a sustained release behaviour of the drug. The authors have shown that the size of nanoparticles has played an important role in the pharmacokinetic profile. In addition, the negative surface charge of nanoparticles and the PEGylation process have been associated with long-circulating capacity. Furthermore, Betancourt et al. (Betancourt et al. 2009) have reported the use of antibodies anchored onto the surface of the nanoparticles to switch these passive-targeting PLGA nanoparticles in active-targeting PLGA nanoparticles.

In some cases, the mechanical and physicochemical properties of these nanoparticles must be improved. For this aim, PGLA has been combined with poloxamer (Barichello et al. 1999). These blends have made possible the development of nanoparticles to be applied by different administration routes (oral and subcutaneous among others) and for carrying both lipophilic drugs as cyclosporine A (Chacón et al. 1999) and hydrophilic ones as vancomycin and phenobarbital (Barichello et al. 1999, Peltonen et al. 2004). Due to that fact, nanoparticles developed with PLA and their derivatives are shown as attractive tools for drug delivery systems design and for improving the release profile of poor-soluble drugs.

8.2.2.3 Alginate

Alginates are polysaccharide-based polymers extracted from brown algae with different properties depending on the species, the region where the algae were found, and time of recollection (Sartuqui et al. 2017). They are composed by anionic linear hetero-polysaccharide formed by two different monomers $(1\rightarrow4)$-β-linked: the β-D-mannuronic acid and the α-L-guluronic acid (pKa = 3.38 and 3.65, respectively). Alginates have two main features: biocompatibility and bio-adhesiveness. Additionally, other properties must be remarked such as water-solubility, biodegradability, their thickening effect, and pH-responsive behaviour. By adding divalent cations such as calcium, they form hydrogels known as "egg-box arrangement" (Grant et al. 1973). Alginates can form hydrogels by interacting with calcium ions of hydroxyapatite and thus develop scaffold systems for bone tissue engineering. Lin and Yeh (Lin and Yeh 2004) have reported that these scaffolds increase cell attachment in the inner portion of the bone matrix. Therefore, as a consequence of their smart behaviour under different pH conditions and due to their capability for developing composites (nano-composites), Benedini et al. (Benedini et al. 2017b) have developed 3D nano scaffolds from synthetic nano-hydroxyapatite and alginates for bone repair. At low ionic strength and neutral pH, the system is shown to be amorphous, but when these conditions are modified, it generates a liquid crystal phase showing different properties, i.e. an increased resistance and a low degradability of the scaffolds. Thus, these biocomposites could additionally be used as a carrier of analgesics, anti-inflammatories, and/or antibiotics because they can change from an amorphous state to liquid crystal phases (inside the composites) by means of modulation of pH and ionic strength of the environment. This smart response makes this device attractive for drug delivery system design.

Many examples of alginate nanoparticles loaded-drugs or nanoparticles based on blends of this polymer have been described in the literature. These particles can be developed by applying different techniques such as ionic gelation, polyelectrolyte complexation, by emulsification or by procedures coupled to nanoemulsions and others (Lopes et al. 2017). Reis et al. (Reis et al. 2008) have reported the development of alginate nanospheres for the oral delivery of peptide-based drugs such as insulin. These authors propose a combination of natural polyelectrolytes, with particles formulated through a nanoemulsion that is gelated by complexation. Albumin-chitosan-coated alginate nanospheres have shown suitable encapsulation efficiency. It has been shown that the coating polymer blend has influenced the insulin release profile and has been crucial to prevent peptic digestion. Therefore, the clinical response by oral administration of this system has demonstrated that alginate nanospheres loaded with insulin and coated by albumin-chitosan have reduced glycemia to ~72% of basal values. Other drugs, with different physicochemical properties, have been loaded on alginate nano/microparticles for different treatments. Aynié et al. (Aynié et al. 1999) have reported a system based on alginate nanoparticles for protecting an antisense-oligonucleotide from degradation in bovine serum medium. This system has modified biodistribution of the oligonucleotide, which was highly accumulated in the lungs, in the liver, and in the spleen after intravenous administration into mice. Therefore, this device has been proposed as a promising carrier for specific delivery in these organs. The use of alginate, blends of this polymer with other polymers as chitosan, or with macromolecules have been suitable for carrying antitumour drugs such as tamoxifen (Martínez et al. 2012) and doxorubicin (Chavanpatil et al. 2007), antibiotics such as tobramycin (Ungaro et al. 2012) and gentamicin (Marie Arockianathan et al. 2012), and also other drugs.

A new generation of vaccines based on particulate systems has been designed and developed to overcome fatal diseases (Malyala and Singh 2010). In this context, Nait Mohamed and Laraba-Djebari (Nait Mohamed and Laraba-Djebari 2016) have reported the use of calcium-alginate nanoparticles as a new promising adjuvant of vaccines for delivering an antidote against scorpion envenomation. In this study, attenuated Androctonus australis hector venom and its toxic fraction were encapsulated into alginate nanoparticles. These nanoparticles (~100 nm diameter) were carried out by ionic gelation (with calcium) and characterized by Fourier transformed infrared spectroscopy, transmission electron microscopy, and scanning electron microscopy. The nanoparticles revealed a protective effect for encapsulating the attenuated antigens. The immunization study was undertaken in rabbits and it showed that nanoparticles allowed an important immunoprotection of all immunized rabbits, with no recorded death.

8.2.3 Drug Delivery Systems Based on Lipid Nanoparticles

Other types of formulations where surfactants play a very important role are lipid nanoparticles and nanostructured lipids. Surfactants are used as stabilizing agents of these nanoparticles because they reduce the interfacial energy between the lipid phase and aqueous solution during their preparation (Leonardi et al. 2014, Naseri et al. 2015).

Lipids have been used as excipients forming vehicles for delivery of poorly soluble drugs in aqueous solutions. This oily phase has been applied by different routes of administration (Rosiaux et al. 2014). The easy manipulation of lipids during their processing and their physicochemical characteristics has allowed the generation of more specific active ingredients delivery systems. The success of a new dosage form is dependent on the offered advantages when it is compared with those pharmaceutical formulations found in the market. There are many pharmaceutical formulations based on lipids, but some drawbacks that new formulation must overcome are related to the release control of the drug. The possibility of controlled drug release from a well-studied formulation such as nano-emulsions is limited due to the small size of the dispersed phase and their liquid state (Magenheim et al. 1993, Mehnert and Mäder 2012). In nano-emulsions, it has been estimated that retarded drug release requires that the drug has a very high log P (P = partition coefficient); therefore, it needs to be very lipophilic. However, these emulsions have some advantages: the main is lack of toxicity. The use of solid lipids instead of liquid oils becomes a very attractive idea to achieve controlled drug release because drug mobility in a solid lipid is lower than in liquid oil. Accordingly, solid lipids have been used for controlling the release of different drugs and consequently, they were thought as a new pharmaceutical formulation (Eldem et al. 1991). Therefore, solid lipid nanoparticles emerged in the 1990s as an alternative in order to overcome several of the disadvantages of other carrier systems such as emulsions, liposomes, and polymeric nanoparticles (Müller et al. 2002). These solid lipid nanoparticles have been preceded by the solid lipid micro-particles which were developed in the 1980s and were prepared by spray drying or spray congealing (Eldem et al. 1991).

The main characteristics of solid lipid nanoparticles such as stability or dissolution of the active agent are intrinsically dependent on their composition. Stability problems of solid lipid nanoparticles are evidenced when solidification followed by crystallization of the lipid from the dispersed phase of nanoparticles leads to the expulsion of the drug. This is because the lipid molecules are progressively crystallizing in more stable forms, yielding an increase in particle size and a decrease in the load capacity (Weber et al. 2014, Westesen et al. 1997). Therefore, a starting point for developing nanoparticles dispersed in aqueous solutions is a concentration range solid lipids from 0.1% to 30% w/w, which are stabilized by a surfactant (or a mixture of surfactants) whose concentration varies between 0.5% and 5% w/w (Joshi and Müller 2009, Müller et al. 2000). One of the most interesting features of solid lipid nanoparticles for pharmaceutical formulation development is their good drug-loading capacity. Another important characteristic (which is shared with liposomes) is their feasibility of encapsulating both hydrophilic and hydrophobic substances with different physicochemical and pharmacological properties. Additionally, it is possible to carry out "stealth" particles capable of avoiding cells of the mononuclear phagocyte system in the liver and the spleen, and the release behaviour of the active agents from the vehicle can be customized (Ghasemiyeh and Mohammadi-Samani 2018, Gordillo-Galeano and Mora-Huertas 2018, Mehnert and Mäder 2012, zur Mühlen et al. 1998).

Some years after the development of solid lipid nanoparticles, the nanostructured lipid carriers have emerged. In these carriers, a portion of solid lipids is replaced by lipid or a mixture of lipids in the liquid state for solving the crystallization and rapid release problems, and also improving the loading capability of solid lipid nanoparticles (Gordillo-

Galeano and Mora-Huertas 2018, Mehnert and Mäder 2012). Both solid lipid nanoparticles and nanostructured lipids have been considered as attractive systems for carrying poor-water soluble drugs because they are developed with biodegradable and biocompatible components. They are easy to scale industrially, can be developed as targeting systems (active or passive), can carry many different types of active agents and can be administered by different routes such as pulmonary, nasal, ocular, intravenous and intrathecal (Attama et al. 2008, Bunjes 2010, Gasco 2007, Gastaldi et al. 2014, Ghasemiyeh and Mohammadi-Samani 2018, Li et al. 2011). Among the drugs that can be transported by these carriers, we find steroidal anti-inflammatories such as prednisolone (zur Mühlen et al. 1998); non-steroidal anti-inflammatories such as ibuprofen (Chantaburanan et al. 2017) and diclofenac (Liu et al. 2011); anti-hyperlipidemic as simvastatin (Tiwari and Pathak 2011); antibiotics as levofloxacin (Islan et al. 2016); antineoplastic such as doxorubicin (Oliveira et al. 2016) and paclitaxel (Miglietta et al. 2000); antiretroviral as stavudine (Shegokar et al. 2011); and other drugs used for gene therapy such as DNA (Carrillo et al. 2013) and interference ARN (siRNA) (Xue and Wong 2011).

The methods of preparing solid lipid nanoparticles and nanostructures lipids, the suitable strategies for improving their features such as loading capability and stability, and methods for characterizing the most relevant parameters such as surface charge, size, and others will not be addressed in this section. For detailed information about these topics, refer to Ghasemiyeh and Mohammadi-Samani (2018), Gordillo-Galeano and Mora-Huertas (2018) and Müller et al. (2000).

8.2.4 Drug Delivery Systems Based on Highly Porous Materials

As we mentioned in previous sub-sections, different types of nanoparticles are employed for carrying drugs to the target site. Another option for this aim is the inorganic porous materials which have been chosen in the last decade due to their unique properties such as high drug-loading capacity, large pore volume, high specific surface area, ordered pore network, chemical and physical stabilities, relatively low toxicity and additionally, easy and cheap production in the laboratory (Baeza et al. 2017, Li et al. 2018b). The morphology of these materials can be tunable depending on the needed requirements. Additionally, the release of the drugs can be controlled producing sustainable release profiles. This feature is sometimes reached by the use of certain polymers generating hybrid materials (Benedini et al. 2017b). In consideration with the size of the carried molecules, their pore size can be tunable and the range is from 2 nm, for small molecules, to 10 or more for bigger ones. The external surface of these systems must be under strict consideration due to which its charge (which is conditioned by its ionization) and the anchored molecules play an important role in the interaction with drugs and also with cells. The aim of the anchored molecules onto the surface of the nanoparticles is based on the possibility to switch the behaviour of these systems from a passive targeting to active targeting and in this way improves their interaction with a specific type of cells. However, due to their rigid nature, porous materials have low penetration in tissues and this feature must be considered for improving many types of treatments (Baeza et al. 2017).

Some of the most relevant porous materials with potential use in human medicine and others currently used are mesoporous silica materials, porous silicon materials, highly porous graphene, and metal-organic nanoparticles. Porous silica nanoparticles are crystalline structures with pores distributed by their surface with diameters between 5 and 20 nm. This feature gives the nanoparticles high loading capacity due to which they show an external surface of 200-500 m^2g^{-1} and a pore volume of 0.5-2 cm^3g^{-1}. Alike to mesoporous materials, porous silica nanoparticles show good biodegradability and biocompatibility (Canham 1995), and do not show immunologic reactions (Anglin et al. 2008, Baeza et al. 2017, Trewyn et al. 2007). One of the main features of these devices is related to the possibility of surface functionalization with active molecules (drugs) or for becoming stealthy nanoparticles. Owing to the negative nature of their surface, electrostatic bonds retain positive molecules but if an active molecule is linked by covalent bonds, it can be released, as the material is degraded. In this context, Xia et al. (Xia et al. 2014) have reported the development of porous silicon nanoparticles as doxorubicin nanocarrier with high stability under physiological conditions, reaching efficient drug loading and controlled release of this therapeutic agent. In this work, bovine serum albumin has been used for improving the efficient loading of doxorubicin, yielding excellent long-term stability under physiological conditions and showing pH-responsive release. Additionally, Xia et al. (2014) have shown that after the interaction between nanoparticles and cells, doxorubicin was released and then it was localized in intracellular nuclei. This fact consequently demonstrates their high efficiency of killing cancer cells.

Such as we mentioned in section 8.2.1, different molecules can coat porous nanoparticles. In that section, we focused on the polymers as a coating. However, porous nanoparticles can be covered by the phospholipids used for forming liposomes. Kong et al. (Kong et al. 2015) have reported a combination of porous silicon nanoparticles and giant liposomes assembled on a microfluidic chip as an advanced platform. These liposomes have the option to be co-loaded with hydrophilic and hydrophobic drugs for their co-delivery. In this context, liposomes have loaded synthesized DNA nanostructures, short gold-nanorods, magnetic nanoparticles, doxorubicin, porous silicon nanoparticles, and the monoclonal antibody erlotinib. Therefore, these giant liposomes show photothermal and magnetic responsiveness, good biocompatibility, high loading capacity, and controllable release (Herranz-Blanco et al. 2015). The silicon nanoparticles can be used to encapsulate hydrophobic therapeutics drugs (in this case erlotinib) within the hydrophilic core of the giant liposomes, endowing high therapeutics loading capacity with tunable ratio and controllable release. Other authors have shown that porous nanomaterials can carry labile therapeutic agents and can be administrated by different routes. Li et al. have reported (Li et al. 2018) the use of porous silicon materials for medical applications focused on drug delivery and immunogenic therapy. One of the major advantages of these porous materials is their excellent biocompatibility and the ability to completely degrade into nontoxic orthosilicic acid (naturally present in the human body). The biodegradability of those materials can be customized by means of the adjustment of porosity and pore size (Salonen et al. 2008, Santos et al. 2010). Silicon porous materials developed for carrying drugs are mainly spherical shaped, micro- and/or nano-sized, easy to fabricate and different routes can dispense them. For example, they can be either applied directly or formulated into micro- or nano-composites to enhance the intestinal absorption of oral delivered small molecules. Among these molecules, we can find the anti-inflammatory, indomethacin or the chemotherapeutic agent, 5-fluorouracil. Additionally, they can guarantee the protection of labile drugs, some peptides, such as glucagon-like peptide-1, peptide tyrosine 3–36, and insulin (protein). These type of devices were also developed for intravenous administration of chemotherapeutic agents such as methotrexate and sorafenib for cancer therapy (Araújo et al. 2015, Liu et al. 2015, Shrestha et al. 2014).

Another type of highly porous materials with high potential for application has been reported by Zare-Zardini et al. (Zare-Zardini et al. 2018). In this work, the authors have shown a new generation of drug delivery systems based on the ginsenoside Rh2, Lysine- and Arginine-treated highly porous graphene for improving the anticancer activity of ginsenoside, for treating ovarian cancer (OVCAR3), breast cancer (MDA-MB), human melanoma (A375) and human mesenchymal stem cells. Thus, ginsenoside-treated graphene oxide, lysine-treated highly porous graphene, arginine-treated highly porous graphene, ginsenoside-treated graphene Lysine and the ginsenoside-treated graphene arginine were synthesized and additionally, their anticancer properties and cytotoxicity were evaluated showing encouraging results.

8.3 Nano-robots: Inspired by Nature and Machines

Nanorobotics has emerged as an application of nanotechnology to the development of nano-devices with a highlighted focus on health care (Saadeh and Vyas 2014). Nano-robots have been investigated, designed and developed for applying in different subgroups within the medical field such as microbiology (Martel 2011), dentistry (Freitas 2000), oncology (Hill et al. 2008, Li et al. 2018a, Tripathi and Kumar 2018), hematology (Freitas 1998), and neurosurgery (Saini and Saini 2010), among others.

Nano-robots or nanobots could be defined as very small devices showing a certain type of "intelligence". Therefore, we could associate this futuristic concept with the submarine which was shrunken to microscopic size ("microbe size") and then injected into the body of the injured scientist in "Fantastic Voyage", one of the most popular books of science fiction written by Isaac Asimov. Thus, nano-robots constitute a heterogeneous group of "smart" structures capable of actuation, sensing, signaling, information processing, manipulation, and behave as a swarm in a much-reduced space. The size of nano-robots is between 1 and 100 nm.

There are two main kinds of nano-robots: assemblers and self-replicators. The first type is simple cell-shaped nanobot formed by molecules or atoms of different types, and is controlled by specific programs. The second one is fundamentally an assembler nano-robot with capability for duplicating itself at a very large and fast rate (Freitas 2005).

From inside the human body, nano-robots could play a critical role in treating different diseases through different actions such as targeting tumours or carrying out organs ablations. However, one of the major challenges which these devices must overcome is related to their own mobility source. For this aim, the miniaturization of a power source onboard with a controllable propulsion and steering system has been one of the drawbacks for the implementation of such mobile robots in different fields. Martel et al. (Martel et al. 2009) have shown that flagellated nano-motors can be combined with the nano-sized magnetosomes of a single magnetotactic bacterium. This integrated device can be used as an effective propulsion system for reaching body zones with difficult accessibility due to which the devices should cross small capillaries. Another important property of these devices is related to their lack of immunologic response. For this aim, surfaces must be carefully developed for achieving a very smooth surface which allows decreasing the likelihood of triggering the body's immune system response (Verma and Chauhan 2014).

Nano-robots based on bacteria must have a smaller size than the upper limit of nanoscale definition (below 500 nm); additionally, they must have a certain speed. Thus, magnetotactic cocci are useful for the intravascular application because they gather these properties. Magnetosomes are responsible for magnetotactic behaviour of these bacteria because they are composed of Fe_3O_4 (magnetite). These bacteria have between 15 and 20 of these pseudo-organelles of 50 nm of diameter (each one) inside an invagination of the cell membrane (Komeili et al. 2006). These highly customizable devices based on bacteria could contain therapeutic agents for different therapies, antibodies or other structures anchored onto their surface for active-targeting; additionally, they could be loaded with sensors for collecting information (Ceyhan et al. 2006, Saadeh and Vyas 2014). This system can be tracked using imaging diagnosis methods such as nuclear magnetic resonance. Through different magnetic field intensities, it is possible to control bacterial behaviour from an external computer (Martel et al. 2009).

There is a nano-robot of 1000 nm with the potential application on emergency transfusions, which works as a blood component to restore primary hemostasis by carrying oxygen to the tissues. It was named respirocyte, and it was designed to fulfill the functions of the erythrocyte. Therefore, their functions are collecting oxygen as it passes through the respiratory system for distribution throughout the bloodstream, collecting carbon dioxide from tissues for release into the lungs, and metabolizing the circulating glucose for obtaining energy. These systems are able to deliver more oxygen (236 times) to the tissues per unit volume than natural erythrocytes. Respirocytes are formed by a computer on board of 58 nm and rotors for loading oxygen and carbon dioxide of 14 nm of diameter. This technology could provide an effective alternative to blood transfusions (Freitas 1998, Saadeh and Vyas 2014).

Nano-robots have been proposed for use in dental care and for improving dental treatments. They can be virtually incorporated to different kind of products used in dental care such as toothpaste or in those used for anaesthesia. Therefore, nano-robots have been proposed for tooth repair, improving the appearance, durability, and decreasing hypersensitivity (Verma and Chauhan 2014). Thus, a suspension containing millions of nano-robots could be buccal administered to the patient and then, they could enter the gingival sulcus, travelling through dental tubules to reach the pulp where they must exert their specific function. In this way, through external control nano-robots could release an analgesic or anaesthetic drug in a highly specific area in proximity to where the dentist will provide care. Additionally, nano-robots could be used in procedures such as root canal fillings or in the treatment of infection by the release of antibiotics locally (Saadeh and Vyas 2014).

Within the neurosurgery field, spinal cord injury and nerve damage are important areas of concern, and they form a significant life-altering event for affected patients. Hence, the restoring of connections of transected axons is the main step to the restoration of the nervous tissue functions. The technical capability of the surgery is limited by the small scale of the structures which must be united. Advancements in technology have led to the development of devices such as a 40 nm diameter nano-knife which has been developed and shown effective for axon-connection restoring surgery (Chang et al. 2007). For this objective, dielectrophoresis has been used. This process involves the use of electrical fields to manipulate polarizable objects in space, which is effective in achieving controlled movement of axons within a surgical field. The next step is to control the reconnection by following the transection of axons and manoeuvering them to the final position by using different methods. Consequently, these nanodevices enable the reconnection of nerves with precision and suitable control (Saadeh and Vyas 2014, Sretavan et al. 2005).

Some nano-robots have been designed (and inspired) by harnessing properties of biological materials such as peptides, proteins, DNAs, and RNAs. In this sense, an autonomous DNA robot was developed and programmed to transport therapeutic agents specifically for tumours treatment. This nano-robot was functionalized on the outside with a DNA aptamer, which binds nucleolin. This protein is specifically expressed on tumour-associated endothelial cells and the core of the nano-robot is loaded with the blood coagulation protease, the thrombin. The nano-robot becomes an active-targeting device because of the aptamer function which binds to nuclein;

additionally, this aptamer triggers the mechanical opening of the nano-robot. Therefore, thrombin is released at the tumour site generating coagulation activity. In mouse models, these nano-robots have produced thrombosis in tumour-associated blood vessels by intravenous administration. The result is the tumour necrosis and inhibition of tumour growth. The nano-robot has proven to be safe and immunologically inert in the animal models (Li et al. 2018a).

8.4 The Upcoming Fantastic Voyage

In this chapter, we have shown the evolution of delivery nano-systems which can be designed with nanometric size and with a certain "intelligence" or a smart response for improving medical treatments. Dosage forms based on nanotechnology have opened the door towards the new era of health care. Systems based on micelles were one of the first nanometric supramolecular systems applied to drug delivery and direct micelles established the first strategies for dissolving poorly-water soluble drugs. Then, liposomes have occupied a preponderant place for carrying active agents with difficulties for reaching a suitable therapeutic response. These systems can carry hydrophobic and hydrophilic drugs due to the properties of their membrane and their aqueous core. Micelles and liposomes can be designed as passive drug-targeting systems due to their size and shell properties. The emergence of PEGylated liposomes has increased the circulation time in blood and consequently, improved the therapeutic response of liposome-based systems. The development of solid lipid nanoparticles and nanostructured lipids has been possible because of the stabilizing properties of surfactants and they have become very attractive formulations for improving the delivery of hydrophobic drugs. The possibility to become these structures into an active-targeting system is an important achievement in addition to their nano-size, good stability features, and their adaptive properties.

Drug delivery nano-systems based on polymers have offered another alternative for delivering different types of active agents. Cyclodextrins, poly-lactic acid (and derivatives) and alginate are suitable molecules for carrying both hydrophilic and hydrophobic drugs, and they show responsive behaviour under different conditions.

Systems based on porous or highly porous materials have many properties shared with those aforementioned systems, nanometric size, possibility to act as an active-targeting system and biocompatibility, but they are rigid systems and this condition straitens some of their potential applications.

Nano-robots, smart self-propulsion nano-devices, have been conceived for improving medical treatments, not only for drug delivery but also as a system for providing therapeutic agents as oxygen or contrast molecules for diagnosis. The application of these devices to medical treatments brings human beings closer to the futuristic concept shown by Isaac Asimov in the "Fantastic Voyage".

References

Akbarzadeh, A., R. Rezaei-Sadabady, S. Davaran, S. W. Joo, N. Zarghami, Y. Hanifehpour, et al. 2013. Liposome: Classification, preparation, and applications. Nanoscale Res. Lett. 8(1): 102.

Alam, M. and C. T. Hartrick. 2005. Extended-release epidural morphine (DepoDur): An old drug with a new profile. Pain Pract. 5(4): 349–353.

Aliabadi, H. M., A. Mahmud, A. D. Sharifabadi and A. Lavasanifar. 2005. Micelles of methoxy poly(ethylene oxide)-b-poly(ε-caprolactone) as vehicles for the solubilization and controlled delivery of cyclosporine A. J. Control Release 104(2): 301-311.

Allen, T. M. and P. R. Cullis. 2013. Liposomal drug delivery systems: From concept to clinical applications. Adv. Drug. Deliv. Rev. 65(1): 36-48.

Anglin, E. J., L. Cheng, W. R. Freeman and M. J. Sailor. 2008. Porous silicon in drug delivery devices and materials. Adv. Drug. Deliv. Rev. 60(11): 1266-1277.

Anirudhan, T. S., P. L. Divya and J. Nima. 2015. Synthesis and characterization of silane coated magnetic nanoparticles/glycidylmethacrylate-grafted-maleated cyclodextrin composite hydrogel as a drug carrier for the controlled delivery of 5-fluorouracil. Mater. Sci. Eng. C Mater. Biol. Appl. 55: 471-481.

Araújo, F., N. Shrestha, M. -A. Shahbazi, D. Liu, B. Herranz-Blanco, E. M. Mäkilä, et al. 2015. Microfluidic assembly of a multifunctional tailorable composite system designed for site specific combined oral delivery of peptide drugs. ACS Nano. 9(8): 8291-8302.

Arima, H., Y. Hayashi, T. Higashi and K. Motoyama. 2015. Recent advances in cyclodextrin delivery techniques. Expert. Opin. Drug. Deliv. 12(9): 1425-1441.

Attama, A. A., S. Reichl and C. C. Müller-Goymann. 2008. Diclofenac sodium delivery to the eye: In vitro evaluation of novel solid lipid nanoparticle formulation using human cornea construct. Int. J. Pharm. 355(1-2): 307-313.

Aulton, M. E. 2004. Pharmaceutics: The Science of Dosage Form Design. Second Ed. Churchill Livingstone, Edinburgh.

Aynié, I., C. Vauthier, H. Chacun, E. Fattal and P. Couvreur. 1999. Spongelike alginate nanoparticles as a new potential system for the delivery of antisense oligonucleotides. Antisense Nucleic Acid Drug. Dev. 9(3): 301-312.

Baeza, A., D. Ruiz-Molina and M. Vallet-Regí. 2017. Recent advances in porous nanoparticles for drug delivery in antitumoral applications: Inorganic nanoparticles and nanoscale metal-organic frameworks. Expert. Opin. Drug Deliv. 14(6): 783-796.

Bangham, A. D. 1972. Lipid bilayers and biomembranes. Annu. Rev. Biochem. 41(1): 753-776.

Barenholz, Y. 2012. Doxil – The first FDA-approved nano-drug: Lessons learned. J. Control Release 160(2): 117-134.

Barichello, J. M., M. Morishita, K. Takayama and T. Nagai. 1999. Encapsulation of hydrophilic and lipophilic drugs in PLGA nanoparticles by the nanoprecipitation method. Drug Dev. Ind. Pharm. 25(4): 471-476.

Bartlett, D. W. and M. E. Davis. 2008. Impact of tumor-specific targeting and dosing schedule on tumor growth inhibition after intravenous administration of siRNA-containing nanoparticles. Biotechnol. Bioeng. 99(4): 975-985.

Bayindir, Z. S. and N. Yuksel. 2010. Characterization of niosomes prepared with various nonionic surfactants for paclitaxel oral delivery. J. Pharm. Sci. 99(4): 2049-2060.

Benedini, L. A., N. Andres and M. L. Fanani. 2017a. Liposomes: From the pioneers to epigenetic therapy. pp. 1—35. *In*: B. Paerson (ed.). From Liposomes: Historical, Clinical and Molecular Perspectives. Nova Science Publishers, N. Y. USA.

Benedini, L., D. Placente, O. Pieroni and P. Messina. 2017b. Assessment of synergistic interactions on self-assembled sodium alginate/nano-hydroxyapatite composites: To the conception of new bone tissue dressings. Colloid Polym. Sci. 295(11): 2109-2113.

Benedini, L., P. V. Messina, R. H. Manzo, D. A. Allemandi, S. D. Palma, E. P. Schulz, et al. 2010. Colloidal properties of amiodarone in water at low concentration. J. Colloid. Interface Sci. 342(2): 407-414.

Betancourt, T., J. D. Byrne, N. Sunaryo, S. W. Crowder, M. Kadapakkam, S. Patel, et al. 2009. PEGylation strategies for active targeting of PLA/PLGA nanoparticles. J. Biomed. Mater. Res. A. 91(1): 263-276.

Bozzuto, G. and A. Molinari. 2015. Liposomes as nanomedical devices. Int. J. Nanomedicine 10: 975-999.

Bunjes, H. 2010. Lipid nanoparticles for the delivery of poorly water-soluble drugs. J. Pharm. Pharmacol. 62(11): 1637-1645.

Canham, L. T. 1995. Bioactive silicon structure fabrication through nanoetching techniques. Adv. Mater. 7(12): 1033-1037.

Carrillo, C., N. Sánchez-Hernández, E. García-Montoya, P. Pérez-Lozano, J. M. Suñé-Negre, J. R. Ticó, et al. 2013. DNA delivery via cationic solid lipid nanoparticles (SLNs). Eur. J. Pharm. Sci. 49(2): 157-165.

Casper, J. M. 1977. Physical chemistry of surfaces (3rd Ed.), Arthur W. Adamson, Wiley-Interscience, New York, 1976, 698 pp. Journal of Polymer Science: Polymer Letters Edition, 15(10): 632-633.

Ceyhan, B., P. Alhorn, C. Lang, D. Schüler and C. M. Niemeyer. 2006. Semisynthetic biogenic magnetosome nanoparticles for the detection of proteins and nucleic acids. Small. 2(11): 1251-1255.

Chacón, M., J. Molpeceres, L. Berges, M. Guzmán and M. R. Aberturas. 1999. Stability and freeze-drying of cyclosporine loaded poly(D,L lactide-glycolide) carriers. Eur. J. Pharm. Sci. 8(2): 99-107.

Challa, R., A. Ahuja, J. Ali and R. K. Khar. 2005. Cyclodextrins in drug delivery: An updated review. AAPS PharmSciTech. 6(2): E329-E357.

Chang, H. -I. and M. -K. Yeh. 2012. Clinical development of liposome-based drugs: Formulation, characterization, and therapeutic efficacy. Int. J. Nanomedicine 7: 49-60.

Chang, W. C., E. A. Hawkes, M. Kliot and D. W. Sretavan. 2007. In vivo use of a nanoknife for axon microsurgery. Neurosurgery 61(4): 683-692.

Chantaburanan, T., V. Teeranachaideekul, D. Chantasart, A. Jintapattanakit and V. B. Junyaprasert. 2017. Effect of binary solid lipid matrix of wax and triglyceride on lipid crystallinity, drug-lipid interaction and drug release of ibuprofen-loaded solid lipid nanoparticles (SLN) for dermal delivery. J. Colloid. Interface Sci. 504: 247-256.

Chavanpatil, M. D., A. Khdair, Y. Patil, H. Handa, G. Mao and J. Panyam. 2007. Polymer-surfactant nanoparticles for sustained release of water-soluble drugs. J. Pharm. Sci. 96(12): 3379-3389.

Daousani, C. and P. Macheras. 2016. Biopharmaceutic classification of drugs revisited. Eur. J. Pharm. Sci. 95: 82-87.

Deamer, D. and A. D. Bangham. 1976. Large volume liposomes by an ether vaporization method. Biochim. Biophys. Acta. 443(3): 629-634.

Deamer, D. W. 2010. From "banghasomes" to liposomes: A memoir of Alec Bangham, 1921-2010. FASEB J. 24(5): 1308-1310.

Dimitriadis, G. J. 1979. Entrapment of plasmid DNA in liposomes. Nucleic Acids Res. 6(8): 2697-2705.

Eldem, T., P. Speiser and A. Hincal. 1991. Optimization of spray-dried and congealed lipid micropellets and characterization of their surface morphology by scanning electron microscopy. Pharm. Res. 8(1): 47-54.

Fahr, A., P. Van Hoogevest, S. May, N. Bergstrand and M. L. S. Leigh. (2005). Transfer of lipophilic drugs between liposomal membranes and biological interfaces: Consequences for drug delivery. Eur. J. Pharm. Sci. 26(3-4): 251-265.

Farah, S., D. G. Anderson and R. Langer. 2016. Physical and mechanical properties of PLA, and their functions in widespread applications – A comprehensive review. Adv. Drug Deliv. Rev. 107: 367-392.

Freitas, R. 2005. A special issue on computational intelligence for bioinformatics. J. Comput. Theor. Nanosci. 2(4): 1-25.

Freitas, R. A. 1998. Exploratory design in medical nanotechnology: A mechanical artificial red cell. Artif. Cells Blood Substit. Immobil. Biotechnol. 26(4): 411-430.

Freitas, R. A. 2000. Nanodentistry. J. Am. Den. Assoc. 131(11): 1559-1565.

Gasco, M. 2007. Lipid nanoparticles: Perspectives and challenges. Adv. Drug Deliv. Rev. 59(6): 377-378.

Gastaldi, L., L. Battaglia, E. Peira, D. Chirio, E. Muntoni, I. Solazzi, et al. 2014. Solid lipid nanoparticles as vehicles of drugs to the brain: Current state of the art. Eur. J. Pharm. and Biopharm. 87(3): 433-444.

Ghasemiyeh, P. and S. Mohammadi-Samani. 2018. Solid lipid nanoparticles and nanostructured lipid carriers as novel drug delivery systems: Applications, advantages and disadvantages. Res. Pharm. Sci. 13(4): 288.

Gordillo-Galeano, A. and C. E. Mora-Huertas. 2018. Solid lipid nanoparticles and nanostructured lipid carriers: A review emphasizing on particle structure and drug release. Eur. J. Pharm. Biopharm. 133: 285-308.

Grant, G. T., E. R. Morris, D. A. Rees, P. J. C. Smith and D. Thom. 1973. Biological interactions between polysaccharides and divalent cations: The egg-box model. FEBS Lett. 32(1): 195-198.

Gregoriadis, G. 1976. The carrier potential of liposomes in biology and medicine. N. Engl. J. Med. 295(14): 765-770.

Gregoriadis, G. 1978. Liposomes in therapeutic and preventive medicine: The development of the drug-carrier concept. Ann. N. Y. Acad Sci. 308(1): 343-370.

Gregoriadis, G. and A. T. Florence. 1993. Liposomes in drug delivery. Drugs 45(1): 15-28.

Hann, I. M. and H. G. Prentice. 2001. Lipid-based amphotericin B: A review of the last 10 years of use. Int. J. Antimicrob. Agents. 17(3): 161-169.

Hattori, Y., Y. Machida, M. Honda, N. Takeuchi, Y. Yoshiike, H. Ohno, et al. 2017. Small interfering RNA delivery into the liver by cationic cholesterol derivative-based liposomes. J. Liposome Res. 27(4): 264-273.

Herranz-Blanco, B., D. Liu, E. Mäkilä, M. -A. Shahbazi, E. Ginestar, H. Zhang, et al. 2015. On-chip self-assembly of a smart hybrid nanocomposite for antitumoral applications. Adv. Funct. Mater. 25(10): 1488-1497.

Hill, C., A. Amodeo, J. V. Joseph and H. R. Patel. 2008. Nano- and microrobotics: How far is the reality? Expert. Rev. Anticancer Ther. 8(12): 1891-1897.

Immordino, M. L., F. Dosio and L. Cattel. 2006. Stealth liposomes: Review of the basic science, rationale, and clinical applications, existing and potential. Int. J. Nanomed. 1(3): 297-315.

Islan, G. A., P. C. Tornello, G. A. Abraham, N. Duran and G. R. Castro. 2016. Smart lipid nanoparticles containing levofloxacin and DNase for lung delivery. Design and characterization. Colloids Surf. B Biointerfaces. 143: 168-176.

Israelachvili, J. N. 2011. Intermolecular and Surface Forces (3rd ed.). Elsevier. Academic Press. Cambridge, Massachusetts, USA.

Israelachvili, J. N., D. J. Mitchell and B. W. Ninham. 1976. Theory of self-assembly of hydrocarbon amphiphiles into micelles and bilayers. J. Chem. Soc. Faraday Trans. 2(72): 1525.

Jiao, J. 2008. Polyoxyethylated nonionic surfactants and their applications in topical ocular drug delivery. Adv. Drug Deliv. Rev. 60(15): 1663-1673.

Jones, M. and J. Leroux. 1999. Polymeric micelles – A new generation of colloidal drug carriers. Eur. J. Pharm. Biopharm. 48(2): 101-111.

Joshi, M. D. and R. H. Müller. 2009. Lipid nanoparticles for parenteral delivery of actives. Eur. J. Pharm. and Biopharm. 71(2): 161-172.

Juliano, R. L., S. Daoud, H. J. Krause and C. W. Grant. 1987. Membrane-to-membrane transfer of lipophilic drugs used against cancer or infectious disease. Ann. N. Y. Acad. Sci. 507: 89-103.

Kastner, E., V. Verma, D. Lowry and Y. Perrie. 2015. Microfluidic-controlled manufacture of liposomes for the solubilisation of a poorly water soluble drug. Int. J. Pharm. 485(1-2).

Kawabata, Y., K. Wada, M. Nakatani, S. Yamada and S. Onoue. 2011. Formulation design for poorly water-soluble drugs based on biopharmaceutics classification system: Basic approaches and practical applications. Int. J. Pharm. 420(1): 1-10.

Khadka, P., J. Ro, H. Kim, I. Kim, J. T. Kim, H. Kim, et al. 2014. Pharmaceutical particle technologies: An approach to improve drug solubility, dissolution and bioavailability. Asian J. Pharm. Sci. 9(6): 304-316.

Khadke, S., P. Stone, A. Rozhin, J. Kroonen and Y. Perrie. 2018. Point of use production of liposomal solubilised products. Int. J. Pharm. 537(1-2): 1-8.

Komeili, A., Z. Li, D. K. Newman and G. J. Jensen. 2006. Magnetosomes are cell membrane invaginations organized by the actin-like protein MamK. Science 311(5758): 242-245.

Kong, F., X. Zhang, H. Zhang, X. Qu, D. Chen, M. Servos, et al. 2015. Inhibition of multidrug resistance of cancer cells by co-delivery of DNA nanostructures and drugs using porous silicon nanoparticles@giant liposomes. Adv. Funct. Mater. 25(22): 3330-3340.

Kumar, G. P. and P. Rajeshwarrao. 2011. Nonionic surfactant vesicular systems for effective drug delivery—An overview. Acta Pharm. Sin. B. 1(4): 208-219.

Kwon, G., M. Naito, M. Yokoyama, T. Okano, Y. Sakurai and K. Kataoka. 1997. Block copolymer micelles for drug delivery: Loading and release of doxorubicin. J. Controlled Release 48(2-3): 195-201.

Langevin, D. 1992. Micelles and microemulsions. Annu. Rev. Phys. Chem. 43(1): 341-369.

Lasic, D. D. 1992. Mixed micelles in drug delivery. Nature 355(6357): 279-280.

Le Garrec, D., S. Gori, L. Luo, D. Lessard, D. C. Smith, M. -A. Yessine, et al. 2004. Poly(N-vinylpyrrolidone)-block-poly(D,L-lactide) as a new polymeric solubilizer for hydrophobic anticancer drugs: In vitro and in vivo evaluation. J. Controlled Release 99(1): 83-101.

Lee, B. K., Y. Yun and K. Park. 2016. PLA micro- and nano-particles. Adv. Drug Deliv. Rev. 107: 176-191.

Leonardi, A., C. Bucolo, G. L. Romano, C. B. M. Platania, F. Drago, G. Puglisi, et al.

2014. Influence of different surfactants on the technological properties and in vivo ocular tolerability of lipid nanoparticles. Int. J. Pharm. 470(1-2): 133-140.

Levine, R. M., T. R. Pearce, M. Adil and E. Kokkoli. 2013. Preparation and characterization of liposome-encapsulated plasmid DNA for gene delivery. Langmuir. 29(29): 9208-9215.

Li, R., J. S. Eun and M. -K. Lee. 2011. Pharmacokinetics and biodistribution of paclitaxel loaded in pegylated solid lipid nanoparticles after intravenous administration. Arch. Pharmacal Res. 34(2): 331-337.

Li, S., Q. Jiang, S. Liu, Y. Zhang, Y. Tian, C. Song, et al. 2018a. A DNA nanorobot functions as a cancer therapeutic in response to a molecular trigger in vivo. Nature Biotechnol. 36(3): 258-264.

Li, W., Z. Liu, F. Fontana, Y. Ding, D. Liu, J. T. Hirvonen, et al. 2018b. Tailoring porous silicon for biomedical applications: From drug delivery to cancer immunotherapy. Adv. Mater. 30(24): 1703740.

Lin, H. -R. and Y. -J. Yeh. 2004. Porous alginate/hydroxyapatite composite scaffolds for bone tissue engineering: Preparation, characterization, and in vitro studies. J. Biomed. Mater. Res. 71B(1): 52-65.

Liu, D., S. Jiang, H. Shen, S. Qin, J. Liu, Q. Zhang, et al. 2011. Diclofenac sodium-loaded solid lipid nanoparticles prepared by emulsion/solvent evaporation method. J. Liposome Res. 13(6): 2375-2386.

Liu, D., F. Yang, F. Xiong and N. Gu. 2016. The smart drug delivery system and its clinical potential. Theranostics. 6(9): 1306-1323.

Liu, D., H. Zhang, E. Mäkilä, J. Fan, B. Herranz-Blanco, C. -F. Wang, et al. 2015. Microfluidic assisted one-step fabrication of porous silicon@acetalated dextran nanocomposites for precisely controlled combination chemotherapy. Biomaterials 39: 249-259.

Liu, R. (Ed.). 2018. Water-Insoluble Drug Formulation, Third Edition. CRC Press, Boca Raton.

Liu, S. -L., X. -S. Sun, X. -Y. Li, Q. -Y. Chen, H. -X. Lin, Y. -F. Wen, et al. 2018. Liposomal paclitaxel versus docetaxel in induction chemotherapy using Taxanes, cisplatin and 5-fluorouracil for locally advanced nasopharyngeal carcinoma. BMC Cancer, 18(1): 1279.

Loftsson, T., P. Jarho, M. Másson and T. Järvinen. 2005. Cyclodextrins in drug delivery. Expert Opin. Drug Delivery 2(2): 335-351.

Lopes, M., B. Abrahim, F. Veiga, R. Seiça, L. M. Cabral, P. Arnaud, et al. 2017. Preparation methods and applications behind alginate-based particles. Expert Opin. Drug Delivery 14(6): 769-782.

Maeda, H., J. Wu, T. Sawa, Y. Matsumura and K. Hori. 2000. Tumor vascular permeability and the EPR effect in macromolecular therapeutics: A review. J. Controlled Release 65(1-2): 271-284.

Magenheim, B., M. Y. Levy and S. Benita. 1993. A new in vitro technique for the evaluation of drug release profile from colloidal carriers – Ultrafiltration technique at low pressure. Int. J. Pharm. 94(1-3): 115-123.

Makadia, H. K. and S. J. Siegel. 2011. Poly Lactic-co-Glycolic Acid (PLGA) as biodegradable controlled drug delivery carrier. Polymers 3(3): 1377-1397.

Malyala, P. and M. Singh. 2010. Micro/nanoparticle adjuvants: Preparation and formulation with antigens. Methods Mol. Biol. 626: 91-101.

Marie Arockianathan, P., S. Sekar, S. Sankar, B. Kumaran and T. P. Sastry. 2012. Evaluation of biocomposite films containing alginate and sago starch impregnated with silver nano particles. Carbohydr. Polym. 90(1): 717-724.

Martel, S. 2011. Flagellated bacterial nanorobots for medical interventions in the human body. pp. 397-416. *In*: Rosen, J., B. Hannaford and R. M. Satava (eds.). Surgical Robotics. Springer, Boston, MA, US.

Martel, S., M. Mohammadi, O. Felfoul, Z. Lu and P. Pouponneau. 2009. Flagellated magnetotactic bacteria as controlled MRI-trackable propulsion and steering systems for medical nanorobots operating in the human microvasculature. Int. J. Robotics Res. 28(4): 571-582.

Martínez, A., M. Benito-Miguel, I. Iglesias, J. M. Teijón and M. D. Blanco. 2012. Tamoxifen-loaded thiolated alginate-albumin nanoparticles as antitumoral drug delivery systems. J. Biomed. Mater. Res. A. 100(6): 1467-1476.

Mathot, F., L. van Beijsterveldt, V. Préat, M. Brewster and A. Ariën. 2006. Intestinal uptake and biodistribution of novel polymeric micelles after oral administration. J. Controlled Release 111(1-2): 47-55.

Mehnert, W. and K. Mäder. 2012. Solid lipid nanoparticles. Adv. Drug Deliv. Rev. 64: 83-101.

Miglietta, A., R. Cavalli, C. Bocca, L. Gabriel and M. Rosa Gasco. 2000. Cellular uptake and cytotoxicity of solid lipid nanospheres (SLN) incorporating doxorubicin or paclitaxel. Int. J. Pharm. 210(1-2): 61-67.

Mohamed, S., N. N. Parayath, S. Taurin and K. Greish. 2014. Polymeric nano-micelles: Versatile platform for targeted delivery in cancer. Ther. Deliv. 5(10): 1101-1121.

Müller, R. H. 1991. Colloidal Carriers for Controlled Drug Delivery and Targeting: Modification, Characterization, and In Vivo Distribution (1st Ed.). CRC Press, Taylor and Francis Group, Stuttgart.

Müller, R. H., M. Radtke and S. A. Wissing. 2002. Solid lipid nanoparticles (SLN) and nanostructured lipid carriers (NLC) in cosmetic and dermatological preparations. Adv. Drug Deliv. Rev. 54: S131-S155.

Müller, R., K. Mäder and S. Gohla. 2000. Solid lipid nanoparticles (SLN) for controlled drug delivery: A review of the state of the art. Eur. J. Pharm. Biopharm. 50(1): 161-177.

Nagarsenker, M. S., L. Amin and A. A. Date. 2008. Potential of cyclodextrin complexation and liposomes in topical delivery of ketorolac: In vitro and in vivo evaluation. AAPS PharmSciTech. 9(4): 1165-1170.

Nahaei, M., H. Valizadeh, B. Baradaran, M. Nahaei, D. Asgari, S. Hallaj-Nezhadi, et al. 2013. Preparation and characterization of chitosan/β-cyclodextrin nanoparticles containing plasmid DNA encoding interleukin-12. Drug Res. 63(01): 7-12.

Nait Mohamed, F. A. and F. Laraba-Djebari. 2016. Development and characterization of a new carrier for vaccine delivery based on calcium-alginate nanoparticles: Safe immunoprotective approach against scorpion envenoming. Vaccine 34(24): 2692-2699.

Naseri, N., H. Valizadeh and P. Zakeri-Milani. 2015. Solid lipid nanoparticles and nanostructured lipid carriers: Structure, preparation and application. Adv. Pharm. Bull. 5(3): 305-313.

Oliveira, M. S., S. V. Mussi, D. A. Gomes, M. I. Yoshida, F. Frezard, V. M. Carregal, et al. 2016. α-Tocopherol succinate improves encapsulation and anticancer activity of doxorubicin loaded in solid lipid nanoparticles. Colloids Surf., B. 140: 246-253.

Olusanya, T. O. B., R. R. Haj Ahmad, D. M. Ibegbu, J. R. Smith and A. A. Elkordy. 2018. Liposomal drug delivery systems and anticancer drugs. Molecules 23(4): E907.

Park, T. G. 1995. Degradation of poly(lactic-co-glycolic acid) microspheres: Effect of copolymer composition. Biomaterials 16(15): 1123-1130.

Patil, Y. P. and S. Jadhav. 2014. Novel methods for liposome preparation. Chem. Phys. Lipids. 177: 8-18.

Peltonen, L., J. Aitta, S. Hyvönen, M. Karjalainen and J. Hirvonen. 2004. Improved entrapment efficiency of hydrophilic drug substance during nanoprecipitation of poly(l)lactide nanoparticles. AAPS PharmSciTech. 5(1): E16.

Placente, D., L. A. Benedini, M. Baldini, J. A. Laiuppa, G. E. Santillán and P. V. Messina. 2018. Multi-drug delivery system based on lipid membrane mimetic coated nano-hydroxyapatite formulations. Int. J. Pharm. 548(1): 559-570.

Pouton, C. W. 2006. Formulation of poorly water-soluble drugs for oral administration: Physicochemical and physiological issues and the lipid formulation classification system. Eur. J. Pharm. Sci. 29(3-4): 278-287.

Puoci, F. (Ed.). 2015. Advanced Polymers in Medicine. Springer International Publishing, Cham.

Quirolo, Z., L. Benedini, M. Sequeira, M. Herrera, T. Veuthey and V. Dodero. 2014. Understanding recognition and self-assembly in biology using the chemist's toolbox: Insight into medicinal chemistry. Curr. Top. Med. Chem. 14(6): 730-739.

Rafiei, P. and A. Haddadi. 2017. Docetaxel-loaded PLGA and PLGA-PEG nanoparticles for intravenous application: Pharmacokinetics and biodistribution profile. Int. J. Nanomed. 12: 935-947.

Rasoulianboroujeni, M., G. Kupgan, F. Moghadam, M. Tahriri, A. Boughdachi, P. Khoshkenar, et al. 2017. Development of a DNA-liposome complex for gene delivery applications. Mater. Sci. Eng. C. 75: 191-197.

Reis, C. P., A. J. Ribeiro, F. Veiga, R. J. Neufeld and C. Damgé. 2008. Polyelectrolyte biomaterial interactions provide nanoparticulate carrier for oral insulin delivery. Drug Deliv. 15(2): 127-139.

Rosiaux, Y., V. Jannin, S. Hughes and D. Marchaud. 2014. Solid lipid excipients – Matrix agents for sustained drug delivery. J. Controlled Release 188: 18-30.

Rowe, R., P. Sheskey and M. Quinn. 2009. Handbook of Pharmaceutical Excipients. Sixth Edition, pp. 549-553.

Ruckmani, K., V. Sankar and M. Sivakumar. 2010. Tissue distribution, pharmacokinetics and stability studies of zidovudine delivered by niosomes and proniosomes. J. Biomed. Nanotechnol. 6(1): 43-51.

Saadeh, Y. and D. Vyas. 2014. Nanorobotic applications in medicine: Current proposals and designs. Am. J. Robot. Surg. 1(1): 4-11.

Saini, R. and S. Saini. 2010. Nanotechnology and surgical neurology. Surg. Neurol. Int. 1: 57.

Salonen, J., A. M. Kaukonen, J. Hirvonen and V. -P. Lehto. 2008. Mesoporous silicon in drug delivery applications. J. Pharm. Sci. 97(2): 632-653.

Santos, H. A., J. Riikonen, J. Salonen, E. Mäkilä, T. Heikkilä, T. Laaksonen, et al. 2010. In vitro cytotoxicity of porous silicon microparticles: Effect of the particle concentration, surface chemistry and size. Acta Biomater. 6(7): 2721-2731.

Sartuqui, J., N. L. D'Elía, A. N. Gravina and L. A. Benedini. 2017. Application of natural, semi-synthetic, and synthetic biopolymers used in drug delivery systems design. pp. 38-65. *In*: J. M. Ruso and P. V. Messina (eds.). Biopolymers for Medical Applications (1st Ed.). CRC Press, Taylor and Francis Group, Boca Raton.

Sha, L., Z. Chen, Z. Chen, A. Zhang and Z. Yang. 2016. Polylactic acid based nanocomposites: Promising safe and biodegradable materials in biomedical field. Int. J. Polymer Sci. 2016: 1-11.

Shegokar, R., K. K. Singh and R. H. Müller. 2011. Production and stability of stavudine solid lipid nanoparticles—From lab to industrial scale. Int. J. Pharm. 416(2): 461-470.

Shelley, H. and R. J. Babu. 2018. Role of cyclodextrins in nanoparticle-based drug delivery systems. J. Pharm. Sci. 107(7): 1741-1753.

Shrestha, N., M. -A. Shahbazi, F. Araújo, H. Zhang, E. M. Mäkilä, J. Kauppila, et al. 2014. Chitosan-modified porous silicon microparticles for enhanced permeability of insulin across intestinal cell monolayers. Biomaterials, 35(25): 7172-7179.

Shuai, X., T. Merdan, A. K. Schaper, F. Xi and T. Kissel. 2004. Core-cross-linked polymeric micelles as paclitaxel carriers. Bioconjugate Chem. 15(3): 441-448.

Sretavan, D. W., W. Chang, E. Hawkes, C. Keller and M. Kliot. 2005. Microscale surgery on single axons. Neurosurgery 57(4): 635-646.

Stapleton, S., C. Allen, M. Pintilie and D. A. Jaffray. 2013. Tumor perfusion imaging predicts the intra-tumoral accumulation of liposomes. J. Controlled Release 172(1): 351-357.

Stone, N. R. H., T. Bicanic, R. Salim and W. Hope. 2016. Liposomal amphotericin B (AmBisome(®)): A review of the pharmacokinetics, pharmacodynamics, clinical experience and future directions. Drugs 76(4): 485-500.

Strieth, S., C. Dunau, U. Michaelis, L. Jäger, D. Gellrich, B. Wollenberg, et al. 2014. Phase I/II clinical study on safety and antivascular effects of paclitaxel encapsulated in cationic liposomes for targeted therapy in advanced head and neck cancer. Head & Neck 36(7): 976-984.

Takagi, T., C. Ramachandran, M. Bermejo, S. Yamashita, L. X. Yu and G. L. Amidon. 2006. A provisional biopharmaceutical classification of the top 200 oral drug products in the United States, Great Britain, Spain, and Japan. Mol. Pharm. 3(6): 631-643.

Tiwari, G., R. Tiwari and A. Rai. 2010. Cyclodextrins in delivery systems: Applications. J Pharm. Bioallied. Sci. 2(2): 72-79.

Tiwari, R. and K. Pathak. 2011. Nanostructured lipid carrier versus solid lipid nanoparticles of simvastatin: Comparative analysis of characteristics, pharmacokinetics and tissue uptake. Int. J. Pharm. 415(1-2): 232-243.

Torchilin, V. P. 2001. Structure and design of polymeric surfactant-based drug delivery systems. J. Controlled Release. 73(2-3): 137-172.

Torchilin, V. P. 2006. Micellar nanocarriers: Pharmaceutical perspectives. Pharm. Res. 24(1): 1-16.

Torchilin, V. P. 2010. Passive and active drug targeting: Drug delivery to tumors as an example. Handb. Exp. Pharmacol. (197): 3-53.

Trewyn, B. G., I. I. Slowing, S. Giri, H. -T. Chen and V. S. -Y. Lin. 2007. Synthesis and functionalization of a mesoporous silica nanoparticle based on the sol-gel process and applications in controlled release. Acc. Chem. Res. 40(9): 846-853.

Tripathi, R. and A. Kumar. 2018. Application of nanorobotics for cancer treatment. Mater. Today 5(3): 9114-9117.

Ungaro, F., I. D'Angelo, C. Coletta, R. d'Emmanuele di Villa Bianca, R. Sorrentino, B. Perfetto, et al. 2012. Dry powders based on PLGA nanoparticles for pulmonary delivery of antibiotics: Modulation of encapsulation efficiency, release rate and lung deposition pattern by hydrophilic polymers. J. Controlled Release 157(1): 149-159.

Verma, S. K. and R. Chauhan. 2014. Nanorobotics in dentistry – A review. Indian J. Dent. 5: 62-70.

Weber, S., A. Zimmer and J. Pardeike. 2014. Solid lipid nanoparticles (SLN) and nanostructured lipid carriers (NLC) for pulmonary application: A review of the state of the art. Eur. J. Pharm. Biopharm. 86(1): 7-22.

Westesen, K., H. Bunjes and M. H. Koch. 1997. Physicochemical characterization of lipid nanoparticles and evaluation of their drug loading capacity and sustained release potential. J. Controlled Release 48(2-3): 223-236.

Xia, B., W. Zhang, J. Shi and S. Xiao. 2014. A novel strategy to fabricate doxorubicin/ bovine serum albumin/porous silicon nanocomposites with pH-triggered drug delivery for cancer therapy in vitro. J. Mater. Chem. B. 2(32): 5280.

Xia, Y., J. Tian and X. Chen. 2016. Effect of surface properties on liposomal siRNA delivery. Biomaterials 79: 56-68.

Xu, P., E. A. Van Kirk, S. Li, W. J. Murdoch, J. Ren, M. D. Hussain, et al. 2006. Highly stable core-surface-crosslinked nanoparticles as cisplatin carriers for cancer chemotherapy. Colloids Surf. B. 48(1): 50-57.

Xu, X., L. Wang, H. -Q. Xu, X. -E. Huang, Y. -D. Qian and J. Xiang. 2013. Clinical comparison between paclitaxel liposome (Lipusu®) and paclitaxel for treatment of patients with metastatic gastric cancer. Asian Pac. J. Cancer Prev. 14(4): 2591-2594.

Xue, H. Y. and H. L. Wong. 2011. Tailoring nanostructured solid-lipid carriers for time-controlled intracellular siRNA kinetics to sustain RNAi-mediated chemosensitization. Biomaterials 32(10): 2662-2672.

Yaksh, T. L., J. C. Provencher, M. L. Rathbun, R. R. Myers, H. Powell, P. Richter, et al. 2000. Safety assessment of encapsulated morphine delivered epidurally in a sustained-release multivesicular liposome preparation in dogs. Drug Deliv. 7(1): 27-36.

Yang, Y., J. Wang, H. Shigematsu, W. Xu, W. M. Shih, J. E. Rothman, et al. 2016. Self-assembly of size-controlled liposomes on DNA nanotemplates. Nature Chem. 8(5): 476-483.

Yu, L. X., G. L. Amidon, J. E. Polli, H. Zhao, M. U. Mehta, D. P. Conner, et al. 2002. Biopharmaceutics classification system: The scientific basis for biowaiver extensions. Pharm. Res. 19(7): 921-925.

Zafar, N., H. Fessi and A. Elaissari. 2014. Cyclodextrin containing biodegradable particles: From preparation to drug delivery applications. Int. J. Pharm. 461(1-2): 351-366.

Zare-Zardini, H., A. Taheri-Kafrani, A. Amiri and A. -K. Bordbar. 2018. New generation of drug delivery systems based on ginsenoside Rh2-, Lysine- and Arginine-treated highly porous graphene for improving anticancer activity. Sci. Rep. 8(1): 586.

Zhang, N., J. Li, W. Jiang, C. Ren, J. Li, J. Xin and K. Li. 2010. Effective protection and controlled release of insulin by cationic beta-cyclodextrin polymers from alginate/chitosan nanoparticles. Int. J. Pharm. 393(1-2): 212-218.

Zhu, L., P. Kate and V. P. Torchilin. 2012. Matrix metalloprotease 2-responsive multifunctional liposomal nanocarrier for enhanced tumor targeting. ACS Nano 6(4): 3491-3498.

zur Mühlen, A., C. Schwarz and W. Mehnert. 1998. Solid lipid nanoparticles (SLN) for controlled drug delivery – Drug release and release mechanism. Eur. J. Pharm. Biopharm. 45(2): 149-155.

Zylberberg, C. and S. Matosevic. 2016. Pharmaceutical liposomal drug delivery: A review of new delivery systems and a look at the regulatory landscape. Drug Deliv. 23(9): 3319-3329.

Diagnostic Test with Targeted Therapy for Cancer: The Theranostic Nanomedicine

"...Original minds are not distinguished by being the first to see a new thing, but instead by seeing the old, familiar thing that is overlooked as something new..."

Friedrich (Wilhelm) Nietzsche (1844-1900)
German-Swiss philosopher and scholar

9.1 A Novel Tactic against Cancer

Theranostic therapy constitutes an encouraging approach that could be defined as a new strategy for both treating and diagnosing some complex diseases through a combination of two well-known concepts: imaging diagnosis and treatment by means of active molecules. Traditionally, a two-steps procedure is carried out for addressing the treatment of a certain disease. First, an imaging diagnosis step and then, based on its results a drug therapy must be agreed. Consequently, theranostic therapy is perceived as a combination of traditional contrast agents and/or therapeutic drugs or molecules with certain properties (magnetism, for example) for the attainment of a one-step diagnosis-treatment process (Sumer and Gao 2008). However, the current relevance of that concept has been accomplished through the simultaneous advance in different fields such as pharmacy, medicine (including imaging diagnosis), chemistry and physics (including nano-science) and also biology. Beyond the prominent advantages, which have naturally arisen from the theranostic concept, there are others based on the advances in nanoscience and nanomedicine, which generate an integration of both disciplines for allowing the conception of a new paradigm, which enables the inclusion of cell-targeting, the generation of ultra-sensitive imaging for diagnosis and consequently, a more selective therapy for patients.

Theranostic nanomedicine could be also proposed as a tool for predicting the treatment response. Through this therapy, the drug delivery, drug release and/or the efficacy of the drug can be monitored through images (Lammers et al. 2011). Another concept, although similar but not equivalent, can emerge from all mentioned above; thus, theranostic could be accurately defined as image-guided drug delivery. This strategy allows us to carry the therapeutic substance towards the target site by means of the application of certain physical properties such as magnetism. Additionally, the active compound can be delivered and its release into that site can also be controlled avoiding or decreasing possible collateral effects in other organs or tissues.

In this chapter, a brief description of the tumour microenvironment, the cell heterogeneity in tumour and the possibility for developing active-targeting and passive-targeting drug delivery systems for application on cancerous tissues will be addressed as an introductory topic. The possibility to develop targeting systems will be analyzed as one of the strengths of this therapy and, finally, the theranostic medicine principles will be described based on the application of three different sources of stimuli such as light, magnetism, and ultrasound.

9.2 Accessing the Tumour, Exploring Its Microenvironment: Passive and Active Tumour Targeting

Poor pharmacokinetics profiles of anti-cancer drug delivery systems and their non-specific distribution through the body of patients leads to severe side effects and toxicity due to which drugs reach both normal and cancer cells similarly. These events, consequently, can lead to the failure of the treatment or a worsening of the health state; this treatment can also lead to the death of the patient. Due to that fact, anti-tumour drug delivery systems could be designed by considering two important features related to their target site: variations of the tumour microenvironment and variations of its angiogenesis. The targeting of the drug delivery system to the specific site of action must be defined depending on the size and surface features of the formulation. In one way, nanometric devices developed through nanotechnology can get into the tumour through its surrounding leaky-vasculature by means of passive penetration because this vasculature shows a raised permeability and an increased retention effect in that site. On the other hand, devices grafted at their surface with a ligand molecule which specifically interacts with those molecules (which interact by means of specific interactions) overexpressed on tumour cells or angiogenic endothelial cells are defined as active targeting devices (Danhier et al. 2010). The challenges of the targeting of nanocarrier systems are related with finding the proper target for a particular disease (complex diseases such as cancer), the correct drug for treating the disease, and how the drug can be carried.

In this section, we describe the structures of the tumour microenvironment, cell heterogeneity and different types of targeting (active and passive) used for improving chemotherapeutic formulations that can be applied in theranostic therapy for cancer.

9.2.1 Tumour Microenvironment

The knowledge about tumour microenvironment leads the researchers to understand and predict how a formulation can behave in that environment and consequently, how an anti-cancer drug system should be designed or which are the necessary features of a contrast agent to overcome the morphologic drawbacks produced for this pathology. Cancerous tissues present some differences in relation to normal ones which encompass vascular abnormalities, decreased pH values, among others. The angiogenesis into a cancerous tissue starts after the growing of the solid tumour from 1 to 2 mm^3 and it is carried out in several steps. Under this condition, the tissue suffers hypoxia that triggers endothelial cell activation and finishes with the vasculature remodeling. Therefore, tumours have the capacity to progress from a non-angiogenic to the angiogenic phenotype. This process is called "angiogenic switch" and plays an important role in the progression of cancer because they provide a supply of oxygen and other nutrients, and allows the dissemination of cancer cells throughout the body (metastasis) (Bergers and Benjamin 2003, Chen and Ni 2014, Naumov et al. 2006).

The use of long-circulating drug delivery systems establishes a promising strategy to overcome the structural vascular abnormalities in some types of cancer. These abnormalities are characterized by a pericyte deficiency, a high proportion of proliferating endothelial cells, and aberrant basement membrane formation. Owing to these modifications, the vascular permeability is increased. Therefore, drug devices, such as nanocarriers (20–200 nm) or nano-sized contrast agents, can get to the interstitial space and accumulate inside the target tissue because of endothelial pores which have a size range between 10 and 1000 nm. Additionally, drugs are retained in the tumour tissue because their drainage is inefficient due to which lymphatic vessels are absent or non-functional in this tissue. This phenomenon has been named the "enhanced permeability and retention effect," and constitutes an important approach of passive targeting (Maeda et al. 2009, Torchilin 2000).

Owing to vascular abnormalities, some changes in blood flow inside the tumour mass are expected. Hence, the transport properties of nano-sized drug delivery systems and contrast agents are affected. Within the tumour, the osmotic pressure is one or two orders of magnitude higher than normal tissues and consequently, this property becomes a barrier for the anti-cancer treatment (Jain 1987).

Another parameter that shows a variation between normal and tumour tissue is the pH. Tumour pH value at the extracellular environment is in the range of 6.0-7.0, and this value is around 7.4 for normal tissues (Cardone et al. 2005). This decrease in pH values in tumour extracellular place is correlated to a cytosolic alkalization in those cells. Hence, the variation of pH in the extracellular and intracellular environment could be used for designing a pH-responsive drug delivery system. This system will keep non-ionized at low pH and consequently, it can cross the cell barriers and then inside cytosol, its ionized molecular form avoids it to get out of cells (Feron 2009).

9.2.2 Cellular Heterogeneity in Cancer

Histological analysis of tumour cells has revealed substantial heterogeneity in their intracellular content, mainly in chromatin, compared to normal tissues. These differences are probably due to variations in genomic and epigenomic profiles, which yield heterogeneous phenotypes. Normal cells sharing tissue with tumour cells represent a small portion for most cancers, and their presence, which is determined by their genomic/epigenomic markers, might not be significantly recognized within tumours. However, the influence of normal cells on tumour cells might partially explain the genomic and epigenomic profiles of tumours which lead to a drastic phenotypic switch in tumour cells (Lelièvre et al. 2014). One of the main recognized sources to generate diversity among tumours is genetic heterogeneity. There are mixtures of point mutations and microsatellite variations as well as chromosomal aberrations that are considered responsible for the genetic modifications. It is well known that due to the genetic heterogeneity, not only differences of tumour tissue among patients are found but also within different regions of a tumour of a certain patient (Boyd et al. 2012). There is a variety of epigenetic modifications (such as histone modifications or DNA methylation) (Benedini et al. 2017) which can be compared with those based on histopathology and genomics (determination of gene sequences and mutations) (Lelièvre et al. 2014). The possibility of identifying these modifications for recognizing different subtypes of a given cancer by means of the identification of different meaningful biomarkers (given in each case) will improve the information, diagnosis and treatment of cancer. In this context, this information is useful for applying a theranostic personalized therapy due to which diagnostic tools, contrast agents and therapeutic agents (chemicals or physicals) should be designed and/or tailored to this information.

9.2.3 Active and Passive Drug Targeting

Passive targeting involves the transport of drugs and other substances through tumour capillary fenestrations towards the extracellular and intracellular tumour space by means of passive diffusion. The diffusion is defined as a process of transport for molecules across the cell membrane, according to a gradient of concentration, and without a contribution of cellular energy. Owing to the interstitial increase of osmotic pressure, the diffusion is the major mode of drug transport (Danhier et al. 2010).

Like we have mentioned in the previous section, selective accumulation of carriers and/or drugs within tumour tissue may be produced by the enhanced permeability and retention effect (Haley and Frenkel 2008). This effect is now becoming a guiding principle in cancer-targeting drug designing because it is applicable to almost all rapidly growing solid tumours (Maeda et al. 2009). Due to that fact, the concentration of anti-cancer drugs within the tumour can be extremely higher than that in normal cells (Iyer et al. 2006). Thus, there are some properties which have an important role in drug delivery systems design. The first one is the size; it must be less than 400 nm to get into the tumour through their vessel fenestrations. For avoiding the capture by the liver, the size must be lower than 100 nm and larger than 10 nm for escaping the kidney's filtration. The second consideration is the charge of the system, which should be neutral for avoiding direct renal elimination. Lastly, the systems should be stealthy for evading the capturing from reticuloendothelial systems (Benedini et al. 2017, Malam et al. 2009).

Active targeting involves the addressing of drug delivery systems towards the active site through ligands anchored at their surface. The ligands should be specifically designed and synthesized for binding the receptors particularly expressed at the target site. For example, a certain type of ligand could be chosen to bind to a receptor homogeneously overexpressed on tumour cells (Adams et al. 2001) or on the inner surface of its vasculature (Gosk et al. 2008) but not expressed by normal cells. Sometimes, the active principle can behave as a therapeutic and also as a ligand molecule. The active targeting can be focused on: active targeting against tumour cells or against its vasculature. The internalization of the active agent to the tumour tissue is the main aim of active targeting against the tumour. This strategy is able to be used for internalization of chemotherapeutic agents and also large macromolecules such as DNA or siRNA (Kirpotin et al. 2006). Through active targeting, tumour cells are killed by the direct effect of the active agent against the tumour tissue almost independently of its vasculature structure (Pastorino et al. 2006). There are many molecules that can be anchored to the surface of nanocarriers such as folic acid (Low and Kularatne 2009) which could be grafted to the surface of the carrier or it could form conjugates with the therapeutic molecule. Transferrin (Daniels et al. 2006) and lectins (Bies et al. 2004) are other molecules used for active targeting. The later generates interaction with carbohydrates (Minko 2004) and the first with a specific receptor. The inhibition or block of the epidermal growth factor receptor can be achieved by means of neutralizing monoclonal antibodies and small-molecule tyrosine kinase inhibitors anchored to the surface of the pharmaceutical formulation. It can also interact specifically with that receptor, which is over-expressed on the surface of certain types of tumour cells (Acharya et al. 2009). However, it has been reported that the use of this strategy provides encouraging clinical responses in only 15% of patients treated (Martinelli et al. 2009). Active targeting against the vasculature of the tumour tissue has some advantages; first, internalization of the drug into the tumour tissue is not necessary. The second is related to the distribution of the drug; hence, there is the possibility of interaction between the drug and the endothelium (target site) immediately after intravenous injection. The third advantage is related to the higher genetic stability of the vasculature cells with respect to the tumour tissue and the last advantage, one of the most important features of the vessels, is the high expression of molecules onto the surface of the endothelium cells. This last can generate interaction with molecules anchored to the surface of the drug delivery system and thus, they can direct the drug to a certain target site expressed on the surface of endothelium. Among the molecules involved in this interaction, there are the vascular endothelial growth factors which are overexpressed in tumour hypoxia states and, therefore, contribute to neovascularization (Shadidi and Sioud 2003). Another molecule is a certain type of integrin which is highly expressed on neovasculature (Desgrosellier and Cheresh 2010, Paolillo et al. 2009). In this sense, the vascular cell adhesion molecule 1 (VCAM-1) is another molecule also expressed in tumours vasculature which is used for directing the treatment (Dienst et al. 2005). Other types of molecules such as zinc endopeptidases (proteins) can also be an objective of the drug delivery system (Genís et al. 2006). Therefore, multifunctional nano-carriers can be designed combining therapeutics and diagnosis strategies based on some of the mentioned features such as long-circulating capacity, response sensibility against temperature variations or pH, and types of targeting (active and passive). In this context, theranostic approaches emerge as a new tool based on two separated concepts, therapeutics and diagnosis. Thus, in the following section of this chapter, different strategies used as theranostic based on nanomedicines for cancer treatment will be described.

9.3 Cancer Theranostic

The progress of both molecular imaging and molecular therapeutics has involved high efforts in multidisciplinary areas. This combination defined as "theranostic" can include: molecular biomarkers and imaging probes for cancer diagnosis, cancer therapy guided by images, and can also involve the development of nanostructures with diagnosis and therapy capabilities. Therefore, cancer theranostics is a combination of strategies for improving the care of patients, decreasing the times between the diagnosis and treatments (when it is possible) and thus, this procedure can be proposed as a personalized treatment for cancer (Chen and Wong 2014). One of the major aims of the theranostic is optimizing the efficiency and safety of the devices used for diagnosis and treatment. In this context, the strengths of this strategy are based on the application of certain properties of the devices such as light responsiveness, sensitivity to the sound or magnetism for the development of systems, which generate an improved diagnosis, and treatment of cancer disease jointly and possibly personalized. This therapy is based on the application of stimuli that can be exogenous or endogenous to reach the release of the content (drug or contrast agent or both) from the nano-carrier. Among endogenous stimuli we can mention some examples: when a formulation with active components go through the cytosol of the cell and reach lysosomes, they can suffer a change on their ionization due to which pH is modified, and this effect can be used for releasing the active compound into the lysosomes (Zhao et al. 2015); enzymatic reactions produced in different organelles is another example (Coll et al. 2011, Luo et al. 2014, Rychak and Klibanov 2014); however, these kinds of processes are not described in this text. We will focus on exogenous stimuli for releasing active components (drugs or contrast agents) from nano-carriers such as light (ultraviolet, visible or near infrared light) (Bao et al. 2012, Zhao et al. 2014), magnetic radiation (Reddy et al. 2012), and ultrasound (Malietzis et al. 2013).

9.3.1 Light

For a long time, the use of light has an important role in medicine. Since antiquity, the application of light as a therapy in medicine has been reported for treating skin diseases, neonatal jaundice, sleep, and psychological disorders, and also cancer (Ackroyd et al. 2001). With advances in the light technology, the development of optical probes/molecules and also with the understanding of the optical properties of tissues, the interest in the use of light for treating different types of cancer has grown considerably. Differences in optical properties of tissue, such as absorption, scattering, polarization and fluorescence, have led to the emergence of several modalities for imaging generation applied to diagnosis including fluorescence imaging and optical tomography, among other techniques (Frangioni 2003, Hilderbrand and Weissleder 2010).

The importance of light responsiveness of a device for application on theranostic therapy lies in its precise capability to be externally modulated (diagnosis agents and/or drug release) by means of light incidence on the target site. Consequently, this fact provides high selectivity and specificity due to the confinement of the light application on the place of the lesion. Therefore, the advantages of devices activated by light are related to the correct rate of release of the active agent (contrast agent or drug) when the device is irradiated and this behaviour improves the possibilities of release within the target site (Chen and Zhao 2018).

Absorption of light by molecules could lead to three types of processes: the first one can be applied to chemotherapeutic drug or another agent. In the first case, photochemical reactions of molecules themselves (photolysis of prodrugs which turned into active drugs) or absorption of light by photosensitive molecules generates physical changes into the formulation and thus, the encapsulated drug is released. For the interaction with other agents, the light can also interact with other molecules such as photosensitizers and these activated molecules can interact with oxygen to generate singlet oxygen for producing the death of the tissue (photodynamic therapy). The second process is produced by the emission of light due to radiative relaxation of molecules (fluorescence), and in the last process, light energy can be transferred to other forms due to nonradiative relaxation of molecules, heat (photothermal therapy). We will focus on photodynamic and photothermal therapies; however, some examples of the other processes are also briefly described.

Neils Finsen developed photodynamic therapy in Denmark. It is a kind of therapy, which uses light properties for treating diseases. In humans, the action of the light would generate oxygen reactive species (singlet oxygen formation) with lethal consequences if certain molecules such as porphyrins did not quench them. Therefore, disturbance in the synthesis of porphyrins causes porphyria. The patients with this disease show skin sensitivity to sunlight due to a pattern of over-production, accumulation, and excretion of intermediaries of haem group biosynthesis. This skin photosensitization is due to porphyrin accumulation and through the consequent photodynamic action (activation by light), the singlet oxygen is produced which induces tissue damage (Ackroyd et al. 2001, Daniell and Hill 1991). This phenomenon has been used for the design of photodynamic therapy. This therapy involves two components, which are non-toxic separately. One of them is the photosensitizing agent (photosensitizer) and the other is a source of light with an appropriated wavelength for activating the photosensitizer (Dolmans et al. 2003).

Chromophores are light-absorbing molecules and they reach states that are more excited by absorption of light. Thus, the excited chromophore returns to the ground (basal) state by the release of energy as fluorescence or heat. When the molecule generates fluorescence, the light released has an equal or longer wavelength than the incident one. However, some chromophores can produce triplet-excited states that are able to transfer the excess of energy to the nearby molecules (or cells). This process has been described as photosensitization (Liu et al. 2014).

Generally, long wavelength light, such as the visible light (400-700 nm) (Chen and Zhao 2018), is used. However, for some specific cases, values of wavelength near ultraviolet (200-400 nm) and near-infrared (700-1000 nm) have been reported (Dolmans et al. 2003). For reaching an improved tissue penetration depth, near-infrared should be used (Zou et al. 2019). For example, the penetration depth after application of near-infrared light (650–900 nm) to skin is up to 2 cm (Chen and Zhao 2018).

The agent frequently used, inherently fluorescent, is a non-toxic drug in the absence of the activating light or a dye. Porphyrins are one of the main photosensitizing agents used in this therapy. They have a porphine structure along with a side chain, which is usually metallic. Porphine is formed by tetrapyrrole backbone connected by bridges of methine in a cyclic configuration (O'Connor et al. 2009). Depending on the specific structure of these agents, the absorption bands are ranged from 600 to 800 nm (Abrahamse and Hamblin 2016). Other tetrapyrrole structures used as photosensitizers are the chlorins which are naturally derived from chlorophyll. Their activation wavelength is between 650 and 700 nm (red light). Examples of these compounds are tetrahydroxyphenylchlorin, benzoporphyrin derivatives and Radachlorin (Biswas et al. 2014, Chan et al. 2010, Wagner et al. 2015). Additionally, bacteriochlorins and phthalocyanines are other types of agents based on pyrrole structure. We can mention synthetic dyes used as a photosensitizer: Phenothiazinium salts, squaraines and others based on boron-dipyrromethene named Bodipy dyes among others (Abrahamse and Hamblin 2016).

The action mechanism of these types of agents could be summarized in the following steps: once into the body of the patient (the agent can be administrated by the known ways, mainly intravenous or topically applied on the skin), the agent, they must be irradiated after a certain period of incubation. Thus, the generation of toxic species of oxygen kills the cells and, consequently, destroys the tissue. This local application avoids collateral damage (Celli et al. 2010, Rai et al. 2010). Mainly, there are two types of reactions: one of them can be defined as a direct reaction. In this reaction, the light-activated photosensitizer transfers its energy directly to the target tissue. This structure could be the cell membrane or another molecule. Therefore, in this transference, the produced radicals interact with oxygen and consequently, oxygenated products are produced. Another reaction is based on the transference of the energy to the oxygen molecules that generate reactive oxygen species, and then, these reactive species interact with the target tissue. It is important to remark that both reactions occur simultaneously and they cannot be carried out in anaerobic conditions or when the tissue has hypoxia (Dolmans et al. 2003, Gomer and Razum 1984).

Photothermal therapy is another technique for cancer treatment based on the local application of light. In this technique, electromagnetic radiation (Visible - Near Infrared light) is applied to the tissue to cause thermal damage (thermal ablation). In photothermal therapy, the energy is converted in heat by means of photo-absorbers structures which causes structural changes in proteins without using a photosensitizing agent for generating oxygen reactive species. Thus, the irradiated tissues are destroyed by the effect of the temperature ranging from 45 °C to 300 °C (Rai et al. 2010). These are the main differences between the previously described technique and the photothermal therapy; however, one of the similarities is the *in situ* irradiation of the target tissue. Owing to this strategy, this technique generates minimal damage in surrounding tissues (Fan et al. 2014) because of high specificity, minimal invasiveness and precise spatial-temporal selectivity (Shanmugam et al. 2014, Zhang et al. 2013, Zou et al. 2016). The light radiation can be also applied for releasing active principles from their pharmaceutical formulation, reaching the sufficient depth because a source of light, Near Infrared, is used. The light responsiveness of drugs can be used as a strategy for releasing these compounds at the target site, at proper timing, in the appropriated dose, depending on the wavelength of the light and its exposure time. Photothermal therapy can be used directly against cancer cells in primary tumours or metastasis (lymph nodes nearby tumour or other metastasis sites) (Zou et al. 2016). The efficacy of this therapy lies on the transformation capability of energy into sufficient heat through photothermal agents. For this aim, nanotechnology plays an important role, and many systems such as photothermal nanotherapeutics agents including noble metal nanostructures, nanocarbons, nanomaterials based on transition metal sulfide/oxides, inorganic nanomaterials (core/shell mesoporous silica nanostructures, carbon nanotubes and nanogold) and organic nanoagents have been widely investigated (Girija and Balasubramanian 2019, Orecchioni et al. 2015, Zou et al. 2016). The application of the different approaches of photothermal therapy depends on the sought aims by therapeutic modality employed. In this sense, the aim of the therapy could be focused only on photothermal treatment for a certain type of cancer (photothermal ablation) or in its possibility to combine photothermal therapy with imaging diagnosis. Additionally, photothermal therapy can be associated with chemotherapy or radiotherapy. In this text, we will focus on its association with imaging diagnosis because this combination belongs to theranostic procedures (Zou et al. 2016). For improving the safety and efficacy of the photothermal ablation, the use of an imaging-guided process is desirable. This guiding can be carried out by means of magnetic resonance imaging, X-ray computed tomography and photoacoustic imaging (Huo et al. 2014, Mehrmohammadi et al. 2013, Wang et al. 2014).

Yu et al. (Yu et al. 2018) have reported the development of a photodynamic therapy-based system formed through multicomponent coordination-driven self-assembly which uses 5,10,15,20- tetra(4-pyridyl) porphyrin, cis-$(PEt_3)_2Pt(OTf)_2$, and disodium terephthalate as the building block. This strategy has been carried out due to which porphyrin always suffers severe $\pi-\pi$ stacking. This fact leads to a significant quench of the excited state. Therefore, a decrease of oxygen singlet is produced with a consequent decrease in quantum yields result. This effect limits the application of porphyrins in photodynamic therapy (Ethirajan et al. 2011). Thus, the intermolecular $\pi-\pi$ stacking is effectively suppressed, resulting in significant enhancement of fluorescence and oxygen singlet generation, which is favourable for near-infrared fluorescence imaging and photodynamic therapy. Additionally, some ions such as Mn and Cu can be hosted into the porphyrins nucleus, called metallacage core, which allows the implementation of highly effective imaging techniques, such as magnetic resonance imaging and positron emission tomography imaging. PEGylation of the metallacage nanoparticles reaches the long-circulating effect. This approach increases the retention effect and provides an active targeting of the device. Therefore, metallacage-loaded system based on nanoparticles has high imaging capability, which allows a precise diagnosis of tumour and real-time monitoring of the delivery, biodistribution, and excretion of the nanoparticles. These nano-devices exhibited excellent anti-metastatic effect and superior anti-tumour performance against tumour models and they achieve ablating the tumours without recurrence after a single treatment. One of the major challenges for tumour phototherapy is based on the development of intelligent photosensitizer agents for responding to tumour-specific signals sensitively and that they can minimize the side effects of the treatment. For this aim, phenyl-based boron dipyrromethene derivatives (BODIPY) have been designed. An important feature of these compounds relies on their absorbance which can be controlled by a tunable penetration depth (He et al. 2017). The possibility of combining photothermal therapy and photodynamic therapy and taking advantage of their synergic activity is desirable. In this context, the type of photosensitizer agent plays an important role. This agent needs to comply with the following feature: its near-infrared absorbance is preferred since it allows an outstanding penetration depth in tissues, and targeting ability (towards tumours) to minimize side effects to normal tissues. Additionally, to obtain an enhanced theranostic efficacy, some functional groups must be introduced to be activated within the environment of lysosomes. Thus, BODIPY derivatives have many advantages such as high fluorescence, anti-photobleaching property, among others. In this context, Zou et al. (Zou et al. 2019) have designed three BODIPY derivatives by Knoevenagel condensation reaction and diethylamino groups have been incorporated to the core of the molecules. An increased number of diethylamino groups confer higher near-infrared absorbance and this feature indicates a potentially adjustable tissue penetration. Additionally, because diethylamino groups function as proton acceptors, once within the lysosome, enhanced photodynamic and photothermal efficacy could be obtained. Therefore, this work shows a new strategy to extend and improve phototherapy through molecular design for a certain wavelength as well as pH-dependent synergistic of both photodynamic therapy and photothermal therapy (Zou et al. 2019).

9.3.2 Magnetic Resonance Imaging

Targeting approaches have emerged before theranostic therapy concepts; however, both have progressed together and nowadays, the first approach is sometimes used for improving the development of the second. Hence, for carrying out a targeting strategy of active molecules for a certain therapy such as cancer, first, the system must find the target structures. This process can be carried out through different techniques such as immunohistochemistry. Because of this finding, a tailored targeting agent should be designed, developed and screened for assessing a new anticancer therapy. The efficacy of targeting can be evaluated through processing imaging procedures such as ultrasound, positron emission tomography, computed tomography or magnetic resonance. Thus, the selection of the imaging procedure should be in concordance with the expected therapeutic response and morphological changes shown within the tumour and in its microenvironment after the application of the device active-targeting is designed (Chen and Ni 2014).

With the advent of new techniques, some unattainable objectives proposed in the past are becoming reality. One of them is the application of the non-invasive methodology to obtain reliable and high-quality images. Another aim is to obtain a system with targeting capacity, which releases the therapeutic compound (a drug or a physical properties-activated active agent) in the objective site. These two requirements can be achieved by applying magnetic resonance to a magnetism-responsive drug delivery system, and as a consequence of this alliance, a theranostic system is conceived (Feng Chen and Ni, 2014). One of the most widely spread techniques for assessing molecular imaging and clinical diagnosis is the magnetic resonance because it is a non-radiation scanning technique with a non-invasive character and it has a capability to show high-resolution images (Wang et al. 2016, Zou et al. 2016). This technique uses a magnetic field and a radiofrequency pulse for generating images of soft tissues (such as brain, heart, ligaments cartilage, etc.) with a high proportion of magnetic sensitive elements such as hydrogen atom. This technique is based on nuclear magnetic resonance principles.

Therefore, the protons of the hydrogen atoms are aligned by a magnetic field and then, through a radiofrequency pulse the protons are excited and rapidly relaxed generating a signal, which is detected by the scanner of the instrument. Thus, the received information is processed and the image is formed (Jeong et al. 2018, Kupersmith et al. 2002, Moser et al. 2007, Selskog et al. 2002). Magnetic resonance images can be divided into T1-weighted and T2-weighted images depending on the passed time after the radiofrequency pulse. The time needed to recover longitudinal magnetization is named relaxation time T1, while that needed to lose transverse magnetization is named relaxation time T2. Both times have dependence with tissue and they form the basis of image contrast among different tissues. For more details about physical principles of this technique, refer to the following reference (Brown et al. 2014).

Like we mentioned above, neovascularization and vascular remodeling emerge because of the needs of the tumour tissue. Accordingly, vascular targeting therapies based on the uses of active molecules can be addressed as encouraging strategy to attack the cancerous tissue. This therapy could be divided into two categories. One of them is based on the inhibition of neovascularization and another on the damage of the established tumour vasculature. Thus, the anti-vascularization effect produced by active molecules can be followed and studied by magnetic resonance (Robinson et al. 2003).

Vascular disrupting agents are used as anti-tumour drugs against the established vasculature of the tumour (Al-Abd et al. 2017). These agents are divided into two categories depending on the target site. Ligand-directed agents cause a vascular disrupting through specific molecules expressed onto the surface of the target tissue, on either the endothelial cell surface or those expressed on the basement membrane of tumour vessels. Generally, ligand-directed agents interact with endothelial cells through antibodies or peptides. Other types of vascular disrupting agents are named as the small molecules. These agents also act on the endothelial cells, but in this case, they play a selective action against the microtubules of the cytoskeleton of those cells shutting down the established tumour vessels and consequently, this fact leads to the death of the cancerous tissue by means of an ischemic process (Dark et al. 1997, Mita et al. 2013, Porcù et al. 2014, Spear et al. 2011, Tozer et al. 2005). Among these agents, we can find two main groups: flavonoids derivatives and tubulin-binding agents. The molecule 5, 6-dimethylxanthenone-4-acetic acid that belongs to the first group is a tricyclic analogous of flavonoid acetic acid and its target site is the actin cytoskeleton. Combretastatin A4 phosphate is the representative of the second group. This small molecule exerts its action against tubulin cytoskeleton. Both types of agents act by exploiting the difference between normal tissue and cancerous by means of either induction of local cytokine production or depolymerization of tubulin. However, the rapid tumour regrowth and eventual relapse of the disease leads to these agents not to be available commercially for clinical applications. It is important to remark that the disruption activity of those two types of molecules have been followed by magnetic resonance and reported by Seshadri and Ciesielski (Seshadri and Ciesielski 2009) and Nielsen et al. (Nielsen et al. 2012), respectively.

Anti-angiogenic agents inhibit the growth of new blood vessels for tumour irrigation. This action is carried out by binding of surface molecules on the endothelial cells or by inhibition of intracellular signaling pathways. There are three types of anti-angiogenic agents that are classified by their action mechanism. The first type acts through direct inhibition of tumour endothelial cells proliferation without consideration of the type of tumour. The second group has indirect action because the inhibition of angiogenic factors (Chen and Ni 2014). The last type has a mixed action affecting the tumour cells and its vessels through a specific interaction, which produces inhibition of kinase, and generates the epidermal growth factor receptor inhibition. Unlike vascular disrupting agents, there are many anti-angiogenic agents approved by Food and Drug Administration (FDA) in the United States (Al-Abd et al. 2017, Alison 2001). Bevacizumab belongs to this family. It is a humanized monoclonal anti-vascular antibody against endothelial growth factor-A. This antibody is applied to several solid tumours such as non-small cell lung cancer, renal cell cancer, colorectal cancer, ovarian cancer, breast cancer, cervical cancer and glioblastoma (Zhou et al. 2012). Another agent is Cabozantinib, an inhibitor of tyrosine kinase receptor used for metastatic medullary thyroid cancer (Khalid et al. 2016).

Vascular disrupting agents and anti-angiogenic agents can be used separately, but neither the firsts nor the seconds are fully effective for cancer treatment. For this reason, using the combination of those two types of agents has been proposed due to which the solid tumours have a high rate of re-growing, and in certain cases, the hypoxic effect produced by vascular disrupting agents may promote tumour angiogenesis through up-regulation mechanism. The therapeutic effects of each agent used separately or in a combination could be monitored through magnetic resonance as imaging biomarkers (Chen and Ni 2014).

Contrast agents are used, for some specific cases, for improving the resolution of magnetic resonance and its sensitivity. Therefore, these agents can generate more accurate diagnosis. Among the contrast agent, super-paramagnetic iron oxide nanoparticles are founded. The magnetic nanoparticles can realign their magnetic moments by means of the application of a time-varying magnetic field (Zhao et al. 2012); therefore, as the excitation frequency of the magnetic field is increased, the applied field will be lagged by the magnetic moments of nanoparticles. This lagged effect depends on the incident angle (Chen and Wong 2014). Because of this effect, a dissipation event is produced which increases the temperature of the magnetic nanoparticles and their surrounding environment, and hence, it can be used as a hyperthermia method for the treatment of certain types of cancer. Using magnetic fluid thermotherapy, different types of cancer can be treated by means of the application of a magnetic field to magnetic fluids containing iron oxide magnetic nanoparticles. These particles delivered into the tumour can produce heat by application of an alternating magnetic field, resulting in hyperthermia or thermoablation of the cancer tissue. Magnetic fields can also be used as a targeting mechanism for the released control of bioactive molecules such as traditional oncologic drugs or those systems based on epigenetic developments.

A promising theranostic device based on magnetism principles has been reported by Efremova et al. (Efremova et al. 2018). They have developed hybrids nanoparticles, which can be used as a platform, comprising the diagnostics function together with the ability for studying the cargo and vehicle functions separately and in conjugation for targeted drug delivery. These multifunctional nanoparticles have been named "all in one" and they enable simultaneous imaging and therapy as well as multimodal imaging with the combination of two or more visualization modalities (Fang et al. 2017). The nanoparticles developed by Efremova et al. (Efremova et al. 2018) show octahedral-like, single-crystalline Fe_3O_4 structure with bulk-like lattice parameters, epitaxially grown on top of Au seeds (Fe_3O_4-Au). This arrangement provides two surfaces for the functionalization chemistry. Additionally, these hybrids nanoparticles show large saturation magnetization values and the presence of the Verwey transition in the temperature dependence of the magnetization, and consequently, bulk-like ferrimagnetic properties. The Fe_3O_4-Au nanoparticles were subsequently covered with a biocompatible polymeric shell and were simultaneously conjugated with two fluorescent dyes or a combination of a drug and a dye. The nano-hybrids were functionalized with covalently attached Sulfo-Cyanine5 NHS ester derivative (a fluorescent dye via thiol-Au bonds) for allowing their tracking. On the other hand, the anticancer drug doxorubicin or Nile Red dye were loaded into the polymeric shell at the Fe_3O_4 surface as a model of hydrophobic drug. Thus, these functionalized Fe_3O_4-Au hybrids were tested in murine breast cancer cell line for stability, *in vitro* toxicity, cell internalization, drug carrier capabilities, and the drug release. After *in vitro* study and the subsequent targeting of fluorescently labeled Fe_3O_4-Au hybrids addressed to the tumours, it was demonstrated that the device revealed high values of passive accumulation in these tissues. Due to that, these novel Fe_3O_4-Au hybrid structures show the combination of therapeutic species and targeting molecules, adding to the synergistic effect of magnetic hyperthermia and photothermal/photodynamic therapy for treatment of cancer diseases and a dual mode contrast agents for the diagnosis through magnetic resonance imaging and computed tomography (Efremova et al. 2018, Tomitaka et al. 2017, Xu et al. 2009). The authors have proposed these hybrids as promising future tools for cancer theranostic therapy.

Multifunctional nanomaterials, combining diagnosis through magnetic resonance imaging and therapeutic properties of doxorubicin for application in cancer research, have been reported by Shi et al. (Shi et al. 2018). Therefore, this technique and the drug have been combined in a nano-platform for improving the contrast, and in this sense, the device exhibited high specificity towards cancer cells compared to normal cells. The results have been evaluated in 2D cultures and within 3D tissue-like biomimetic matrices. These nano-platforms were prepared from hollow silica nanoparticles coated with MnO_2 nanosheets and conjugated with the AS1411 aptamer. This is a nucleolin aptamer, which is wrapped around water-soluble carbon dots and used as a probe for the detection of several types of cancer cells (Motaghi et al. 2017). Therefore, in this case, it is used as a targeting agent. This device shows an encouraging release of the drug and an enhancement of the magnetic resonance imaging signal (Shi et al. 2018); therefore, it can be considered as an auspicious device for use in theranostic therapy.

9.3.3 Ultrasound

The description of theranostic therapy based on sound application is addressed in this section. Particularly, ultrasound is used for diagnosis and therapy. This oscillating sound pressure wave has a greater frequency than human beings can hear. This technique also uses high echogenicity agents such as gas-filled micro-bubbles. The enhancement of ultrasound backscatter or reflection of the ultrasound waves can be carried out by means of contrast agents which provide a high difference of echogenicity of ultrasound and a sonogram with an increased contrast is obtained (Postema 2011). Thus, this improved ultrasound can be used for studying blood perfusion and its flow. From this concept emerges the high-intensity focused ultrasound which is used as a therapeutic application of ultrasound for destroying cancerous tissues by heating (Malietzis et al. 2013). Many ultrasound-responsive nanocarriers based on traditional pharmaceutical formulations such as liposomes, polymeric nanoparticles, microbubbles among other systems, have been designed for delivering chemotherapy agents within tumour tissues and in this way improve the pharmacological response (Quirolo et al. 2014). These systems, developed with an average diameter between 0.8 and 10 μm, are loaded with active principles but they also contain a small percentage of gas. This co-encapsulation yields in acoustically active particles allowing their contraction and expansion in response to ultrasound waves; therefore, the expansion of the system through the application of ultrasound waves will produce the breakdown of the system, at a certain point, and will permit the release of its content locally (Sirsi and Borden 2012).

Ultrasound has many advantages over other imaging diagnosis modalities, which include its low cost, broad availability, portability, patients are not exposed to ionizing radiation, and it is a minimally invasive procedure. Additionally, it has deep penetration and high-resolution real-time view of the observed organ (Hannah et al. 2014). The intensity of the ultrasound is easily tunable from low intensities, which are used for diagnosis, to high-intensity used for tumour treatment. The intensity range is about 720 mW/cm^2 to 10^5 W/cm^2, respectively (Kiessling et al. 2014).

This technique can be used as stimuli for causing damage to a drug delivery system and generate the accurate release of its content within the target tissue. Ultrasound stimuli are used as theranostic therapy; however, for the development of the suitable ultrasound-triggered release system, a sufficient concentration of ultrasound contrast agent in the formulation must be considered and it should reach high encapsulation efficiency. There are many systems used as ultrasound contrast agents: microbubbles with perfluorocarbons are one of them; however, they have several problems related to their chemical modification feasibility and drug loading capabilities. There are other agents such as liposomes, nanoparticles, and micelles which have many advantages against microbubbles (more stable and easily modifiable) but their imaging contrast produced is not so suitable due to their small size (Benedini et al. 2014, Liang et al. 2015, Quirolo et al. 2014).

Through the uses of nanoparticles, it is possible to generate a strong ultrasound contrast. However, for this purpose, a co-encapsulating procedure must be carried out for establishing a mixture between gas-loaded echogenic nanoparticles and drug-loaded micelles. This aim is attainable by carrying out gas bubbles arrangement around nanocarriers, and thus, physically entrapped echo enhancers can be successfully generated. This fabrication method could produce some complications which would generate low drug-loading nanocarriers (Gao et al. 2005, Rapoport et al. 2007). In this context, Chen et al. (Chen et al. 2017) have reported a stable theranostic drug carrier system with sufficient ultrasound contrast and the ability to release large amounts of therapeutics under ultrasound treatment based on polyelectrolyte multilayer microcapsules. This system is proposed as an alternative to overcome the low loading capacity, the low capability for composition tunability and the lack of adjustable physical-chemical properties of other systems (Gao et al. 2015). Chen et al. (Chen et al. 2017) proposed nano-sized particles (<50 nm) composed by ultra-thin multilayer shells which are assembled through layer-by-layer deposition of polymers onto suitable templates. These microcarriers based on tannic acid, poly (N-vinylpyrrolidone) possess a high ultrasound-imaging contrast, and they can deliver encapsulated therapeutics agents under both low-intensity ultrasound irradiation and high-intensity.

9.4 A Flourishing Platform for Cancer Diagnosis and Therapy

As we have mentioned in the previous chapters, the use of nanomaterials for the construction of emergent and improved devices applied to medicine has substantially grown in the last 20 years. This growth has been followed by the development of a strategy in which diagnosis and therapeutic are combined for obtaining a one-step discipline: theranostic. This therapy has been mainly focused on the treatment of one of the major causes of deaths in the United States of America: cancer (Fan et al. 2014). The concept of theranostic therapy applied for cancer includes the use of molecular biomarkers and imaging probes for diagnosis, cancer therapy guided by images, and also it can involve the development of nanostructures with diagnosis and therapy capabilities. Owing to the nature and prognostic of cancer disease, this is an illness which generates a suffering state by itself. For this reason, the application of a strategy, which allows diagnosis and treatment at the same step, and/or monitoring the progress of the disease, is an encouraging option. The diagnosis of diseases is commonly carried out by clinical revisions and by the uses of devices that employed physical properties to obtain information about the body of the patient. Contrast agents are usually used in cancer diagnosis for improving the resolution of the image and the type of agent depends on the applied technique. Theranostic therapy makes possible the option for including a therapeutic agent added to the contrast one. For cancer treatment, this therapeutic agent is a chemotherapy drug and this therapy allows the following of this drug through the body until its target site. Chemotherapy drugs have poor pharmacokinetic profiles and high toxicity and hence, it is desirable that the drug reaches the target site and produces a low frequency of adverse effects. Consequently, this fact can be carried out by means of application targeting strategy (active or passive and/or by application of physical stimuli). Cancer tissues have some morphological difference with respect to the normal ones. These features, described in this chapter, are highly important because of the understanding of these differences allows the possibility to design active and passive targeting systems for application in theranostic therapy which contributes to the specificity of the system.

Different theranostic therapies, which are based on the application of certain properties of the devices such as light responsiveness, sensitivity to the sound or magnetism, generate an improved diagnosis, and additionally, customized the treatment of cancer disease. Finally, one of the major strengths of this therapy lies in the precise external modulation of diagnosis elements, drug release, and the availability of device tracking when a certain stimulus affects the target site. Multifunctional nanoparticles with combined diagnostic and therapeutic functions show great promise towards personalized nanomedicine, and particularly to the cancer treatment. However, attaining consistently high performance of these functions *in vivo* in one single nanoconstruct remains extremely challenging and must be the purpose of forthcoming investigations.

References

Abrahamse, H. and M. R. Hamblin. 2016. New photosensitizers for photodynamic therapy. Biochemical J. 473(4): 347-364.

Acharya, S., F. Dilnawaz and S. K. Sahoo. 2009. Targeted epidermal growth factor receptor nanoparticle bioconjugates for breast cancer therapy. Biomaterials 30(29): 5737-5750.

Ackroyd, R., C. Kelty, N. Brown and M. Reed. 2001. The history of photodetection and photodynamic therapy. Photochem. Photobiol. 74(5): 656-669.

Adams, G. P., R. Schier, A. M. McCall, H. H. Simmons, E. M. Horak, R. K. Alpaugh, et al. 2001. High affinity restricts the localization and tumor penetration of single-chain Fv antibody molecules. Cancer Res. 61(12): 4750-4755.

Al-Abd, A. M., A. J. Alamoudi, A. B. Abdel-Naim, T. A. Neamatallah and O. M. Ashour. (2017). Anti-angiogenic agents for the treatment of solid tumors: Potential pathways, therapy and current strategies – A review. J. Adv. Res. 8(6): 591-605.

Alison, M. R. 2001. Cancer. pp. 1-8. *In*: F. Rose and K. Osborne (eds.). Encyclopedia of Life Sciences. John Wiley & Sons Ltd., Chichester, UK.

Bao, C., M. Jin, B. Li, Y. Xu, J. Jin and L. Zhu. 2012. Long conjugated 2-nitrobenzyl derivative caged anticancer prodrugs with visible light regulated release: Preparation and functionalizations. Org. Biomol. Chem. 10(27): 5238-5244.

Benedini, L. A., N. Andres and M. L. Fanani. 2017. Liposomes: From the pioneers to epigenetic therapy. pp. 1-35. *In*: B. Paerson (ed.). From Liposomes: Historical, Clinical and Molecular Perspectives. Nova Science Publishers, N. Y. USA.

Benedini, L., S. Antollini, M. L. Fanani, S. Palma, P. Messina and P. Schulz. 2014. Study of the influence of ascorbyl palmitate and amiodarone in the stability of unilamellar liposomes. Mol. Membr. Biol. 31(2-3): 85-94.

Bergers, G. and L. E. Benjamin. 2003. Tumorigenesis and the angiogenic switch. Nat. Rev. Cancer. 3(6): 401-410.

Bies, C., C. -M. Lehr and J. F. Woodley. 2004. Lectin-mediated drug targeting: History and applications. Adv. Drug Deliv. Rev. 56(4): 425-435.

Biswas, R., J. H. Moon and J. -C. Ahn. 2014. Chlorin e6 derivative radachlorin mainly accumulates in mitochondria, lysosome and endoplasmic reticulum and shows high affinity toward tumors in nude mice in photodynamic therapy. Photochem. Photobiol. 90(5): 1108-1118.

Boyd, L. K., X. Mao and Y. -J. Lu. 2012. The complexity of prostate cancer: Genomic alterations and heterogeneity. Nat. Rev. Urol. 9(11): 652-664.

Brown, R., E. Haacke, Y. -C. Cheng, M. Thompson and R. Venkatesan. 2014. Magnetic Resonance Imaging (R. W. Brown, Y. -C. N. Cheng, E. M. Haacke, M. R. Thompson and R. Venkatesan, Eds.) (2nd ed.). John Wiley & Sons Ltd., Chichester, UK.

Cardone, R. A., V. Casavola and S. J. Reshkin. 2005. The role of disturbed pH dynamics and the Na^+/H^+ exchanger in metastasis. Nat. Rev. Cancer 5(10): 786-795.

Celli, J. P., B. Q. Spring, I. Rizvi, C. L. Evans, K. S. Samkoe, S. Verma, et al. 2010. Imaging and photodynamic therapy: Mechanisms, monitoring, and optimization. Chem. Rev. 110(5): 2795-2838.

Chan, W. M., T. -H. Lim, A. Pece, R. Silva and N. Yoshimura. 2010. Verteporfin PDT for non-standard indications – A review of current literature. Graefes Arch. Clin. Exp. Ophthalmol. 248(5): 613-626.

Chen, F. and Y. Ni. 2014. Magnetic resonance imaging of cancer therapy. pp. 95-126. *In*: Chen, X. and S. Wong (eds.). Cancer Theranostics. Elsevier. Academic Press, Cambridge, Massachusetts, USA.

Chen, H. and Y. Zhao. 2018. Applications of light-responsive systems for cancer theranostics. ACS Appl. Mater. Interfaces. 10(25): 21021-21034.

Chen, J., S. Ratnayaka, A. Alford, V. Kozlovskaya, F. Liu, B. Xue, et al. 2017. Theranostic multilayer capsules for ultrasound imaging and guided drug delivery. ACS Nano 11(3): 3135-3146.

Chen, X. and S. T. C. Wong. 2014. Cancer theranostics. pp. 3-8. *In*: Chen X. and S. Wong (eds.). Cancer Theranostics. Elsevier. Academic Press, Cambridge, Massachusetts, USA.

Coll, C., L. Mondragón, R. Martínez-Máñez, F. Sancenón, M. D. Marcos, J. Soto, et al. 2011. Enzyme-mediated controlled release systems by anchoring peptide sequences on mesoporous silica supports. Angew. Chem., Int. Ed. 50(9): 2138-2140.

Danhier, F., O. Feron and V. Préat. 2010. To exploit the tumor microenvironment: Passive and active tumor targeting of nanocarriers for anti-cancer drug delivery. J. Controlled Release 148(2): 135-146.

Daniell, M. D. and J. S. Hill. 1991. A history of photodynamic therapy. Aust. N. Z. J. Surg. 61(5): 340-348.

Daniels, T. R., T. Delgado, G. Helguera and M. L. Penichet. 2006. The transferrin receptor. Part II: Targeted delivery of therapeutic agents into cancer cells. Clin. Immunol. 121(2): 159-176.

Dark, G. G., S. A. Hill, V. E. Prise, G. M. Tozer, G. R. Pettit and D. J. Chaplin. 1997. Combretastatin A-4, an agent that displays potent and selective toxicity toward tumor vasculature. Cancer Res. 57(10): 1829-1834.

Desgrosellier, J. S. and D. A. Cheresh. 2010. Integrins in cancer: Biological implications and therapeutic opportunities. Nat. Rev. Cancer, 10(1): 9-22.

Dienst, A., A. Grunow, M. Unruh, B. Rabausch, J. E. Nör, J. W. U. Fries and C. Gottstein. 2005. Specific occlusion of murine and human tumor vasculature by VCAM-1-targeted recombinant fusion proteins. J. Natl. Cancer Inst. 97(10): 733-747.

Dolmans, D. E. J. G. J., D. Fukumura and R. K. Jain. 2003. Photodynamic therapy for cancer. Nat. Rev. Cancer. 3(5): 380-387.

Efremova, M. V., V. A. Naumenko, M. Spasova, A. S. Garanina, M. A. Abakumov, A. D. Blokhina, et al. 2018. Magnetite-Gold nanohybrids as ideal all-in-one platforms for theranostics. Sci. Rep. 8(1): 11295.

Ethirajan, M., Y. Chen, P. Joshi and R. K. Pandey. 2011. The role of porphyrin chemistry in tumor imaging and photodynamic therapy. Chem. Soc. Rev. 40(1): 340-362.

Fan, Z., P. P. Fu, H. Yu and P. C. Ray. 2014. Theranostic nanomedicine for cancer detection and treatment. J. Food Drug Anal. 22(1): 3-17.

Fang, S., J. Lin, C. Li, P. Huang, W. Hou, C. Zhang, et al. 2017. Dual-stimuli responsive nanotheranostics for multimodal imaging guided trimodal synergistic therapy. Small, 13(6): 1602580.

Feron, O. 2009. Pyruvate into lactate and back: From the Warburg effect to symbiotic energy fuel exchange in cancer cells. Radiother. Oncol. 92(3): 329-333.

Frangioni, J. V. 2003. In vivo near-infrared fluorescence imaging. Curr. Opin. Chem. Biol. 7(5): 626-634.

Gao, H., D. Wen and G. B. Sukhorukov. 2015. Composite silica nanoparticle/polyelectrolyte microcapsules with reduced permeability and enhanced ultrasound sensitivity. J. Mater. Chem. B. 3(9): 1888-1897.

Gao, Z. -G., H. D. Fain and N. Rapoport. 2005. Controlled and targeted tumor chemotherapy by micellar-encapsulated drug and ultrasound. J. Controlled Release. 102(1): 203-222.

Genís, L., B. G. Gálvez, P. Gonzalo and A. G. Arroyo. 2006. MT1-MMP: Universal or particular player in angiogenesis? Cancer Metastasis Rev. 25(1): 77-86.

Girija, A. R. and S. Balasubramanian. 2019. Theragnostic potentials of core/shell mesoporous silica nanostructures. Nanotheranostics, 3(1): 1-40.

Gomer, C. J. and N. J. Razum. 1984. Acute skin response in albino mice following porphyrin photosensitization under oxic and anoxic conditions. Photochem. Photobiol. 40(4): 435-439.

Gosk, S., T. Moos, C. Gottstein and G. Bendas. 2008. VCAM-1 directed immunoliposomes selectively target tumor vasculature in vivo. Biochim. Biophys. Acta. 1778(4): 854-863.

Haley, B. and E. Frenkel. 2008. Nanoparticles for drug delivery in cancer treatment. Urol. Oncol. 26(1): 57-64.

Hannah, A., G. Luke, K. Wilson, K. Homan and S. Emelianov. 2014. Indocyanine green-loaded photoacoustic nanodroplets: Dual contrast nanoconstructs for enhanced photoacoustic and ultrasound imaging. ACS Nano. 8(1): 250-259.

He, H., S. Ji, Y. He, A. Zhu, Y. Zou, Y. Deng, et al. 2017. Photoconversion-tunable fluorophore vesicles for wavelength-dependent photoinduced cancer therapy. Adv. Mater. 29(19): 1606690.

Hilderbrand, S. A. and R. Weissleder. 2010. Near-infrared fluorescence: Application to in vivo molecular imaging. Curr. Opin. Chem. Biol. 14(1): 71-79.

Huo, D., J. He, H. Li, A. J. Huang, H. Y. Zhao, Y. Ding, et al. 2014. X-ray CT guided fault-free photothermal ablation of metastatic lymph nodes with ultrafine HER-2 targeting W18O49 nanoparticles. Biomaterials, 35(33): 9155-9166.

Iyer, A. K., G. Khaled, J. Fang and H. Maeda. 2006. Exploiting the enhanced permeability and retention effect for tumor targeting. Drug Discov. Today. 11(17-18): 812-818.

Jain, R. K. 1987. Transport of molecules in the tumor interstitium: A review. Cancer Res. 47(12): 3039-3051.

Jeong, Y., H. S. Hwang and K. Na. 2018. Theranostics and contrast agents for magnetic resonance imaging. Biomater. Res. 22: 20.

Khalid, E. B., E. -M. E. -K. Ayman, H. Rahman, G. Abdelkarim and A. Najda. 2016. Natural products against cancer angiogenesis. Tumour Biol. 37(11): 14513-14536.

Kiessling, F., S. Fokong, J. Bzyl, W. Lederle, M. Palmowski and T. Lammers. 2014. Recent advances in molecular, multimodal and theranostic ultrasound imaging. Adv. Drug Deliv. Rev. 72: 15-27.

Kirpotin, D. B., D. C. Drummond, Y. Shao, M. R. Shalaby, K. Hong, U. B. Nielsen, et al. 2006. Antibody targeting of long-circulating lipidic nanoparticles does not increase tumor localization but does increase internalization in animal models. Cancer Res. 66(13): 6732-6740.

Kupersmith, M. J., T. Alban, B. Zeiffer and D. Lefton. 2002. Contrast-enhanced MRI in acute optic neuritis: Relationship to visual performance. Brain 125(4): 812-822.

Lammers, T., S. Aime, W. E. Hennink, G. Storm and F. Kiessling. 2011. Theranostic Nanomedicine. Acc. Chem. Res. 44(10): 1029-1038.

Lelièvre, S. A., K. B. Hodges and P. -A. Vidi. 2014. Application of theranostics to measure and treat cell heterogeneity in cancer. pp. 493-516. *In*: Chen, X. and S. Wong (eds.). Cancer Theranostics. Elsevier. Academic Press, Cambridge, Massachusetts, USA.

Liang, X., J. Gao, L. Jiang, J. Luo, L. Jing, X. Li, et al. 2015. Nanohybrid liposomal cerasomes with good physiological stability and rapid temperature responsiveness for high intensity focused ultrasound triggered local chemotherapy of cancer. ACS Nano. 9(2): 1280-1293.

Liu, T. W., E. Huynh, T. D. MacDonald and G. Zheng. 2014. Porphyrins for imaging, photodynamic therapy, and photothermal therapy. pp. 229-254. *In*: Chen, X. and S. Wong (eds.). Cancer Theranostics Elsevier. Academic Press, Cambridge, Massachusetts, USA.

Low, P. S. and S. A. Kularatne. 2009. Folate-targeted therapeutic and imaging agents for cancer. Curr. Opin. Chem. Biol. 13(3): 256-262.

Luo, Z., Y. Hu, K. Cai, X. Ding, Q. Zhang, M. Li, et al. 2014. Intracellular redox-activated anticancer drug delivery by functionalized hollow mesoporous silica nanoreservoirs with tumor specificity. Biomaterials, 35(27): 7951-7962.

Maeda, H., G. Y. Bharate and J. Daruwalla. 2009. Polymeric drugs for efficient tumor-targeted drug delivery based on EPR-effect. Eur. J. Pharm. Biopharm.71(3): 409-419.

Malam, Y., M. Loizidou and A. M. Seifalian. 2009. Liposomes and nanoparticles: Nanosized vehicles for drug delivery in cancer. Trends Pharmacol. Sci. 30(11): 592-599.

Malietzis, G., L. Monzon, J. Hand, H. Wasan, E. Leen, M. Abel, et al. 2013. High-intensity focused ultrasound: Advances in technology and experimental trials support enhanced utility of focused ultrasound surgery in oncology. Br. J. Radiol. 86(1024): 20130044.

Martinelli, E., R. De Palma, M. Orditura, F. De Vita and F. Ciardiello. 2009. Anti-epidermal growth factor receptor monoclonal antibodies in cancer therapy. Clin. Exp. Immunol. 158(1): 1-9.

Mehrmohammadi, M., S. Joon Yoon, D. Yeager and S. Y. Emelianov. 2013. Photoacoustic imaging for cancer detection and staging. Curr. Mol. Imaging. 2(1): 89-105.

Minko, T. 2004. Drug targeting to the colon with lectins and neoglycoconjugates. Adv. Drug Deliv. Rev. 56(4): 491-509.

Mita, M. M., L. Sargsyan, A. C. Mita and M. Spear. 2013. Vascular-disrupting agents in oncology. Expert Opin. Invest. Drugs. 22(3): 317-328.

Moser, T., J. -C. Dosch, A. Moussaoui and J. -L. Dietemann. 2007. Wrist ligament tears: Evaluation of MRI and combined MDCT and MR arthrography. AJR Am. J. Roentgenol. 188(5): 1278-1286.

Motaghi, H., M. A. Mehrgardi and P. Bouvet. 2017. Carbon dots-AS1411 aptamer nanoconjugate for ultrasensitive spectrofluorometric detection of cancer cells. Sci. Rep. 7(1): 10513.

Naumov, G. N., L. A. Akslen and J. Folkman. 2006. Role of angiogenesis in human tumor dormancy: Animal models of the angiogenic switch. Cell Cycle. 5(16): 1779-1787.

Nielsen, T., L. Bentzen, M. Pedersen, T. Tramm, P. F. J. W. Rijken, J. Bussink, et al. 2012. Combretastatin A-4 phosphate affects tumor vessel volume and size distribution as assessed using MRI-based vessel size imaging. Clin. Cancer Res. 18(23): 6469-6477.

O'Connor, A. E., W. M. Gallagher and A. T. Byrne. 2009. Porphyrin and nonporphyrin photosensitizers in oncology: Preclinical and clinical advances in photodynamic therapy. Photochem. Photobiol. 85(5): 1053-1074.

Orecchioni, M., R. Cabizza, A. Bianco and L. G. Delogu. 2015. Graphene as cancer theranostic tool: Progress and future challenges. Theranostics, 5(7): 710-723.

Paolillo, M., M. A. Russo, M. Serra, L. Colombo and S. Schinelli. 2009. Small molecule integrin antagonists in cancer therapy. Mini. Rev. Med. Chem. 9(12): 1439-1446.

Pastorino, F., C. Brignole, D. Di Paolo, B. Nico, A. Pezzolo, D. Marimpietri, et al. 2006. Targeting liposomal chemotherapy via both tumor cell-specific and tumor vasculature-specific ligands potentiates therapeutic efficacy. Cancer Res. 66(20): 10073-10082.

Porcù, E., R. Bortolozzi, G. Basso and G. Viola. 2014. Recent advances in vascular disrupting agents in cancer therapy. Future Med. Chem. 6(13): 1485-1498.

Postema, M. 2011. Contrast-enhanced and targeted ultrasound. World J Gastroenterol. 17(1): 28-41.

Quirolo, Z., L. Benedini, M. Sequeira, M. Herrera, T. Veuthey and V. Dodero. 2014. Understanding recognition and self-assembly in biology using the chemist's toolbox. Insight into medicinal chemistry. Curr. Top. Med. Chem. 14(6): 730-739.

Rai, P., S. Mallidi, X. Zheng, R. Rahmanzadeh, Y. Mir, S. Elrington, et al. 2010. Development and applications of photo-triggered theranostic agents. Adv. Drug Deliv. Rev. 62(11): 1094-1124.

Rapoport, N., Z. Gao and A. Kennedy. 2007. Multifunctional nanoparticles for combining ultrasonic tumor imaging and targeted chemotherapy. J. Natl. Cancer Inst. 99(14): 1095-1106.

Reddy, L. H., J. L. Arias, J. Nicolas and P. Couvreur. 2012. Magnetic nanoparticles: Design and characterization, toxicity and biocompatibility, pharmaceutical and biomedical applications. Chem. Rev. 112(11): 5818-5878.

Robinson, S. P., D. J. O. McIntyre, D. Checkley, J. J. Tessier, F. A. Howe, J. R. Griffiths, et al. 2003. Tumour dose response to the antivascular agent ZD6126 assessed by magnetic resonance imaging. Br. J. Cancer, 88(10): 1592-1597.

Rychak, J. J. and A. L. Klibanov. 2014. Nucleic acid delivery with microbubbles and ultrasound. Adv. Drug Deliv. Rev. 72: 82-93.

Selskog, P., E. Heiberg, T. Ebbers, L. Wigstrom and M. Karlsson. 2002. Kinematics of the heart: Strain-rate imaging from time-resolved three-dimensional phase contrast MRI. IEEE Trans. Med. Imaging. 21(9): 1105-1109.

Seshadri, M. and M. J. Ciesielski. 2009. MRI-based characterization of vascular disruption by 5,6-dimethylxanthenone-acetic acid in gliomas. J. Cereb. Blood Flow Metab. 29(8): 1373-1382.

Shadidi, M. and M. Sioud. 2003. Selective targeting of cancer cells using synthetic peptides. Drug Resist. Updat. 6(6): 363-371.

Shanmugam, V., S. Selvakumar and C. -S. Yeh. 2014. Near-infrared light-responsive nanomaterials in cancer therapeutics. Chem. Soc. Rev. 43(17): 6254-6287.

Shi, Y., F. Guenneau, X. Wang, C. Hélary and T. Coradin. 2018. MnO_2-gated nanoplatforms with targeted controlled drug release and contrast-enhanced MRI properties: From 2D cell culture to 3D biomimetic hydrogels. Nanotheranostics, 2(4): 403-416.

Sirsi, S. R. and M. A. Borden. 2012. Advances in ultrasound mediated gene therapy using microbubble contrast agents. Theranostics, 2(12): 1208-1222.

Spear, M. A., P. LoRusso, A. Mita and M. Mita. 2011. Vascular disrupting agents (VDA) in oncology: Advancing towards new therapeutic paradigms in the clinic. Curr. Drug Targets. 12(14): 2009-2015.

Sumer, B. and J. Gao. 2008. Theranostic nanomedicine for cancer. Nanomedicine, 3(2): 137-140.

Tomitaka, A., H. Arami, A. Raymond, A. Yndart, A. Kaushik, R. D. Jayant, et al. 2017. Development of magneto-plasmonic nanoparticles for multimodal image-guided therapy to the brain. Nanoscale, 9(2): 764-773.

Torchilin, V. P. 2000. Drug targeting. Eur. J. Pharm. Sci. 11(2): S81-91.

Tozer, G. M., C. Kanthou and B. C. Baguley. 2005. Disrupting tumour blood vessels. Nat. Rev Cancer. 5(6): 423-435.

Wagner, A., U. W. Denzer, D. Neureiter, T. Kiesslich, A. Puespoeck, E. A. J. Rauws, et al. 2015. Temoporfin improves efficacy of photodynamic therapy in advanced biliary tract carcinoma: A multicenter prospective phase II study. Hepatology. 62(5): 1456-1465.

Wang, S., Q. Zhang, X. F. Luo, J. Li, H. He, F. Yang, et al. 2014. Magnetic graphene-based nanotheranostic agent for dual-modality mapping guided photothermal therapy in regional lymph nodal metastasis of pancreatic cancer. Biomaterials. 35(35): 9473-9483.

Wang, S., Q. Zhang, P. Yang, X. Yu, L. -Y. Huang, S. Shen and S. Cai. (2016). Manganese oxide-coated carbon nanotubes as dual-modality lymph mapping agents for photothermal therapy of tumor metastasis. ACS Appl. Mater. Interfaces. 8(6): 3736-3743.

Xu, C., B. Wang and S. Sun. 2009. Dumbbell-like $Au-Fe_3O_4$ nanoparticles for target-specific platin delivery. J. Am. Chem. Soc. 131(12): 4216-4217.

Yu, G., S. Yu, M. L. Saha, J. Zhou, T. R. Cook, B. C. Yung, et al. 2018. A discrete organoplatinum(II) metallacage as a multimodality theranostic platform for cancer photochemotherapy. Nat. Commun. 9(1): 4335.

Zhang, Z., J. Wang and C. Chen. 2013. Near-infrared light-mediated nanoplatforms for cancer thermo-chemotherapy and optical imaging. Adv. Mater. 25(28): 3869-3880.

Zhao, L., J. Peng, Q. Huang, C. Li, M. Chen, Y. Sun, et al. 2014. Near-infrared photoregulated drug release in living tumor tissue via yolk-shell upconversion nanocages. Adv. Funct. Mater. 24(3): 363-371.

Zhao, Q., L. Wang, R. Cheng, L. Mao, R. D. Arnold, E. W. Howerth, et al. 2012. Magnetic nanoparticle-based hyperthermia for head & neck cancer in mouse models. Theranostics, 2(1): 113-121.

Zhao, Y., Z. Luo, M. Li, Q. Qu, X. Ma, S. -H. Yu and Y. Zhao. 2015. A preloaded amorphous calcium carbonate/doxorubicin@silica nanoreactor for pH-responsive delivery of an anticancer drug. Angew. Chem., Int. Ed. 54(3): 919-922.

Zhou, H., N. O. Binmadi, Y. -H. Yang, P. Proia and J. R. Basile. 2012. Semaphorin 4D cooperates with VEGF to promote angiogenesis and tumor progression. Angiogenesis, 15(3): 391-407.

Zou, J., P. Wang, Y. Wang, G. Liu, Y. Zhang, Q. Zhang, et al. 2019. Penetration depth tunable BODIPY derivatives for pH triggered enhanced photothermal/photodynamic synergistic therapy. Chem. Sci. 10(1): 268-276.

Zou, L., H. Wang, B. He, L. Zeng, T. Tan, H. Cao, et al. 2016. Current approaches of photothermal therapy in treating cancer metastasis with nanotherapeutics. Theranostics. 6(6): 762-772.

Nanotechnology Application in Tissue Regeneration and Regenerative Medicine: Smart Tools in Modern Healthcare

"… In fact, when you combine stem cell technology with the technology known as tissue engineering you can actually grow up entire organs, so as you suggest that sometime in the future you get in an auto accident and lose your kidney, we would simply take a few skin cells and grow you up a new kidney. In fact, this has already been done …"

Dr. Robert Lanza, Head of Astellas Global Regenerative Medicine, and Chief Scientific Officer of the Astellas Institute for Regenerative Medicine and Adjunct Professor at Wake Forest University School of Medicine He was selected for the 2014 TIME 100 list of the 100 Most Influential People in the World (Lanza 2019)

10.1 Can Tools Operating at the Nano-size Restore Injured or Ageing Tissue?

"…By using our novel nanochip technology, injured or compromised organs can be replaced…" "…We have shown that skin is a fertile land where we can grow the elements of any organ that is declining…" said Chandan Sen, from the Ohio State University, to The Guardian (Davis 2017).

As we mentioned in chapter 3, nano-size is the scale at which biological processes befall. The machinery inside cells operates on a nano-metric scale and they behave *in vivo* in response to the biological signals they receive from the surrounding environment, which is structured by nanometre-scaled components. Likewise, the organs owe their specific function to their structure, based on a hierarchical organization that goes from the nano-scale to the macro-scale. In this way, it is logical to think that operating at the same scale at which biological processes occur, these can be manipulated for the benefit of health. Opposite to the urodele amphibians commonly referred to as salamanders, the human body has a low regenerative potential. Traditionally, transplantation of intact tissues and organs has been the bedrock to replace damaged and diseased parts of the body. However, and despite the rising success and greater improvement in post-transplant outcome, the unavailability of adequate organs for transplantation to meet the existing demand has resulted in organ shortage crises. In this sense, millions of people worldwide would benefit greatly if tissues and organs could be replaced on demand. This organ shortage crisis has deprived thousands of patients of a new and better quality of life and has caused a substantial increase in the cost of alternative medical care such as dialysis (Abouna 2008). The total cost to society in terms of caring for patients with failing organs and debilitating diseases is enormous (Stephan 2017). Scientists and clinicians, motivated by the need to develop safe and reliable sources of tissues and organs, have been improving therapies and technologies that can regenerate tissues and in some cases create new tissues altogether. In this context, tissue engineering and regenerative medicine disciplines arise (Berthiaume et al. 2011). The spirit of these emerging multidisciplinary fields is the development of biological substitutes that restore, maintain, or improve damaged tissue and organ functionality. Nanotechnology may have the necessary tools to attain such goals since only nano-scale materials can mimic properties of natural tissues (Engel et al. 2008, Zhang and Webster 2009). Consequently, regenerative medicine and tissue engineering has experienced great progresses along with the emergence of nanotechnology. It has been shown that increased tissue regeneration can be achieved on almost any surface by employing novel nano-textured surface features (Dalby et al. 2014, Xu et al. 2016a, Sartuqui et al. 2018). Various nano-structured polymers (Armentano et al. 2018, Tandon et al. 2018) and metals (alloys) (Awad et al. 2017, Maher et al. 2017) have been investigated for their bio- and cyto-compatibility properties and numerous studies have reported that nanotechnology accelerates various regenerative therapies, such as those for the bone (Gong et al. 2015, Tang et al. 2016), vascular (Wang et al. 2017a), heart (Marchesan et al. 2016), cartilage (Cross et al. 2016), bladder (Lin et al. 2015) and nerve tissue (Xu et al. 2016b).

Current technologically advanced nano-devices have shown promising results *in vitro*, and some of them have also been successfully tested *in vivo* with animal models, thus paving the way toward a personalized medicine and transforming the fundamentals of diagnosis, treatment, and prevention of any diseases (please refer to chapters 4–9). The advanced applications of this approach to regenerative medicine will undoubtedly improve the present issue of personal and global health care, and its economic and social burden. In this chapter, we will explore how the architecture of native extracellular matrices can be mimicked at the nanoscale level and therefore provides the primary base for the regeneration of new tissue and the creation of artificial organs. Several of the latest nanotechnology findings in regenerative medicine and tissue engineering will be described, as well as their relative levels of success. The aspects related to the nano materials currently used for tissue regeneration will be discussed and it will furthermore point out the most promising strategies in the field of cell therapies, hard tissue applications, orthoepy and implantology with an outlook to future research in the related areas.

10.2 Tissue Engineering and Regenerative Medicine

We will start with a brief description of the scientific areas "Tissue engineering" and "Regenerative medicine" to introduce the reader toward the actions and objectives of these disciplines in order to strengthen their understanding and later, to perceive how nanotechnology directly impacts their fundamental pillars. Tissue engineering and/or regenerative medicine are fields of life science employing both engineering and biological principles to create new tissues and organs and to promote their regeneration after any damage, disorder or illness. Although regenerative medicine and tissue engineering are frequently cited as a whole, and in some cases the mistake of considering them as the same discipline is made, they are actually two separate areas of study that within bioengineering and medicine are working together to find new and effective treatments for a host of challenging diseases and ailments (Mao 2007). Through communion of selected scientific subjects, including molecular biology, genetics, immunology and biochemistry, regenerative medicine is focused on replacing or repairing tissue and organs that have been damaged during diseases, trauma, genetic, chromosomal or age disorders, by stimulating the body's own repair mechanisms. In order to reduce demand for organs by using the precepts of regenerative medicine, researchers have to first understand the abilities and machineries of stem cells, how they attach, spread, grow, differentiate and die, and particularly how these cells interact together to replace damaged tissue. Currently, a plethora of applications for regenerative medicines are being found. Treatments include neurological, cardiovascular, orthopedic and musculoskeletal therapies.

Similar to regenerative medicine, the closely related field of tissue engineering focuses on developing tissue outside of the body and promises to deliver "off-the-self" organs grown from patients' own stem cells to improve supply (Cassidy 2014). Investigators fuse the values of engineering and life sciences to attain synthetic and naturally three-dimensional 3D materials to produce functional tissues in the lab and then apply them to the affected areas. As the regenerative medicine is reliant on the abilities and mechanism of stem cells, tissue engineering is subject to the construction of biodegradable structures that hold implanted cells in place until they develop into integrated and functional tissue (scaffolds) and the use of growth factors or physical stimulus that replicate natural conditions found in the body. The most commonly cited example of tissue engineering is artificial skin. But, the field is rapidly expanding from the relatively mature skin substitution idea to the ongoing proof-of-concept of cartilage replacement to early stage of liver surrogate. A good amount of thoughtful work has also yielded prototypes of other tissue substitutes like nerve conduits, blood vessels, and bioartificial liver, kidney, and even heart tissue are being designed (Berthiaume et al. 2011, Dzobo et al. 2018). Nevertheless, the movement forward to clinical product, however, has been slow. Some historical highlights related to tissue engineering and regenerative medicine are shown in Box 10.1.

Box 10.1. Chronological developments concerning tissue engineering and regenerative medicine. Information extracted from references (Vacanti 2006, Berthiaume et al. 2011, Baptista and Atala 2014)

- 3000 B.C.: The *Sushruta Samhita*, an ancient Sanskrit text on medicine of ancient India, described the first skin grafting.
- 500–400 B.C.: Genesis I:1 "The Lord, breathed a deep sleep on the man and while he was asleep he took out one of his ribs and closed up its place with flesh. The Lord God then built up into a woman the rib that he had taken from the man", may be interpreted as the oldest written reference to "tissue engineering".
- 1438–1440: The famous painting by Fra Angeliac entitled, "Healing of Justinian" depicting the brothers Saints Damien and Cosmos Transplanting a Homograft limb onto a wounded soldier is often referred to as the first historical reference to "tissue engineering".
- 1794: Autologous skin grafting was achieved in Europe by Bunger, Reverdin, and Baronio .
- 1881: Cadaveric skin allograft by Girdner (Girdner 1881).
- 1910: R.G. Harrison first reached the groundbreaking achievement of *in vitro* cell cultivation, demonstrating an active growth of cells in culture.
- 1944: Clinically refrigerated skin allografts by Webster (Webster 1944).
- 1949: Human spermatozoa cryopreservation at subzero temperatures (-79°C) developed by Polge (Polge et al. 1949).
- 1952: Skin cryopreservation developed by Billingham.
- 1957-1959: Dr. Don Thomas team treated for the first time leukemic patients with allogeneic marrow cells in Seattle.
- 1962: Ivalon sponge developed as "synthetic substitute for skin" by Chardack .
- 1970's: W. T. Green, a paediatric orthopedic surgeon working at the Children's Hospital, M.D., undertook a number of experiments to generate new cartilage using chondrocytes seeded onto spicules of bone and implanted in nude mice. Although he was unsuccessful, Green correctly concluded that it would one day be possible to generate new tissue, and even organs, by seeding cells onto appropriate scaffolds.
- 1976: Drs. Burke and Yannas of the Massachusetts General Hospital and M.I.T. discovered the first scaffold with regenerative activity. They worked together making both *in vitro* and *in vivo* experiments to generate a tissue-engineered skin substitute using a collagen matrix to support the growth of dermal fibroblasts.
- 1977-1979: Dr. Howard Green transferred sheets of keratinacytes onto burn patients, while Dr. Eugene Bell seeded collagen gels with fibroblasts, referring to them as contracted collagen gels.
- 1979: Human epidermal cells cultured in the absence of dermal components or medium supplements (Eisinger et al. 1979).

- 1981: Developed of a composite living skin equivalent to Bell´s collagen gels; it was later commercialized as Apligraf by Organogenesis. Apligraf, derived from neonatal foreskin and bovine type I collagen, was the first bi-layered living skin equivalent approved in the US and other countries for use in venous ulcers (Trent and Kirsner 1998).
- 1986: Creation of longterm functional replacement of skin membrane composed by a silicone elastomer, covalently cross-linked network of bovine hide collagen and glycosaminoglycan (GAG) Collagen-glycosaminoglycans (Yannas et al. 1984). Later, it was commercialized as Integra® Dermal Regeneration Template by Integra Lifesciences.
- 1988: The term "tissue engineering" was officially coined at a National Science Foundation workshop (Akter 2016). Genzyme Tissue Repair began marketing Epicel, cell transplantation in synthetic biodegradable polymers, as an unregulated product. The product had been considered a banked human tissue until 1996 when FDA announced that manipulated autologous cell-based products used for structural repair or reconstruction required regulatory oversight.
- 1994: Brittberg proposed a chondrocyte culture and transplantation methodology to repair full-thickness defects of articular cartilage in the knee that have a poor capacity for healing; later, it was commercialized as Carticel by Genzyme (Brittberg et al. 1994). Drs. Charles A. and Joseph P. Vacanti conceived and founded, in Boston, the Tissue Engineering Society (TES); it was officially incorporated in the state of Massachusetts on January 8, 1996.
- 1997: Dr. Minora Ueda of Nagoya University established a large tissue engineering effort in Japan, and organized the first meeting of the Japanese Tissue Engineering Society in Nagoya.
- 1999: Genzyme applied for regulatory approval of Epicel.
- 2003: Computer-aided jet-based 3D tissue engineering was proposed as an organ printing method (Mironov et al. 2003).
- 2004: Polymeric networks microarray was developed for seeding embryonic stem cells (ES's) (Anderson et al. 2004).
- 2007: First experimental attempt, performed on a rat model, to create a conduit urinary diversion with a tissue engineering approach (Adamowicz et al. 2017). FDA's Center for Devices and Radiologic Health (CDRH) approved Epicel under the HDE regulatory statute.
- 2008: Engineered trachea from decellularized matrix seeded with human cells was derived from stem cells (Komura et al. 2008).
- 2010: Engineered tissue trachea transplantation (Baiguera et al. 2010).
- 2011: Chen et al. reported "essential 8" (E8), a xeno-free and completely defined culture medium, for human pluripotent stem cells hPSC cultivation (Chen et al. 2011).
- 2012: Chen and coworkers developed a promising method to generate cell banks of several hESC lines (Chen et al. 2012).

(Contd.)

Box 10.1. (*Contd.*)

- 2013: Lead regulatory responsibility for the Epicel HDE was transferred to the Center for Biologics Evaluation and Research (CBER) based on an assessment of the primary mode of action under the Combination Products regulations. This change was part of a transfer of oversight responsibilities for certain wound care products containing live cells from the CDRH to CBER.
- 2014: 3D Microarray chip containing 36 different scaffolds was implanted in Wistar rats (Oliveira et al. 2014). Epicel ownership was transferred from Genzyme to Vericel.
- 2016: FDA approved a Paediatric Labeling supplement, which specified use in both adult and paediatric patients, added paediatric labeling information, and granted an exemption from the profit prohibition. A 3D bioprinting system to produce tissue constructs on a human scale with structural integrity (Kang et al. 2016).
- 2017: First Annual Review of Paediatric Safety for Epicel was presented.
- 2018: Hasan Erbil Abaci et al. proposed a biomimetic approach for generation of human hair follicles (HFs) within human stem cells (HSCs) by recapitulating the physiological 3D organization of cells in the HF microenvironment using 3D-printed molds (Abaci et al. 2018). Functional skeletal muscle tissues from human pluripotent stem cells (hPSCs) was reported for the first time (Rao et al. 2018).

10.3 Impact of Nanotechnology on the Stem Cells Therapy: Nano-features as Stem Cell Regulators

The use of nanotechnologies to manipulate and track stem cells holds a great promise for the study of their biology and, even more, to influence throughout their expansion, differentiation, and transplantation the improvement of current stem cells approaches (Ferreira et al. 2008). In chapter 5, we have discussed how nanotechnology's emerging platforms could be useful in measuring, understanding, and manipulating stem cells and, through them, the body function codified into genes. Some examples include magnetic nanoparticles and quantum dots for stem cell labelling and *in vivo* tracking; lipid/protein nanoparticles, carbon nanotubes, liposomes, liquid crystals and polyplexes for the intracellular delivery of genes/oligonucleotides. Beyond these applications, it is now known and accepted that mesenchymal stem cells' (MSCs) fate is largely determined by their mechanosensitivity to biophysical signals from the cellular microenvironment, i.e. the topographical features and the collection of extracellular matrix (ECM) components present in their niches. At the nano-scale, where features and individual cell receptors have the similar size, nano-topography can be manipulated to induce alterations in cell adhesion. This would imply that it might therefore be possible to target receptor-driven pathways and to manage the response of the cells to stimulate tissue regeneration (Dalby et al. 2014).

To achieve the major clinical goal of repairing central nervous system, including brain and spinal cord injury, enough quantity and quality of neuron cells are needed; this is a challenge because of their weak capability to proliferate. To conquer this obstacle, the researchers put their eyes on neural natural niches where stem cells are exposed to biophysical cues that guided their sophisticated behaviour, specifically those that are subjected to nano-topographical and bioelectric stimuli. In this way, Xiaodi Zhang and co-workers designed a nano-striped structural array exhibiting a pattern ranged from 200 to 500 nm (ridge, groove, and height) based on piezoelectric poly-vinylidene fluoride (PVDF). They demonstrated that the combination of a piezoelectric and nano-scaled interfaces can provide a real assistance for the neural differentiation of stem cells (Zhang et al. 2019b). The local piezoelectric potential near the cell membrane provides a continuous electric stimulation for the living cells, in response to migration and traction which is affected by the nano-topography, that increases the expressions of neuronal lineage marker (Tuj-1), Glial fibrillary acidic protein (GFAP), and Microtubule-associated protein 2 (MAP-2). Weiqiang Chen et al. (Chen et al. 2018) also took advantage of the fact that nano-topographic cues can provide potent regulatory signals to mediate human pluripotent stem cell (hPSCs) behaviours. They demonstrated that the neuro-epithelial conversion and the motor neuron (MN) progenitor differentiation of hPSCs can be influenced by nano-engineered vitronectin-coated glass surfaces. The authors claimed that hPSCs, after sensing substrate nano-topography, translate biophysical signal through a regulatory network involving cell adhesion, the actin CSK, and Hippo/YAP signalling to mediate neural differentiation.

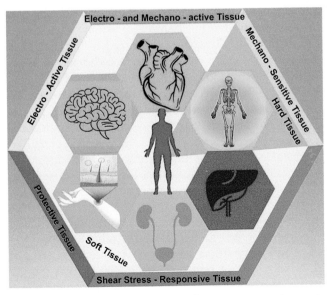

Figure 10.1. Current applications of tissue regeneration and regenerative medicine in the human body.

Real tissues exhibit definite architectural shapes (pits, folds and branching points) that may create "hot spots" of tensional stress due to localized cell distortion. Important groundwork towards the regeneration of neural tissue also resulted from the development of methods to confine neuronal adhesion at specific locations to match predefined network topologies and connectivity that their niche environments (Brusatin et al. 2018, Schulte et al. 2018). The phenomenon in which the cells retain information from historical stimuli, the so-called "time-retention effect", is another fact to take into consideration at the time to modulate stem cell differentiation through a biophysical stimulus. Accordingly, some approaches analysed the effects of time dependent retention of the nano-topographical cues to properly modulate the stem cell fate (Yang et al. 2019). Using a stress-responsive and tuneable nano-wrinkle topography, Seungwon S. Yang et al. (Yang et al. 2019) confirmed that longer periods of exposure of human neural stem cells (hNSCs) to the nano-wrinkle substrates were associated with higher levels of neuronal gene expression and that they could retain the nanotopographical stimuli depending on the dosing time during differentiation.

10.4 Regulation of Tissue Engineering through Surface Nano-topography: Physical Cues to Attain Regeneration

10.4.1 Nano-size Technology Attending Orthopaedic Medicine

Bone is a mechanosensitive tissue, which is continuously exposed to mechanical forces while walking or exercising and simultaneously acts as a damper, absorbing sudden mechanical variations. Such kind of tissue is composed by hierarchical structures organized from nano- to macro-scale, see chapter 3. Therefore, the construction of artificial nano-topographical substrates providing physiologically relevant mechanical stimuli and operating by induction of physical cues to regulate stem cells behaviour can afford an accurate approach to address the problems associated to orthopaedic implants performance and novel tissue engineering tactics to bone repair/regeneration (Laurencin et al. 2009, Kim et al. 2013a).

The burden of musculoskeletal trauma is worldwide; with an ageing population and growing culture for active lifestyles, this trend is expected to continue. Extremely affecting deprived persons who have the least access to quality orthopaedic trauma care, currently, bone pathologies like fractures, osteoarthritis, osteoporosis or cancers are becoming increasingly common and are a source of extra global disability-adjusted life years (DALYs) compared to the combination of HIV/AIDs, malaria, and tuberculosis (Zhang and Webster 2009, Conway et al. 2017). Then, orthopaedic trauma care is essential, and must be a priority in the horizontal development of global health systems (Agarwal-Harding et al. 2016). In the search of a better design of orthopaedic implant devices, one approach that has received much attention is the simulation of their metallic nano-surface roughness; progresses in nanotechnology exposed new promises to yield biomimetic surfaces that resemble the cell *in vivo* growth environment at a nanoscale level (Wang et al. 2015). In this manner, thin nano-porous and highly adherent layers of anodised surfaces formed on titanium and its alloys are being developed as coatings for metallic surgical implants (Briggs et al. 2004, Krząkała et al. 2013, Zhang et al. 2019a). Nano-topography enhances, among others, tissue organization, facilitate active self-assembly of ECM molecules to further mediate cell attachment, cell-to-cell communication, orientation, differentiation (Das et al. 2007, Kim et al. 2013b, Sun et al. 2015) and stem cell fate regulation (Khang et al. 2012). Organized nanostructured surfaces have been also shown to modulate protein adsorption and integrin binding (Castro-Raucci et al. 2016); they have shown influences on the orientation of focal adhesion, fibrinogenesis (McNamara et al. 2011, Wang et al. 2017a) and also on the induction of blood clot formation followed by osseointegration (Traini et al. 2018).

Additionally, to host tissue cellular stimulation, surface modification of a material is a promising avenue to resolve the global problem of implant infections that is frequently associated with their placement surgery (Vasilev et al. 2009). Antibacterial surfaces in nature is a popular strategy; biomimicry of cicada, dragonfly and butterfly wings, taro and lotus leaves, catkin packages, gecko and shark skin micro/nano-structures inspired scientist to design antibacterial, antifouling and self-cleaning orthopaedic devices (Zhang et al. 2006, Hasan and Chatterjee 2015, Andrés et al. 2018, Ye et al. 2019). Along this line of investigation, for example, Chan and co-workers (Chan et al. 2017) reported laser-induced nano-spiky surfaces resembling cicada wings printed on commercially pure (CP) Ti (Grade 2), Ti6Al4V (Grade 5) and CoCrMo alloy implant materials. After treatment, the metal surfaces showed a marked reduction in adhesion followed by the death of the *Staphylococcus aureus* strain, a bacterium commonly implicated in orthopaedic implant infections. Inspired also by dragonfly wings, Bhadra et al. (Bhadra et al. 2015) fabricated on titanium surfaces antibacterial nano-patterned arrays using a simple hydrothermal etching process. In such conditions, nano-topography provides the metallic titanium surface with a selective bactericidal activity, eliminating almost 50% of *Pseudomonas aeruginosa* and about 20% of the *Staphylococcus aureus*.

The effect on nanotopographical cues that increase roughness and wettability of metallic surfaces promote osteogenic differentiation of stem cells and eventually provide antibacterial properties. These modifications increase the production of osteoblastic factors, bone formation and a reduction of potential risk of infection. However, there is a third effect that we should mention: their effect on immune cells. The initial host response to an implantable material is primarily controlled by macrophages and the factors they secrete in response to the injury caused by surgery and the material cues, then the control of inflammation to direct immune response will enhance the success of implanted materials. Kelly M. Hotchkiss and co-workers

(Hotchkiss et al. 2016) demonstrated that surface roughness and wettability exerted an effect on the activation and production of inflammatory factors by macrophages. The authors claimed that a mishmash of topographical selected roughness/ patterns and hydrophilicity may interact synergistically to yield a microenvironment suitable to macrophage activation similar to the anti-inflammatory M2-like, reducing healing times and increasing osseointegration. Although we have at the present an excellent starting point, there is still a fair amount of research to be done before the effective application of nano-textured orthopaedic implants. Together, cooperative interplay between surface nanotopography and chemistry can put forth immunomodulatory effects. Christo et al. (Christo et al. 2016) demonstrated that surface nanotopography enhanced the matrix metalloproteinase – 9 production from primary neutrophils, and a reduction of pro-inflammatory cytokine secretion from primary macrophages.

The absence of large-scale, low cost and fast fabrication methods of uniform nanostructures is one of the causes. Moreover, researchers need to optimise nano-topographies to inhibit bacteria adhesion and growth while simultaneously promote osteoblast metabolic activity and bone regrowth. This tactic could be much more industrialised in the future because a passive strategy, based on physical stimuli, has the advantage to be free from any chemical toxicity or contamination risk, which is quite interesting even if it has the disadvantage of the absence of an asset and strong action.

10.4.2 Matrix Nano-topography as a Regulator of Electro-active Tissues

Electro-active tissues, unlike mechano-responsive tissues, are affected by electrical stimuli; as indicated by its name, "electro-active" tissues respond to pulsatile or abrupt electrical stimuli (Kim et al. 2013a). Representative par excellence of the electroactive tissue, the brain tissue processes information and stimulation received from the eyes, ears and skin through the relay of signals by neurons. Contrasting with neurons that are only electrically active, other tissues like the skeletal muscle and heart tissue are both mechano- and electro-active. Nevertheless, and saving their differences, all respond to nanoscale stimuli. Human induced pluripotent stem cells (hiPSCs) possess inefficient neural differentiation. hiPSCs ruling processes have been mainly examined through the assessment of biochemical aspects; nevertheless, the present directed differentiation protocols for hPSCs that completely rely on biochemical factors remain suboptimal. These facts motivate new investigations regarding the influence of biophysical cues throughout cellular surroundings, in particular the roll of nano-topography on hiPSC fate decision (Song et al. 2016, Kim et al. 2017b, Chen et al. 2018). The roles of geometry and dimensions of nano-topography in neural lineage commitment of hiPSCs, including self-renewal and differentiation, were analysed with emphasis on the effects of geometry, feature size and height of nano-topography. Lattices composed of hexagonally arranged nano-pillars (500 nm in diameter), each having a height of 150 or 560 nm, brought neuronal differentiation in concert with dual Smad inhibitors. Nano-pillars of 560 nm height reduced cell proliferation, enhanced cytoplasmic localization of Yes-associated protein, and promoted neuronal differentiation (up to 60% βIII-tubulin + cells) compared with the flat control (Song et al. 2016). Weiqiang Chen and co-workers (Chen et al. 2018) applied a method developed in their laboratory, based on reactive-ion etching (RIE) to generate random nanoscale structures on glass surfaces. Using such nano-engineered topographic substrates, the authors demonstrated, with high precision and reproducibility, that the early neuroepithelial conversion and motorized neuron progenitor differentiation of hiPSCs could be promoted through a regulatory signalling network that involved all together cell adhesion, actomyosin cytoskeleton and the Hippo/YAP signalling (Chen et al. 2018).

Nanophotography can be combined with additional stimuli, such as bioelectricity, in order to stimulate cellular responses. Following this line of work, Zhang et al. (Zhang et al. 2019b) designed piezoelectric polyvinylidene fluoride surfaces exhibiting nano-stripe array structures. The obtained material, which can generate a surface piezoelectric potential up to millivolt as a function of cell movement and traction, had a profound influence on cell proliferation, adhesion, and differentiation. The results revealed that there was a synergic effect after the mishmash of piezoelectricity and nanotopography that stimulated neuron-like differentiation of bone marrow-derived mesenchymal stem cells (rbMSCs) and that effect was superior to the attained using nano-topography alone. Synergistic effects of biophysical and biochemical cues are also used to attain a better performance to neural stem cells reaction. Pre-coating nano-surfaces with 3,4-dihydroxy-l-phenylalanine (DOPA), Kisuk Yang et al. (Yang et al. 2015) achieved a facilitated immobilization of poly-l-lysine and fibronectin on biodegradable poly(lactic-co-glycolic acid) (PLGA) substrates via bio-inspired catechol chemistry. The addition of nerve growth factor further directed cellular alignment along the patterned grooves by contact guidance, leading to enhanced focal adhesion, skeletal protein reorganization, and a superior neuronal differentiation of hiNSCs that showed highly extended neurites of cell bodies and enhanced expression of Tuj1 and MAP2 neuronal markers (Yang et al. 2015). Nano-topography can also increase the number of axon collateral branches and promote their growth. A study performed by J. Seo and co-workers, from the Department of Chemistry Center for Cell-Encapsulation Research KAIST, the Department of Applied Chemistry of Kyung Hee University and the Department of Chemistry Chungnam National University of South Korea, revealed that the nanostructured surface stimulated lateral filopodia at the axon shaft via cytoskeletal changes, leading to the formation of axonal branches (Seo et al. 2018).

The prominent examples mentioned above highlight the fact that neural cells use their intracellular machineries to actively recognize surface nano-topography. In this way, it can be assumed, then, that at a specific range of the nano-scale, there is a superior differentiation of stem cells into neurons and a progressive acceleration of neuronal communication. Surface nano-topography, therefore, is deeply involved in the brain renewal development; advance insights into the field of nano–neuron interfaces would allow a better approach of unsolved biological problems and of the increased engineering needs regarding electroactive tissue regeneration.

10.4.3 Nano-scale Topography in Wound Healing

Protective tissue, the skin, from a structural point of view, is a set of three layers of organized extracellular matrix (ECM) fibres, majorly, of collagen, fibronectin and keratin: epidermis, dermis and hypodermis (Kim et al. 2013a). The outermost of the three layers, the epidermis, is composed mostly of desmosomes, a specialized cell designed for cell-to-cell adhesion and found in tissue that experiences intense mechanical shear stress. The fundamental structural unit of desmosomes are known as desmosome-intermediate filament complexes (DIFC), which is a network of cadherin proteins, linker proteins and keratin intermediate filaments (Garrod and Chidgey 2008). The DIFCs can be divided into three regions, all of them exhibiting nanostructure: (i) the extracellular core region, or desmoglea, ~34 nm in length, (ii) the outer dense plaque, which is about 15–20 nm in length, and (iii) the inner dense plaque, about 15–20 nm in length. In addition to the own internal nanostructure, desmosomes assembly is organized at the nano-scale. The extracellular space of viable epidermis contains desmosomes arranged in a characteristic extracellular crosswise of about five nm periodicity. The extracellular space between viable and cornified epidermis contains transition desmosomes at different stages of reorganization interconnected between ~9 and ~48 nm thick regions (Al-Amoudi et al. 2005).

The dermis, or corium, is the intermediate layer between epidermis and subcutaneous tissues or hypodermis. There, collagen, fibronectin and keratin fibres of few 10 nm in diameter moulded in aligned fibrils of 60–120 nm diameters with a length of few micrometers and arranged in a basket-weave meshwork organization, are responsible for mechanical support and cellular function regulation (Kim et al. 2013a). In addition, a macroscopically directional anisotropy of collagen fibres, which is called Langer's line (Langer 1978), was present. Then, nano-topography, intense anisotropic mechanical properties and the response to high mechanical stress are the fundamental characteristics of skin tissue function; this has inspired *in vitro* and *in vivo* wound healing studies (Hamdan et al. 2017, Castano et al. 2018).

The nano-topography of a wound dressing scaffold has been the focus of interest of researchers aiming at minimizing scar formation while maintaining a high rate of wound closure (Hamdan et al. 2017). Hong-Nam Kim and co-workers (Kim et al. 2012) designed a synthetic extracellular matrix (ECM) scaffold in the form of uniformly-spaced nano-grooved surfaces and analysed the effect of nano-topography densities (vertically or paralleled arrangements of nano-groves of 550 nm width with different gaps of 550, 1100, and 2750 nm and spacing ratio of 1:1, 1:2 and 1:5, respectively) in dermal wound healing. The authors claimed that the wound was guided by the nano-topographical cues in the absence of growth factors; nano-patterns of 1:2 spacing ratio, similar to the natural organization of collagen fibres, yielded the best wound healing performance in terms of migration speed. Later experiments (Kim et al. 2017b) demonstrated a maximum biphasic trend of fibroblasts body and nucleus elongation at ridge and groove widths of 500–600 nm density, whereas maximum migration speed was observed at 300–400 nm with a monotonic decrease upon increasing the piece size. Similar studies performed on a nano-designed polyurethane acrylate grooved matrix coated with gelatine validated that the migration of human mesenchymal stem cells (hMSCs) at certain groove distances is significantly higher than on smooth surfaces (Kim et al. 2013b).

A feasible, cheap and perhaps the most efficient method to introduce active nano-topography in ECM-like scaffolds is the electrospinning. In this way, it is highly used to make submicronic and nano-metric structures for skin regeneration. Following this line of work and using a single or two-step modified electrospinning set-up, Jeong In Kim and Cheol Sang Kim (Kim and Kim 2018) created an adjustable-angled pattern polycaprolactone (PCL)/collagen nano-fibrous mats. The authors stated that the nano-topographical pattern provided a hierarchical cellular control by influencing fibroblast morphology and guiding the formation of integrin. The cellular infiltration of membrane was tested, *in vivo*, to evaluate wound healing progression as a function of depth; results showed a positive effect on neo-vascularization, re-epithelization, and the acceleration of wound healing compared to control group.

Summarizing, the reparation of tissue matrices and wound healing could be controlled via physical cues mediated by feature-size-dependent cell/nano-architecture interactions; then, it must be considered within the scaffolding construction strategies at the same level than chemical and biological signals.

10.4.4 Nano Topography-guided Mechano and Electroactive Tissue Regeneration

One of the major causes of morbidity and mortality worldwide are cardiovascular diseases (CVDs), leading to myocardial infarction and heart failure. The biological system by himself cannot restore the capabilities of a damaged heart and the current treatment involves complicated pharmacological and surgical interventions. At that point, less aggressive and more cost-effective approaches are sought to repair tissues after an injury (Moorthi et al. 2017). The scientific world fixed its gaze on cell therapy. However, human induced pluripotent stem cell-derived cardiomyocytes (hiPSC-CMs) show immature phenotypes in contractile properties, electrophysiology, metabolism, structure, and protein isoform expression, thus greatly limiting their application in regenerative medicine, disease modelling, and drug screening (Lundy et al. 2013, Leung 2016). A deep analysis of heart tissue reveals that it is active both mechanically and electrically and, that in order to correct mechanical contraction and electrical conduction, the sets of unidirectionally aligned sarcomeres, the basic functional unit of the muscles, are essential (Kim et al. 2013b). To manipulate such responses and to induce similar physiological cellular organization *in vitro*, topographic cues can be used (Carson et al. 2016). Sarcomere length is linearly correlated to force spawned from cell contractions and can be associated to the Frank-Starling law on the nanoscale (Konhilas et al. 2002). Then, 1.8–2.0 μm sarcomere lengths correspond to adult cardiomyocytes, while fewer dimensions are associated to immature human pluripotent stem cell-derived cardiomyocytes (~1.5 μm) (Lundy et al. 2013). Based on observations of the oriented myocardial extracellular matrix (ECM) fibres found *in vivo*, Carson and co-workers (Carson et al. 2016) designed a nano-grooved surface array integrated with a self-assembling chimeric peptide containing the Arg-Gly-Asp (RGD) to study the influence of several nanoscale arrangements on the active growth of hiPSC-CMs. The authors found that nano-topographical feature size affects the cellular response in a biphasic manner, with an improved progress (a maximum sarcomere length in cells) attained on grooves widths of about 700–1000 nm. Literature information sustains that cellular development in response to dynamic biophysical cues is essential (Mengsteab et al. 2016, Moorthi et al. 2017, Cui et al. 2019). Mengsteab et al. (Mengsteab et al. 2016) worked on dynamic anisotropic nano-topographic cues, mimicking the cardiac ECM structure, using thermally induced shape memory polymer, poly(e-caprolactone). The aim of the study was to evaluate the special effects of a 90° transition in substrate pattern orientation on the contractile direction and structural organization of cardiomyocyte sheets. Cardiomyocyte sheets cultured on poly(e-caprolactone) exhibited anisotropic contractions before heat-induced shape transition; after 48 h of stimulation, they significantly reoriented their direction of contraction and exhibited a bimodal distribution, with peaks at ~45°. Electroactive and nano-topographical stimuli were combined to attain a better native myocardial extracellular matrix (ECM) mimetic scaffold. Tsui et al. (Tsui et al. 2018) constructed bioinspired cardiac scaffolds from electroconductive acid-modified silk fibroin–poly(pyrrole) (AMSF + PPy) substrates patterned with nano-ridges and nano-grooves to enhance the structural and functional properties of hiPSC-CMs. The authors stated that anisotropic topographical cues increased cellular organization and sarcomere development, while electroconductive cues promoted a significant improvement in the expression and polarization of connexin 43 (Cx43), a critical regulator of cell–cell electrical coupling. In addition, they claimed that there was a synergic effect between the biomimetic topography and the electro-conductivity properties. This synergic influence augmented the expression of genes, encoding key proteins involved in the regulation of contractile and electrophysiological functions of mature human cardiac tissue.

The above described examples and others that can be seen in literature extensively support the use of biomimetic topographies and synergic effects, combining physical and electrical stimulus to orient hiPSC-CMs and to drive the development of cellular structure and function in engineered cardiac tissues. However, optimal dimensions for nanoscale topographies in terms of promoting stem cell derived cardiomyocyte development are yet to be established and require further investigation to translate results from bench-to-bedside.

10.4.5 Shear Stress-sensitive Tissues Regeneration

Within *"shear stress-sensitive"* classification are included those particular tissues that are incessantly exposed to fluidic conditions such as the kidney, liver and vascular systems. The vast capacity of stem cells therapies for developing novel kidney disease treatments, like renal bioengineering and drug development programmes, is frustrated because of the current inability to direct their differentiation to specialised renal cells and to fulfil the variety of kidney functions (MacGregor-Ramiasa et al. 2017). Novel nano-rough surface substrates were used as a useful platform for guiding the fate kidney stem cell *in vitro*. (MacGregor-Ramiasa et al. 2017, Hopp et al. 2019). Silver-coated catheters, wound dressings and a range of other devices are currently used on current medical treatments; they are also tested to induce mouse kidney-derived stem cells differentiation (Chowdhury et al. 2018). The combined stiffness + nano-topology regenerative approaches are also considered to optimally and permanently restore the functions of damaged or diseased liver tissue (Perez et al. 2017). The number of reports exploiting micro- to nano-topography to attain tissue regeneration is superior concerning to vascular systems and blood vessels. The performance of synthetic vascular grafts is estimated by their ability to maintain the cellular behaviours *in vitro* similar to those *in vivo*. Numerous researches have been performed to assess the effects of micro- and nano-topographical scaffolds on the performances of the endothelial cells (ECs) and the possibility of their use in vascular grafts (Chong et al. 2015, Nakayama et al. 2015, Taskin et al. 2017). Proliferation and migration are other vital functions which may be affected by physical cues and that promote the development and wound healing of blood vessels (Kim et al. 2016, Lee et al. 2016). Foreign-body reaction can also be avoided by the nano-topography in order to improve tissue repair (Wang et al. 2016). Angiogenesis is another important behaviour because it is the initial process for the formation of new blood vessels (Kim et al. 2016).

In the current scenario, nano-topography cues to stimulate *"shear-stress tissues"* are being investigated in the preclinical stage but are not yet in clinical trials. Although many radical treatments are proposed, more advanced systems must be developed with improved efficacy to attain the next-generation of solutions for the treatment of liver, vascular or kidney diseases.

10.5 Promises for a Better-quality Tissue Repair and a Decreased Reliance on Transplantations

Would it not be huge to grow a new hand or a new kidney from their own stem cells? It would be fantastic to replace damaged cells with a cocktail of new and healthy ones, to give a weakened heart the ability to repair itself or, better yet, growing organs in the laboratory.

That potential of regenerative medicine and tissue engineering, cutting-edge fields that pursue the treatment of disease and injury through original methods to replace, regenerate and revitalise tissues, is vast. Actually, scientists are working to make treatments available for clinical use, and, to some extent, they are getting it. Since the origin of the field several decades ago, a number of regenerative and tissue engineering therapies, including those designed for wound healing and orthopaedics applications, have received Food and Drug Administration (FDA) approval and are now commercially available (Sampogna et al. 2015). For example, the GRAFTJACKET® Regenerative Tissue Matrix (RTM) that is a trademark of Wright Medical Technology, Inc. and licensed to KCI USA, Inc. Products processed by LifeCell Corporation Scaffolds are designed to provide repair or replacement of damaged or inadequate integumental tissue, such as diabetic foot ulcers, venous leg ulcers, pressure ulcers, or for other homologous uses of human integument (https://www.mykci.com/products/graftjacket-regenerative-tissue-matrix). In the same line, there is PuraPly® Antimicrobial (PuraPly® AM) by Organogenesis, which is a FDA 510(k)-cleared Class II medical device. The wound matrix is indicated for the management of a variety of acute and chronic wound types: pressure, venous, diabetic and chronic vascular ulcers; tunnelled and undermined wounds, trauma and draining wounds; and first- and second-degree burns (https://organogenesis.com/products/puraply-antimicrobial-wound-care.html). There are also products to repair peripheral injured nerves like those offered by Axogen® (https://www.axogeninc.com/patients/), or Stryker (https://www.stryker.com/us/en/trauma-and-extremities/products/nerve-repair.html). Tissue engineers have reached the construction of real organs in the laboratory. A series of child and teenage patients displaying a birth defect called spina bifida, have received urinary bladders grown from their own cells (https://www.bbc.com/news/business-45470799).

However, despite the progress made and the potential of market expansion (https://www.marketsandmarkets.com/Market-Reports/regenerative-medicine-market-65442579.html), several of the new applications cited in this chapter will continue in the planning or proof-of-concept stage, as preclinical investigation or under regulatory evaluation. They failed to reach the marketplace because of the costs involved, ethical or theological concerns, or because of the complex and tedious processes regarding postmarked studies. Patient testimonials and other marketing provided by clinics may be misleading and that is to the detriment of scientific progress because it raises unspeakable doubts and fears. In addition, there are many scientifically challenges involved in tissue engineering and regenerative medicine. The organ-building approach requires precise cells to induce reparation and sometimes the cells from the patient's own organ might not be available or they do not have functionality. Scientific community is exploring genetic reprogramming to turn blood or skin cells into the appropriate cells for growing organ (Mertens et al. 2016, Tanabe et al. 2018), and looking for stem cells from bone marrow or body fat that could be nudged into becoming the right kinds of cells for particular organs (Li and Belmonte 2016, Scudellari 2016).

Notwithstanding all the drawbacks and current challenges, there is reason for optimism; the position of personalized medicine in tissue engineering and other cell-based therapies has been accepted, envisioning the development of customized approaches. As more and more is investigated about the design of biomimetic scaffolds tissue, about cell-specific function *in vitro* or about tissue regeneration *in vivo* upon implantation, which are much improved due to nano-topographical guidance and better cell-to-cell communications, there will be a significant decrease in health care and social co-lateral costs related to ineffective or inadequate approaches.

To address this fruitful area, engineers and scientists will require all of their ability to manufacture and distribute complex products in order to reach the larger number of patients who might potentially profit from bioengineered therapeutics. Effective communication and in-depth education will also be necessary to provide all the necessary information to the public, in order to minimize fears and avoid misunderstandings.

References

Abaci, H. E., A. Coffman, Y. Doucet, J. Chen, J. Jacków, E. Wang, et al. 2018. Tissue engineering of human hair follicles using a biomimetic developmental approach. Nat. Commun. 9(1): 5301.

Abouna, G. M. 2008. Organ shortage crisis: Problems and possible solutions. Transplantation Proc. 40(1): 34-38.

Adamowicz, J., M. Pokrywczynska, S. B. Van, T. Kloskowski and T. Drewa. 2017. Concise review: Tissue engineering of urinary bladder; we still have a long way to go? Stem Cells Transl. Med. 6(11): 2033-2043.

Agarwal-Harding, K. J., A. von Keudell, L. G. Zirkle, J. G. Meara and G. S. Dyer. 2016. Understanding and addressing the global need for orthopaedic trauma care. JBJS 98(21): 1844-1853.

Akter, F. 2016. What is tissue engineering? pp. 1-2. *In*: F. Akter (ed.). Tissue Engineering Made Easy. Academic Press, Cambridge, Massachusetts, USA.

Al-Amoudi, A., J. Dubochet and L. Norlén. 2005. Nanostructure of the epidermal extracellular space as observed by cryo-electron microscopy of vitreous sections of human skin. J. Invest. Dermatol. 124(4): 764-777.

Anderson, D. G., S. Levenberg and R. Langer. 2004. Nanoliter-scale synthesis of arrayed biomaterials and application to human embryonic stem cells. Nat. Biotechnol. 22(7): 863-866.

Andrés, N. C., J. M. Sieben, M. Baldini, C. H. Rodríguez, Á. Famiglietti and P. V. Messina. 2018. Electroactive Mg2+-Hydroxyapatite Nanostructured Networks against Drug-Resistant Bone Infection Strains. ACS Appl. Mater. Interfaces. 10(23): 19534-19544.

Armentano, I., L. Tarpani, F. Morena, S. Martino, L. Latterini and L. Torre. 2018. Nanostructured biopolymer-based materials for regenerative medicine applications. Curr. Org. Chem. 22(12): 1193-1204.

Awad, N. K., S. L. Edwards and Y. S. Morsi. 2017. A review of TiO_2 NTs on Ti metal: Electrochemical synthesis, functionalization and potential use as bone implants. Mater. Sci. Eng. C. 76: 1401-1412.

Baiguera, S., M. A. Birchall and P. Macchiarini. 2010. Tissue-engineered tracheal transplantation. Transplantation 89(5): 485-491.

Baptista, P. M. and A. Atala. 2014. Regenerative medicine: The hurdles and hopes. Transl. Res. 163(4): 255-258.

Berthiaume, F., T. J. Maguire and M. L. Yarmush. 2011. Tissue engineering and regenerative medicine: History, progress, and challenges. Annu. Rev. Chem. Biomol. Eng. 2(1): 403-430.

Bhadra, C. M., V. Khanh Truong, V. T. H. Pham, M. Al Kobaisi, G. Seniutinas, J. Y. Wang, et al. 2015. Antibacterial titanium nano-patterned arrays inspired by dragonfly wings. Scie. Rep. 5: 16817.

Briggs, E. P., A. R. Walpole, P. R. Wilshaw, M. Karlsson and E. Pålsgård. 2004. Formation of highly adherent nano-porous alumina on Ti-based substrates: A novel bone implant coating. J. Mater. Sci.: Mater. Med. 15(9): 1021-1029.

Brittberg, M., A. Lindahl, A. Nilsson, C. Ohlsson, O. Isaksson and L. Peterson. 1994. Treatment of deep cartilage defects in the knee with autologous chondrocyte transplantation. N. Engl. J. Med. 331(14): 889-895.

Brusatin, G., T. Panciera, A. Gandin, A. Citron and S. Piccolo. 2018. Biomaterials and engineered microenvironments to control YAP/TAZ-dependent cell behaviour. Nat. Mater. 17(12): 1063-1075.

Carson, D., M. Hnilova, X. Yang, C. L. Nemeth, J. H. Tsui, A. S. T. Smith, et al. 2016. Nanotopography-induced structural anisotropy and sarcomere development in human cardiomyocytes derived from induced pluripotent stem cells. ACS Appl. Mater. Interfaces. 8(34): 21923-21932.

Cassidy, J. W. 2014. Nanotechnology in the regeneration of complex tissues. Bone Tissue Regen. Insights 5: 25-35.

Castano, O., S. Perez-Amodio, C. Navarro-Requena, M. Á. Mateos-Timoneda and E. Engel. 2018. Instructive microenvironments in skin wound healing: Biomaterials as signal releasing platforms. Adv. Drug. Deliv. Rev. 129: 95-117.

Conway, D. J., R. Coughlin, A. Caldwell and D. Shearer. 2017. The institute for global orthopedics and traumatology: A model for academic collaboration in orthopedic surgery. Front. Public. Health. 5: 146-146.

Cross, L. M., A. Thakur, N. A. Jalili, M. Detamore and A. K. Gaharwar. 2016. Nanoengineered biomaterials for repair and regeneration of orthopedic tissue interfaces. Acta Biomater. 42: 2-17.

Cui, C., J. Wang, D. Qian, J. Huang, J. Lin, P. Kingshott, et al. 2019. Binary colloidal crystals drive spheroid formation and accelerate maturation of human-induced pluripotent stem cell-derived cardiomyocytes. ACS Appl. Mater. Interfaces. 11(4): 3679-3689.

Chan, C. -W., L. Carson, G. C. Smith, A. Morelli and S. Lee. 2017. Enhancing the antibacterial performance of orthopaedic implant materials by fibre laser surface engineering. Appl. Surf. Scie. 404: 67-81.

Chen, G., D. R. Gulbranson, Z. Hou, J. M. Bolin, V. Ruotti, M. D. Probasco, et al. 2011. Chemically defined conditions for human iPSC derivation and culture. Nat. Meth. 8(5): 424.

Chen, V. C., S. M. Couture, J. Ye, Z. Lin, G. Hua, H. -I. P. Huang, et al. 2012. Scalable GMP compliant suspension culture system for human ES cells. Stem Cell Res. 8(3): 388-402.

Chen, W., S. Han, W. Qian, S. Weng, H. Yang, Y. Sun, et al. 2018. Nanotopography regulates motor neuron differentiation of human pluripotent stem cells. Nanoscale 10(7): 3556-3565.

Chong, D., L. -A. Turner, N. Gadegaard, A. Seifalian, M. Dalby and G. Hamilton. 2015. Nanotopography and plasma treatment: Redesigning the surface for vascular graft endothelialisation. Eur. J. Vasc. Endovasc. Surg. 49(3): 335-343.

Chowdhury, N. R., I. Hopp, P. Zilm, P. Murray and K. Vasilev. 2018. Silver nanoparticle modified surfaces induce differentiation of mouse kidney-derived stem cells. RSC Adv. 8(36): 20334-20340.

Christo, S. N., A. Bachhuka, K. R. Diener, A. Mierczynska, J. D. Hayball and K. Vasilev. 2016. The role of surface nanotopography and chemistry on primary neutrophil and macrophage cellular responses. Adv. Healthcare Mater. 5(8): 956-965.

Dalby, M. J., N. Gadegaard and R. O. C. Oreffo. 2014. Harnessing nanotopography and integrin-matrix interactions to influence stem cell fate. Nat. Mater. 13: 558-569.

Das, K., S. Bose and A. Bandyopadhyay. 2007. Surface modifications and cell-materials interactions with anodized Ti. Acta Biomater. 3(4): 573-585.

Davis, N. 2017. Nanochip could heal injuries or regrow organs with one touch, say researchers. from https://www.theguardian.com/science/2017/aug/07/nanochip-could-heal-injuries-or-regrow-organs-with-one-touch-say-researchers.

Dzobo, K., N. E. Thomford, D. A. Senthebane, H. Shipanga, A. Rowe, C. Dandara, et al. 2018. Advances in regenerative medicine and tissue engineering: Innovation and transformation of medicine. Stem Cells Int. 2018: 2495848.

Eisinger, M., J. S. Lee, J. Hefton, Z. Darzynkiewicz, J. Chiao and E. De Harven. 1979. Human epidermal cell cultures: Growth and differentiation in the absence of differentiation in the absence of dermal components or medium supplements. Proc. Natl. Acad. Sci. U.S.A. 76(10): 5340-5344.

Engel, E., A. Michiardi, M. Navarro, D. Lacroix and J. A. Planell. 2008. Nanotechnology in regenerative medicine: The materials side. Trends Biotechnol. 26(1): 39-47.

F Trent, J. and R. S. Kirsner. 1998. Tissue engineered skin: Apligraf, a bi-layered living skin equivalent. Int. J. Clin. Pract. 52(6): 408-413.

Ferreira, L., J. M. Karp, L. Nobre and R. Langer. 2008. New Opportunities: The Use of Nanotechnologies to Manipulate and Track Stem Cells. Cell Stem Cell 3(2): 136-146.

Garrod, D. and M. Chidgey. 2008. Desmosome structure, composition and function. Biochim. Biophys. Acta, Biomembr. 1778(3): 572-587.

Girdner, J. H. 1881. Skin-grafting with grafts taken from the dead subject. Med. Rec. (1866-1922). 20(5): 119.

Gong, T., J. Xie, J. Liao, T. Zhang, S. Lin and Y. Lin. 2015. Nanomaterials and bone regeneration. Bone Res. 3: 15029.

Hamdan, S., I. Pastar, S. Drakulich, E. Dikici, M. Tomic-Canic, S. Deo, et al. 2017.

Nanotechnology-driven therapeutic interventions in wound healing: Potential uses and applications. ACS Central Scie. 3(3): 163-175.

Hasan, J. and K. Chatterjee. 2015. Recent advances in engineering topography mediated antibacterial surfaces. Nanoscale 7(38): 15568-15575.

Hopp, I., M. N. MacGregor, K. Doherty, R. M. Visalakshan, K. Vasilev, R. L. Williams, et al. 2019. Plasma polymer coatings to direct the differentiation of mouse kidney-derived stem cells intopodocyte and proximal tubule-like cells. ACS Biomater. Sci. Eng. 5(6): 2834-2845.

Hotchkiss, K. M., G. B. Reddy, S. L. Hyzy, Z. Schwartz, B. D. Boyan and R. Olivares-Navarrete. 2016. Titanium surface characteristics, including topography and wettability, alter macrophage activation. Acta Biomater. 31: 425-434.

In Kim, J. and C. S. Kim. 2018. Harnessing nanotopography of PCL/collagen nanocomposite membrane and changes in cell morphology coordinated with wound healing activity. Mater. Sci. Eng. C. 91: 824-837.

Kang, H. -W., S. J. Lee, I. K. Ko, C. Kengla, J. J. Yoo and A. Atala. 2016. A 3D bioprinting system to produce human-scale tissue constructs with structural integrity. Nat. Biotechnol. 34(3): 312-319.

Khang, D., J. Choi, Y. -M. Im, Y. -J. Kim, J. -H. Jang, S. S. Kang, et al. 2012. Role of subnano-, nano- and submicron-surface features on osteoblast differentiation of bone marrow mesenchymal stem cells. Biomaterials 33(26): 5997-6007.

Kim, H. K., E. Kim, H. Jang, Y. -K. Kim and K. Kang. 2017. Neuron-material nanointerfaces: Surface nanotopography governs neuronal differentiation and development. Chem. Nano. Mat. 3(5): 278-287.

Kim, H. N., Y. Hong, M. S. Kim, S. M. Kim and K. -Y. Suh. 2012. Effect of orientation and density of nanotopography in dermal wound healing. Biomaterials 33(34): 8782-8792.

Kim, H. N., A. Jiao, N. S. Hwang, M. S. Kim, D. H. Kang, D. -H. Kim and K. -Y. Suh. 2013. Nanotopography-guided tissue engineering and regenerative medicine. Adv. Drug Deliv. Rev. 65(4): 536-558.

Kim, J., W. -G. Bae, Y. J. Kim, H. Seonwoo, H. -W. Choung, K. -J. Jang, et al. 2017. Directional matrix nanotopography with varied sizes for engineering wound healing. Adv. Healthcare Mater. 6(19): 1700297.

Kim, J., H. N. Kim, K. -T. Lim, Y. Kim, H. Seonwoo, S. H. Park, et al. 2013. Designing nanotopographical density of extracellular matrix for controlled morphology and function of human mesenchymal stem cells. Scie. Rep. 3: 3552.

Kim, T. H., S. H. Kim and Y. Jung. 2016. The effects of nanotopography and coculture systems to promote angiogenesis for wound repair. Nanomedicine 11(22): 2997-3007.

Komura, M., H. Komura, Y. Kanamori, Y. Tanaka, K. Suzuki, M. Sugiyama, et al. 2008. An animal model study for tissue-engineered trachea fabricated from a biodegradable scaffold using chondrocytes to augment repair of tracheal stenosis. J. Pediatr. Surg. 43(12): 2141-2146.

Konhilas, J. P., T. C. Irving and P. P. De Tombe. 2002. Frank-Starling law of the heart and the cellular mechanisms of length-dependent activation. Pflugers Arch. 445(3): 305-310.

Krząkała, A., A. Kazek-Kęsik and W. Simka. 2013. Application of plasma electrolytic oxidation to bioactive surface formation on titanium and its alloys. RSC Adv. 3(43): 19725-19743.

Langer, K. 1978. Zur Anatomie und Physiologie der Haut. Über die Spaltbarkeit der Cutis. Sitzungsbericht der Mathematisch-naturwissenschaftlichen Classe der Wiener Kaiserlichen Academie der Wissenschaften Abt. 44 (1861). English Translation: Br. J. Plast. Surg. 31: 3-8.

Lanza, R. (2019). from http://www.robertlanza.com/.

Laurencin, C. T., S. G. Kumbar and S. P. Nukavarapu. 2009. Nanotechnology and orthopedics: A personal perspective. Wiley Interdiscip. Rev.: Nanomed. Nanobiotechnol. 1(1): 6-10.

Lee, R., S. Das, M. Hourwitz, X. Sun, C. Parent, J. Fourkas, et al. 2016. Nanotopography guides and directs cell migration in amoeboid and epithelial cells. APS Meeting Abstracts.

Leung, W. W. -Y. 2016. Effects of Nanotopography on Structural Maturation and Differentiation of Human Induced Pluripotent Stem Cell-Derived Cardiomyocytes. https://digital.lib.washington.edu/researchworks/handle/1773/38066.

Li, M. and J. C. I. Belmonte. 2016. Looking to the future following 10 years of induced pluripotent stem cell technologies. Nat. Protoc. 11(9): 1579-1585.

Lin, H. -K., S. V. Madihally, B. Palmer, D. Frimberger, K. -M. Fung and B. P. Kropp. 2015. Biomatrices for bladder reconstruction. Adv. Drug Deliv. Rev. 82-83: 47-63.

Lundy, S. D., W. -Z. Zhu, M. Regnier and M. A. Laflamme. 2013. Structural and functional maturation of cardiomyocytes derived from human pluripotent stem cells. Stem Cells Dev. 22(14): 1991-2002.

MacGregor-Ramiasa, M., I. Hopp, A. Bachhuka, P. Murray and K. Vasilev. 2017. Surface nanotopography guides kidney-derived stem cell differentiation into podocytes. Acta Biomater. 56: 171-180.

Maher, S., G. Kaur, L. Lima-Marques, A. Evdokiou and D. Losic. 2017. Engineering of micro- to nanostructured 3d-printed drug-releasing titanium implants for enhanced osseointegration and localized delivery of anticancer drugs. ACS Appl. Mater. Interfaces 9(35): 29562-29570.

Mao, J. J. 2007. Translational Approaches in Tissue Engineering and Regenerative Medicine. Artech House. Norwood, Massachusetts, USA.

Marchesan, S., S. Bosi, A. Alshatwi and M. Prato. 2016. Carbon nanotubes for organ regeneration: An electrifying performance. Nano Today 11(4): 398-401.

McNamara, L. E., T. Sjöström, K. E. Burgess, J. J. Kim, E. Liu, S. Gordonov, et al. 2011. Skeletal stem cell physiology on functionally distinct titania nanotopographies. Biomaterials 32(30): 7403-7410.

Mengsteab, P. Y., K. Uto, A. S. T. Smith, S. Frankel, E. Fisher, Z. Nawas, et al. 2016. Spatiotemporal control of cardiac anisotropy using dynamic nanotopographic cues. Biomaterials 86: 1-10.

Mertens, J., M. C. Marchetto, C. Bardy and F. H. Gage. 2016. Evaluating cell reprogramming, differentiation and conversion technologies in neuroscience. Nat. Rev. Neuroscie 17(7): 424-437.

Mironov, V., T. Boland, T. Trusk, G. Forgacs and R. R. Markwald. 2003. Organ printing: Computer-aided jet-based 3D tissue engineering. Trends Biotechnol. 21(4): 157-161.

Moorthi, A., Y. -C. Tyan and T. -W. Chung. 2017. Surface-modified polymers for cardiac tissue engineering. Biomater. Scie 5(10): 1976-1987.

MS Castro-Raucci, L., M. S. Francischini, L. N. Teixeira, E. P. Ferraz, H. B. Lopes, P. T. de Oliveira, et al. 2016. Titanium with nanotopography induces osteoblast differentiation by regulating endogenous bone morphogenetic protein expression and signaling pathway. J. Cell Biochem. 117(7): 1718-1726.

Nakayama, K. H., P. A. Joshi, E. S. Lai, P. Gujar, L. -M. Joubert, B. Chen and N. F. Huang 2015. Bilayered vascular graft derived from human induced pluripotent stem cells with biomimetic structure and function. Regener. Med. 10(6): 745-755.

Oliveira, M. B., M. P. Ribeiro, S. P. Miguel, A. I. Neto, P. Coutinho, I. J. Correia and J. F. Mano. 2014. In vivo high-content evaluation of three-dimensional scaffolds biocompatibility. Tissue Eng., Part C. 20(11): 851-864.

Perez, R. A., C. -R. Jung and H. -W. Kim. 2017. Biomaterials and culture technologies for regenerative therapy of liver tissue. Adv. Healthcare Mater. 6(2): 1600791.

Polge, C., A. U. Smith and A. S. Parkes. 1949. Revival of spermatozoa after vitrification and dehydration at low temperatures. Nature 164(4172): 666.

Rao, L., Y. Qian, A. Khodabukus, T. Ribar and N. Bursac. 2018. Engineering human pluripotent stem cells into a functional skeletal muscle tissue. Nat. Commun. 9(1): 126.

Sampogna, G., S. Y. Guraya and A. Forgione. 2015. Regenerative medicine: Historical roots and potential strategies in modern medicine. J. Microsc. Ultrastruct. 3(3): 101-107.

Sartuqui, J., C. Gardin, L. Ferroni, B. Zavan and P. V. Messina. 2018. Nanostructured hydroxyapatite networks: Synergy of physical and chemical cues to induce an osteogenic fate in an additive-free medium. Mater. Today Commun. 16: 152-163.

Scudellari, M. 2016. A decade of iPS cells. Nature 534(7607): 310-313.

Schulte, C., J. Lamanna, A. S. Moro, C. Piazzoni, F. Borghi, M. Chighizola, et al. 2018. Neuronal cells confinement by micropatterned cluster-assembled dots with mechanotransductive nanotopography. ACS Biomater. Sci. Eng. 4(12): 4062-4075.

Seo, J., J. Kim, S. Joo, J. Y. Choi, K. Kang, W. K. Cho and I. S. Choi. 2018. Nanotopography-promoted formation of axon collateral branches of hippocampal neurons. Small 14(33): 1801763.

Song, L., K. Wang, Y. Li and Y. Yang. 2016. Nanotopography promoted neuronal differentiation of human induced pluripotent stem cells. Colloids Surf., B. 148: 49-58.

Stephan, A. 2017. Organ shortage: Can we decrease the demand? Experimental and clinical transplantation. Official Journal of the Middle East Society for Organ Transplantation 15(Suppl 1): 6-9.

Sun, Y.-S., J.-F. Liu, C.-P. Wu and H.-H. Huang. 2015. Nanoporous surface topography enhances bone cell differentiation on Ti–6Al–7Nb alloy in bone implant applications. J. Alloys Compd. 643: S124-S132.

Tanabe, K., C. E. Ang, S. Chanda, V. H. Olmos, D. Haag, D. F. Levinson, et al. 2018. Transdifferentiation of human adult peripheral blood T cells into neurons. Proc. Natl. Acad. Sci. U.S.A. 115(25): 6470-6475.

Tandon, B., A. Magaz, R. Balint, J. J. Blaker and S. H. Cartmell. 2018. Electroactive biomaterials: Vehicles for controlled delivery of therapeutic agents for drug delivery and tissue regeneration. Adv. Drug Delivery Rev. 129: 148-168.

Tang, D., R. S. Tare, L.-Y. Yang, D. F. Williams, K.-L. Ou and R. O. Oreffo. 2016. Biofabrication of bone tissue: Approaches, challenges and translation for bone regeneration. Biomaterials. 83: 363-382.

Taskin, M. B., D. Xia, F. Besenbacher, M. Dong and M. Chen. 2017. Nanotopography featured polycaprolactone/polyethyleneoxide microfibers modulate endothelial cell response. Nanoscale. 9(26): 9218-9229.

Traini, T., G. Murmura, B. Sinjari, G. Perfetti, A. Scarano, C. D'Arcangelo and S. Caputi. 2018. The surface anodization of titanium dental implants improves blood clot formation followed by osseointegration. Coatings 8(7): 252.

Tsui, J. H., N. A. Ostrovsky-Snider, D. M. P. Yama, J. D. Donohue, J. S. Choi, R. Chavanachat, et al. 2018. Conductive silk–polypyrrole composite scaffolds with bioinspired nanotopographic cues for cardiac tissue engineering. J. Mater. Chem. 6(44): 7185-7196.

Vacanti, C. A. 2006. The history of tissue engineering. J. Cell. Mol. Med. 10(3): 569-576.

Vasilev, K., J. Cook and H. J. Griesser. 2009. Antibacterial surfaces for biomedical devices. Expert Rev. Med. Devices. 6(5): 553-567.

Wang, G., S. Moya, Z. Lu, D. Gregurec and H. Zreiqat. 2015. Enhancing orthopedic implant bioactivity: Refining the nanotopography. Nanomedicine 10(8): 1327-1341.

Wang, K., X. He, W. Linthicum, R. Mezan, L. Wang, Y. Rojanasakul, et al. 2017a. Carbon nanotubes induced fibrogenesis on nanostructured substrates. Environ. Scie. Nano. 4(3): 689-699.

Wang, K., W. -D. Hou, X. Wang, C. Han, I. Vuletic, N. Su, et al. 2016. Overcoming foreign-body reaction through nanotopography: Biocompatibility and immunoisolation properties of a nanofibrous membrane. Biomaterials 102: 249-258.

Wang, Z., F. Wen, P. N. Lim, Q. Zhang, T. Konishi, D. Wang, et al. 2017b. Nanomaterial scaffolds to regenerate musculoskeletal tissue: Signals from within for neovessel formation. Drug Discovery Today. 22(9): 1385-1391.

Webster, J. P. 1944. Refrigerated skin grafts. Ann Surg. 120(4): 431- 48.

Xu, D., L. Fan, L. Gao, Y. Xiong, Y. Wang, Q. Ye, et al. 2016a. Micro-nanostructured polyaniline assembled in cellulose matrix via interfacial polymerization for applications in nerve regeneration. ACS Appl. Mater. Interfaces 8(27): 17090-17097.

Xu, J. -y., X. -s. Chen, C. -y. Zhang, Y. Liu, J. Wang and F. -l. Deng. 2016b. Improved bioactivity of selective laser melting titanium: Surface modification with micro-/ nano-textured hierarchical topography and bone regeneration performance evaluation. Mater Sci. Eng. C Mater. Biol. Appl. 68: 229-240.

Yang, K., E. Park, J. S. Lee, I. -S. Kim, K. Hong, K. I. Park, et al. 2015. Biodegradable nanotopography combined with neurotrophic signals enhances contact guidance and neuronal differentiation of human neural stem cells. Macromol Biosci. 15(10): 1348 - 1356.

Yang, S. S., J. Cha, S. -W. Cho and P. Kim. 2019. Time-dependent retention of nanotopographical cues in differentiated neural stem cells. ACS Biomater. Sci. Eng. https://doi.org/10.1021/acsbiomaterials.8b01057

Yannas, I., J. Burke, M. Warpehoski, P. Stasikelis, E. Skrabut, D. Orgill, et al. 1984. Prompt, longterm functional replacement of skin. Trans. Am. Soc. Artif. Intern. Organs. 27: 19-23.

Ye, J., J. Deng, Y. Chen, T. Yang, Y. Zhu, C. Wu, et al. 2019. Cicada and catkin inspired dual biomimetic antibacterial structure for the surface modification of implant material. Biomater. Scie. 7(7): 2826-2832.

Zhang, G., J. Zhang, G. Xie, Z. Liu and H. Shao. 2006. Cicada wings: A stamp from nature for nanoimprint lithography. Small 2(12): 1440-1443.

Zhang, L. and T. J. Webster. 2009. Nanotechnology and nanomaterials: Promises for improved tissue regeneration. Nano Today 4(1): 66-80.

Zhang, M., X. Wang, X. Huang, Y. Wang, R. Hang, X. Zhang, et al. 2019a. A high current anodization to fabricate a nano-porous structure on the surface of Ti-based implants. J. Mater. Sci. Mater. Med. 30(1): 2.

Zhang, X., X. Cui, D. Wang, S. Wang, Z. Liu, G. Zhao, et al. 2019b. Piezoelectric nanotopography induced neuron-like differentiation of stem cells. Adv. Funct. Mater. 29(22): 1900372.

Nano-scaled Biotechnology: The Upcoming Medicine

"... we will develop new scientific tools and create new technologies for the diagnosis and treatment of diseases and medical conditions. We will develop new tools which will allow us to develop a better understanding of how cells function, and misfunction, at the molecular level. Research will also focus on the development of new diagnostic and therapeutic strategies, for example, for the early detection and treatment of cancer...."

Peter Searson, Director, Johns Hopkins Institute for NanoBioTechnology
From the interview with News-Now, Nanomedicine,
Editor Rocky Rawstern

11.1 The Power of the Small

The raid of nanotechnology in the arena of medicine renovated the tactics of detection and treatment of a wide range of medical disorders and ailments. Throughout this book, we analysed specific examples that showed the magnificence of nanotechnology and we discovered how the manipulation of matter at the atomic level has the potential to fight diseases and to restore health. Nanoscale materials, exhibiting reduced dimensions and increased surface areas that cause them to have greater chemical reactivity, biological activity and catalytic behaviour in addition to a better penetration inside biological membranes and better access to cells, tissues and organs of living organisms, has revolutionised healthcare. The use of nano-tools in medical care includes different nanotechnology-based carriers like liposomes, dendrimers, polymeric micelles, carbon dots, quantum dots, carbon nanotubes, magnetic nanoparticles, silica nanoparticles, silver nanoparticles and gold nanoparticles. They transform pharmaceutical field and bio-imaging detection resolving problems of inefficacy or nonspecific effects (Patil et al. 2019, Zhu et al. 2019). The creation of nano-bots, nano-machines, nano-swimmers, nano-engines and nano-rockets theoretically can provide an accurate and effective identification and annihilation of cancer cells (Liu et al. 2019, Thorat et al. 2019). They can guide biochemical reactions by positioning reactive molecules with atomic precision (Nedorezova et al. 2019) and integrate into our systems to monitor and measure blood chemistry, notifying when something is out of order, detect temperature, pH or pressure variation, inflammation and more (Erkoc et al. 2019, Varshney et al. 2019). Another crucial nanoscale biological target is DNA. The application of nanotechnology in gene therapy provides non-viral structures that are as effective as viral systems in gene transfection (Fan et al. 2019). Then, tissue engineering and regenerative medicine are fascinating prospects; advances in nanotechnology in recent years have added a new dimension to their scientific approaches. The ability of nanomaterials to operate at the same scale as biological processes have meant that their application in tissue engineering and regenerative medicine has generated a real possibility of functional replacement of damaged organs and tissues (Yang et al. 2019). For example, nanotechnology offers hope to people suffering from diabetes, atherosclerosis, Alzheimer's, Parkinson's, brain injuries, tumours and neurological disorders. Nano-constructs could deliver neuroprotective molecules directly to the brain to recover or protect nerve cells from damage or degeneration. Nanotechnology has also been emerging in this field in the form of nano-engineered scaffolds that could one-day result in a tool for rewiring intricate tissue networks like bone, cartilage, skin, muscle and cardio-vascular frameworks and much more.

Finally, and not least important, nanotechnology has the potential to have a real impact on the fight against poverty-related diseases. For example, the development of new nanosensors capable of detecting tuberculosis and tropical diseases such as malaria and dengue eliminates the need for various stages of labeling and amplification generally associated with traditional diagnostic processes, making the results available in minutes and with a lower operating cost.

With this scenario, the scope and opportunities of the global market of nano-tools applied to healthcare and medical treatment expects that the worldwide nanotechnology industry will grow to reach $261,063 million (USD) by 2023 (Allied Market Research 2019, Gavrilović and Maksimović 2019). It is a necessity that the advantages of working with nanomedicine benefit as many patients as possible. Investigations focused on prevalent illnesses causing the most deaths worldwide, according to the World Health Organization (WHO 2018), highly validate the use of nanomedicine to tackle large public health issues (Pautler and Brenner 2010), to ensure that treatments reach all, to decrease poverty and inequality (Woodson 2016). Consequently, a robust government initiative toward the improvement of healthcare facilities through insurance schemes stimulates the growth of nanomedicine market (Allied Market Research 2019). The expansions of some markets, like that of India, not only has government support, but is also reinforced by the high adoption of innovative technologies by a great number of pharmaceutical companies. Numerically speaking, the nanomedicines market in India is expected to grow at the highest CAGR of 13.4% from 2016 to 2023 (Allied Market Research 2019).

It can be projected that following scientific research, investment in nanomedicine will continue at an accelerated rate, which will cause these products, in the not too distant future, to revolutionize medical care in all disciplines and specialties.

11.2 Risks, Doubts, and Ethical Concerns about Nanomedicine

Many successful nano-scaled materials tested in laboratories belong to the group of nanomedicines, and every day more and more are being added to that list. However, and compared to the list of products created, very few are being applied over the world.

What is the reason for this? Is it due to its complexity or, its confusing and unspecific regulatory approval pathways? Or, perhaps, is it due to ethical concerns, or a lack of education regarding such technology? We think that it is a matter of the convergence of all those multiple aspects.

Along with innovation and technological latent revolution, the application of nanomedicine is also associated to profound social and environmental impacts that raises various ethical concerns (Resnik and Tinkle 2007, Kazemi et al. 2014). The first level of concerns is related to the scientific uncertainty rooted in the continuous difference, and the restrictions that these impose, between the experimental research scenarios and the operational settings at the biological environments where nanomedicines are applied. The second dimension is related to the fragmentation of the regulatory framework. At present, there are poorly defined and diverse, if not divergent, regulatory strategies to deal with the problems that must be regulated. A third aspect refers to the opinions and attitudes of social stakeholders on the manipulation of nanotechnology: civic, business, religious, scientists, clinicians, health authorities and governance framework (Arnaldi and Muratorio 2013).

At present, there is a plenty of ethical interrogations encompassing nanomedicine, most of them are not new; however, the substantial ethical issues relating to nanomedicine that are only assessed involve risk calculation, risk management, and risk communication in clinical trials (Resnik and Tinkle 2007, Kazemi et al. 2014, Marques et al. 2019, Nasrollahzadeh and Sajadi 2019).

If we want to promote the use of nanotechnology in medicine, in addition to achieving experimental advances, it is highly necessary to be proactive to address its ethical, social and regulatory aspects and to educate people about its benefits and risks. This is the key point to gain and maintain public support (Feather and Aznar 2010, Winkelmann and Bhushan 2016, Zingg and Fischer 2019) and to minimize the bad reputation and the fear of using nanotechnology in public health, thus avoiding a negative stakeholder's reaction.

11.3 Immortality and Human Perfection

Among the questions that play a decisive role in the application of nanomedicine, we will deal with especially two: the possibility of attaining immortality and radical improvement of human abilities.

If you go back in time, the life expectancy, in general, was a lot shorter. The first question to answer is… is aging an ailment?

Aging is not that much different from any other physical disorder because it implies bone density deterioration, crumpled and drooping skin, sarcopenia, hormonal imbalances, slow wound healing, the loss of cells that we need and the accumulation of senescent cells which we do not need, loss of cognitive ability and memory. Added to them aging have other problems such as protein crosslinks, which cause arterial stiffening and intra and extracellular aggregates formation which are basically garbage which accumulate over time as a side effect of the bodies day-to-day metabolism and increasingly lead to infirmity. Finally, there is an increased likelihood of chronic disease and eventually death.

After establishing that aging is a disease, we asked the second question, could aging be cured?

Our cells are progressively damaged due to the influence of several factors, like reactive oxygen species and free radicals, which cause mutations in both the mitochondrial and nuclear DNA. With aging, the capacity of the natural machinery for repairing these damages diminishes and cells mortality augments, this being the primary reason for aging and death. Currently, many examples, some of which are provided in this book, demonstrate that nanotechnology can be applied to solve the above-mentioned conditions through the application of radical methodologies that would be able to repair single cells at the molecular level, and then nanotechnology should be able to repair the damage created by aging.

This brings us to our final questions: could nanotechnology replace senescent cells or reprogram cells so they do not senescence, which would keep the body from aging? So if you never get sick and never get old could you live forever?

Some scientists believe that it is possible, according to Ray Kurzweil in his book "The Singularity is Near: When Humans Transcend Biology" (Kurzweil 2005) "... *Nanotechnology will enable the design of nanobots: robots designed at the molecular level, measured in microns (millionths of a meter), such as "respirocytes" (mechanical red-blood cells). Nanobots will have myriad roles within the human body, including reversing human aging (to the extent that this task will not already have been completed through biotechnology, such as genetic engineering)...*"

Evidence shows that nanotechnological advances may come to treat and solve many of the current ailment, infections, trauma and disorder conditions; however, nothing guarantees that new and unknown diseases will affect humanity. These may be treatable or not, but in any case as it was done historically, potential future solutions will be reached. According to our vision, the quality and extension of life expectancy will be overcome every day, but we do not believe that immortality will ever be achieved.

While immortality is a bit further, the improvement of the human race through nanotechnology could be just around the corner. Can nanotechnology enhance humans?

The relationship between technology and its use to "aid" or "enhance" human capability is controversial. Where does human enhancement against true pathological conditions or disabilities end? Technological advances will imminently provide various devices that will be crossing point with the human body in various ways to deal with disease: enhancing vision, cognitive functions or improving a person's ability to move independently (Kim et al. 2013). The notion of human "aid" or "enhancement" as a way to treat disorders and illness can be transformed to improve human weaknesses and also propose to "optimize" human beings' physical, cognitive and psychological abilities (Barfield 2019, Gaspar and Giger 2019, Pariseau-Legault et al. 2019). A scientific-philosophical theoretical debate has emerged among "posthumanists" who are in favour of the use of technology on humanity for non-therapeutic purposes (Milburn 2002), "transhumanist" who believes that human nature can and should be transcended with the aid of technological change (Bostrom 2005) and "bioconservatives" who reject enhancement technology to consider them "dehumanizing" (Kotze 2018, Parra-Sáez 2019).

The truth is that nanotechnology will have an enormous influence thanks to its potentials; the human desire for enhancement contrasts with its moral and religious beliefs resulting in a conflict between them. Thus, one of the most important aspects that future nanotechnology research should rest on is to propose different methods to distinguish between "therapy" and "improvement". This is a necessity in order to handle nano-scaled biotechnological treatments and products in a cleverer and more responsible manner. Especially so that the recipients of this technology can decide whether to accept it without prejudice and alarm or not.

11.4 Final Remarks

Nanomedicine is now widely seen as having huge potential to bring benefits to many areas of research and application, and are attracting rapidly increasing investments from governments and from businesses in many parts of the world. Current investigation confirms that nanotechnologies applied to healthcare clearly offer exciting possibilities that could benefit society as a whole. However, there is also a clear necessity of profound clarification in areas of ambiguity and appropriate regulation to ensure that nanotechnologies develop and should be applied in a safe and socially acceptable way. Dynamic, proactive, and socially responsible research will drive nanomedicine as it plays an increasingly integral and transformative role in medicine and public health in the 21st century. The most important potential interventions that are likely to be developed with nanotechnology lie in the potential for repairing our bodies at the molecular and cellular level, which brings with it the questions of the increase of human longevity and supremacy. This latent, and increasingly closer, technological improvement carries with it ethical and moral alarms that we will be forced to face. We will be at a crossroads and we will have to choose between the virtuous and the demonic side of nanotechnology, or maybe we can reach an agreement and exploit the best side of each one.

Nanotechnology is driving a great expansion of medicine both at a scientific and practical level from which nobody wants to be left out. Despite the many obstacles that have slowed the accessibility of nanomaterials to the market, they are expected to become an essential and vital part of our conventional therapies.

References

Allied Market Research. 2019. Nanomedicine Market by Modality (Treatments and Diagnostics), Application (Drug Delivery, Diagnostic Imaging, Vaccines, Regenerative Medicine, Implants, and Others), and Indication (Oncological Diseases, Infectious Diseases, Cardiovascular Diseases, Orthopedic Disorders, Neurological Diseases, Urological Diseases, Ophthalmological Diseases, Immunological Diseases, and Others) - Global Opportunity Analysis and Industry Forecast, 2017-2023. from https://www.alliedmarketresearch.com/nanomedicine-market.

Arnaldi, S. and A. Muratorio. 2013. Nanotechnology, uncertainty and regulation: A guest editorial. NanoEthics 7(3): 173-175.

Barfield, W. 2019. The process of evolution, human enhancement technology, and cyborgs. Philosophies 4(1): 10.

Bostrom, N. 2005. Transhumanist values. J. Phil. Res. 30(Supplement): 3-14.

Erkoc, P., I. C. Yasa, H. Ceylan, O. Yasa, Y. Alapan and M. Sitti. 2019. Mobile microrobots for active therapeutic delivery. Adv. Ther. 2(1): 1800064.

Fan, B., X. Yang, X. Li, S. Lv, H. Zhang, J. Sun, et al. 2019. Photoacoustic-imaging-guided therapy of functionalized melanin nanoparticles: Combination of photothermal ablation and gene therapy against laryngeal squamous cell carcinoma. Nanoscale 11(13): 6285-6296.

Feather, J. L. and M. F. Aznar. 2010. Nanoscience education, workforce training, and K-12 resources. CRC Press. Boca Raton, Florida, USA.

Gaspar, R. and J. C. Giger. 2019. Emerging technologies, emerging risks: Current approaches on the future risks of human enhancement technologies. Human Behav. Emerging Technol. 1(1): 67-68.

Gavrilović, Z. and M. Maksimović. 2019. Nanomedical Devices as a Tool for Consumer Research. *In*: Badnjevic, A. R. Škrbić and L. Gurbeta Pokvić (eds.). International Conference on Medical and Biological Engineering. Springer Nature Switzerland.

Kazemi, A., M. Majidinia and A. A. Jamali. 2014. The question of ethics in nanomedicine. J. Clin. Res. Bioeth. 5(4): 1.

Kim, T. -i., J. G. McCall, Y. H. Jung, X. Huang, E. R. Siuda, Y. Li, et al. 2013. Injectable, cellular-scale optoelectronics with applications for wireless optogenetics. Science 340(6129): 211-216.

Kotze, M. 2018. The theological ethics of human enhancement: Genetic engineering, robotics and nanotechnology. In die Skriflig. 52: 1-8.

Kurzweil, R. 2005. The Singularity is Near: When Humans Transcend Biology. Penguin Group, New York, USA.

Liu, D., V. García-López, R. S. Gunasekera, L. G. Nilewski, L. B. Alemany, A. Aliyan, et al. 2019. Near infrared light activates molecular nanomachines to drill into and kill cells. ACS Nano 13(6): 6813-6823.

Marques, M. R. C., Q. Choo, M. Ashtikar, T. C. Rocha, S. Bremer-Hoffmann and M. G. Wacker. 2019. Nanomedicines – Tiny particles and big challenges. Adv. Drug. Deliv. Rev. S0169-409X(19)30066-3

Milburn, C. 2002. Nanotechnology in the age of posthuman engineering: Science fiction as science. Configurations 10(2): 261-295.

Nasrollahzadeh, M. and S. M. Sajadi. 2019. Chapter 7: Risks of Nanotechnology to Human Life. Interface Science and Technology. M. Nasrollahzadeh, S. M. Sajadi, M. Sajjadi, Z. Issaabadi and M. Atarod. Elsevier 28: 323-336.

Nedorezova, D. D., A. F. Fakhardo, D. V. Nemirich, E. A. Bryushkova and D. M. Kolpashchikov. 2019. Towards DNA nanomachines for cancer treatment: Achieving selective and efficient cleavage of folded RNA. Angew. Chem. Int. Ed. Engl. 131(14): 4702-4706.

Pariseau-Legault, P., D. Holmes and S. J. Murray. 2019. Understanding human enhancement technologies through critical phenomenology. Nurs. Philos. 20(1): e12229.

Parra-Sáez, J. 2019. Human Perfection and Contemporary Enhancement Technologies. Handbook of Research on Industrial Advancement in Scientific Knowledge, IGI Global: 74-94.

Patil, A., V. Mishra, S. Thakur, B. Riyaz, A. Kaur, R. Khursheed, et al. 2019. Nanotechnology derived nanotools in biomedical perspectives: An update. Curr. Nanosci. 15(2): 137-146.

Pautler, M. and S. Brenner. 2010. Nanomedicine: Promises and challenges for the future of public health. Int. J. Nanomedicine 5: 803-809.

Resnik, D. B. and S. S. Tinkle. 2007. Ethics in nanomedicine. Nanomedicine (London, England), 2(3): 345-350.

Thorat, N. D., H. Townely, G. Brennan, A. K. Parchur, C. Silien, J. Bauer, et al. 2019. Progress in remotely triggered hybrid nanostructures for next-generation brain cancer theranostics. ACS Biomater. Sci. Eng. 5(6): 2669-2687.

Varshney, N., A. Patel, Y. Deng, W. Haselmayr, P. K. Varshney and A. Nallanathan. 2019. Abnormality Detection inside Blood Vessels with Mobile Nanomachines. IEEE Transactions on Molecular, Biological and Multi-Scale Communications. 1-1.

W.H.O. 2018. The top 10 causes of death. from https://www.who.int/en/news-room/fact-sheets/detail/the-top-10-causes-of-death.

Winkelmann, K. and B. Bhushan (eds.). 2016. Global Perspectives of Nanoscience and Engineering Education. Springer International Publishing, Switzerland.

Woodson, T. S. 2016. Public private partnerships and emerging technologies: A look at nanomedicine for diseases of poverty. Res. Policy 45(7): 1410-1418.

Yang, Y., A. Chawla, J. Zhang, A. Esa, H. L. Jang and A. Khademhosseini. 2019. Applications of nanotechnology for regenerative medicine; healing tissues at the nanoscale. pp. 485-504. *In*: Atala A., R. Lanza, A. G. Mikos and R. Nelson (eds.). Principles of Regenerative Medicine. Academic Press. Cambridge, Massachusetts, USA.

Zhu, W., J. Guo, Y. Ju, R. E. Serda, J. G. Croissant, J. Shang, et al. 2019. Drug delivery: Modular metal-organic polyhedra superassembly: From molecular-level design to targeted drug delivery. Adv. Mater. 31(12): 1970082.

Zingg, R. and M. Fischer. 2019. The consolidation of nanomedicine. Wiley Interdiscip. Rev. Nanomed. Nanobiotechnol. e1569.

Index